Applied Electrochemistry
Principles, Practices And Applications

Applied Electrochemistry
Principles, Practices And Applications

Contributors :
Bao-Lin Xiao,
Ying-Xue Zhao, *et al.*

KOROS PRESS LIMITED
London, UK

Applied Electrochemistry : *Principles, Practices and Applications*
Contributors : Bao-Lin Xiao *and* Ying-Xue Zhao, *et al.*

Published by Koros Press Limited

www.korospress.com

United Kingdom

Copyright 2016
Printed in 2017 for Sale in the Indian Subcontinent

Applied Electrochemistry : *Principles, Practices and Applications*

ISBN: 978-1-78163-554-4

British Library Cataloguing in Publication Data
A CIP record for this book is available from the British Library

Exclusively distributed by CBS Publishers & Distributors Pvt. Ltd.

Sales & Distribution Rights only for India, Pakistan, Bangladesh, Sri Lanka, Nepal and Bhutan.This book is not to be sold outside these territories.

PREFACE

Electrochemistry provides the opportunity to run chemical redox reactions directly with electrons as reaction partners and even against the free energy gradient via external electrical energy input into the reaction system. These unique features create a wide range of industrial applications of electrochemical reactions like fluorene and chlorine production, electrowinning of metals, electromachining and electropolishing, electrodialysis, electrophoresis, and other electrochemical separation technologies. Electrochemistry takes advantage of thermodynamic driving forces for chemical conversion in technologies like batteries, rechargeable batteries, fuel cells, and of course, corrosion processes. Electrochemistry will play a key role in any future sustainable energy system, both for energy storage and energy conversion. These fascinating opportunities have given rise to widespread interest in electrochemical technologies among non-specialists, like engineers of all disciplines, life scientists, economists, and even politicians.

In the preparation of text standard works and review by renowned author have been freely consulted and the reference given chapter wise. At the end of the book will be found useful by those who wish to make a more detailed study of the topics discussed.

This page left intentionally blank.

CONTENTS

This page left intentionally blank.

LIST OF CONTRIBUTORS

Jun Hong

School of Life Sciences, Henan University, JinMing Road, Kaifeng 475000, China; E-Mails: 66yingxue@163.com (Y.-X.Z.); arixxl@163.com (B.-L.X.)

Ying-Xue Zhao

School of Life Sciences, Henan University, JinMing Road, Kaifeng 475000, China; E-Mails: 66yingxue@163.com (Y.-X.Z.); arixxl@163.com (B.-L.X.)

Bao-Lin Xiao

School of Life Sciences, Henan University, JinMing Road, Kaifeng 475000, China; E-Mails: 66yingxue@163.com (Y.-X.Z.); arixxl@163.com (B.-L.X.)

Ali Akbar Moosavi-Movahedi

Institute of Biochemistry and Biophysics, University of Tehran, Enquelab Avenue, P.O. Box 13145-1384, Tehran, Iran; E-Mail: hadi@ibb.ut.ac.ir

Hedayatollah Ghourchian

Institute of Biochemistry and Biophysics, University of Tehran, Enquelab Avenue, P.O. Box 13145-1384, Tehran, Iran; E-Mail: hadi@ibb.ut.ac.ir

Nader Sheibani

Department of Ophthalmology and Visual Sciences, University of Wisconsin, 600 Highland Avenue, K6/456 CSC, Madison, WI 53792-4673, USA; E-Mail: nsheibanikar@wisc.edu

This page left intentionally blank.

Chapter 1

FUEL CELL

INTRODUCTION

A **fuel cell** is a device that converts the chemical energy from a fuel into electricity through a chemical reaction with oxygen or another oxidizing agent. Hydrogen is the most common fuel, but hydrocarbons such as natural gas and alcohols like methanol are sometimes used. Fuel cells are different from batteries in that they require a constant source of fuel and oxygen/air to sustain the chemical reaction; however, fuel cells can produce electricity continuously for as long as these inputs are supplied.

The first fuel cells were invented in 1838. The first commercial use of fuel cells came more than a century later in NASA space programs to generate power for probes, satellites and space capsules. Since then, fuel cells have been used in many other applications. Fuel cells are used for primary and backup power for commercial, industrial and residential buildings and in remote or inaccessible areas. They are also used to power fuel-cell vehicles, including forklifts, automobiles, buses, airplanes, boats, motorcycles and submarines.

There are many types of fuel cells, but they all consist of an anode, a cathode and an electrolyte that allows charges to move between the two sides of the fuel cell. Electrons are drawn from the anode to the cathode through an external circuit, producing direct current electricity. As the main difference among fuel cell types is the electrolyte, fuel cells are classified by the type of electrolyte they use followed by the difference in startup time ranging from 1 sec for PEMFC to 10 min for SOFC. Fuel cells come in a variety of sizes. Individual fuel cells produce relatively small electrical potentials, about 0.7 volts, so cells are "stacked", or placed in series, to increase the voltage and meet an application's requirements. In addition to electricity, fuel cells produce water, heat and, depending on the fuel source, very small amounts of nitrogen dioxide and other emissions. The energy efficiency of a fuel cell is generally between 40–60%, or up to 85% efficient in co-generation if waste heat is captured for use.

The fuel cell market is growing, and Pike Research has estimated that the stationary fuel cell market will reach 50 GW by 2020.

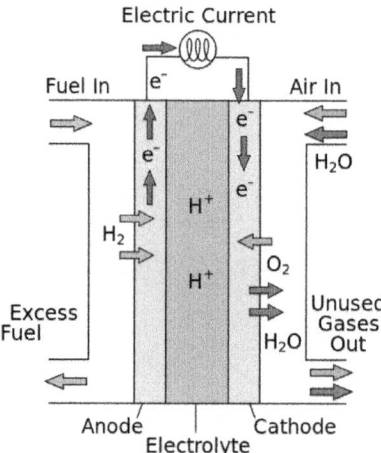

Fig. : Scheme of a proton-conducting fuel cell.

HISTORY

The first references to hydrogen fuel cells appeared in 1838. In a letter dated October 1838 but published in the December 1838 edition of *The London and Edinburgh Philosophical Magazine and Journal of Science*, Welsh physicist and barrister William Grove wrote about the development of his first crude fuel cells. He used a combination of sheet iron, copper and porcelain plates, and a solution of sulphate of copper and dilute acid. In a letter to the same publication written in December 1838 but published in June 1839, German physicist Christian Friedrich Schönbein discussed the first crude fuel cell that he had invented. His letter discussed current generated from hydrogen and oxygen dissolved in water. Grove later sketched his design, in 1842, in the same journal. The fuel cell he made used similar materials to today's phosphoric-acid fuel cell.

In 1939, British engineer Francis Thomas Bacon successfully developed a 5 kW stationary fuel cell. In 1955, W. Thomas Grubb, a chemist working for the General Electric Company (GE), further modified the original fuel cell design by using a sulphonated polystyrene ion-exchange membrane as the electrolyte. Three years later another GE chemist, Leonard Niedrach, devised a way of depositing platinum onto the membrane, which served as catalyst for the necessary hydrogen oxidation and oxygen reduction reactions. This became known as the "Grubb-Niedrach fuel cell". GE went on to develop this technology with NASA and McDonnell Aircraft, leading to its use during Project Gemini. This was the first commercial use of a fuel cell. In 1959, a team led by Harry Ihrig built a 15 kW fuel cell tractor for Allis-Chalmers, which was demonstrated across the U.S. at state fairs. This system used potassium hydroxide as the electrolyte and compressed hydrogen and oxygen as the reactants. Later in 1959, Bacon and his colleagues demonstrated a practical five-kilowatt unit capable of powering a welding machine. In the 1960s, Pratt and Whitney licensed Bacon's U.S. patents for use in the U.S. space program to supply electricity and drinking water (hydrogen and oxygen being readily available from the spacecraft tanks). In 1991, the first hydrogen fuel cell automobile was developed by Roger Billings.

UTC Power was the first company to manufacture and commercialize a large, stationary fuel cell system for use as a co-generation power plant in hospitals, universities and large office buildings. UTC Power continues to be the sole supplier of fuel cells to NASA for use in space vehicles, having supplied fuel cells for the Apollo missions, and the Space Shuttle program, and is developing fuel cells for cell phone towers and other applications.

TYPES OF FUEL CELLS; DESIGN

Fuel cells come in many varieties; however, they all work in the same general manner. They are made up of three adjacent segments : the anode, the electrolyte, and the cathode. Two chemical reactions occur at the interfaces of the three different segments. The net result of the two reactions is that fuel is consumed, water or carbon dioxide is created, and an electric current is created, which can be used to power electrical devices, normally referred to as the load.

At the anode a catalyst oxidizes the fuel, usually hydrogen, turning the fuel into a positively charged ion and a negatively charged electron. The electrolyte is a substance specifically designed so ions can pass through it, but the electrons cannot. The freed electrons travel through a wire creating the electric current. The ions travel through the electrolyte to the cathode. Once reaching the cathode, the ions are reunited with the electrons and the two react with a third chemical, usually oxygen, to create water or carbon dioxide.

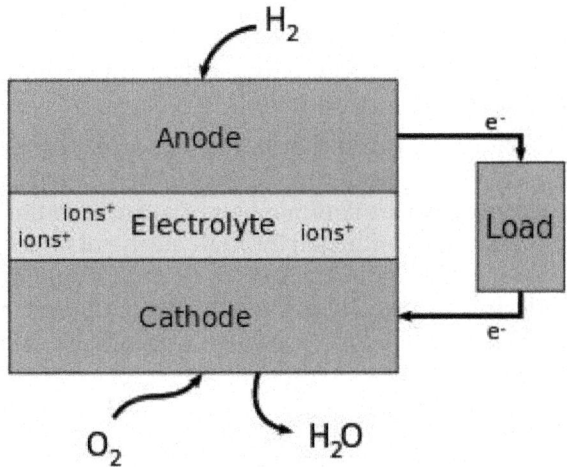

Fig. : A block diagram of a fuel cell.

The most important design features in a fuel cell are :

- The electrolyte substance. The electrolyte substance usually defines the *type* of fuel cell.
- The fuel that is used. The most common fuel is hydrogen.
- The anode catalyst breaks down the fuel into electrons and ions. The anode catalyst is usually made up of very fine platinum powder.
- The cathode catalyst turns the ions into the waste chemicals like water or carbon dioxide. The cathode catalyst is often made up of nickel but it can also be a nanomaterial-based catalyst.

A typical fuel cell produces a voltage from 0.6 V to 0.7 V at full rated load. Voltage decreases as current increases, due to several factors :

- Activation loss
- Ohmic loss (voltage drop due to resistance of the cell components and inter-connections)
- Mass transport loss (depletion of reactants at catalyst sites under high loads, causing rapid loss of voltage).

To deliver the desired amount of energy, the fuel cells can be combined in series and parallel circuits to yield higher voltage, and parallel-channel of configurations allow a higher current to be supplied. Such a design is called a *fuel cell stack*. The cell surface area can be increased, to allow stronger current from each cell. In the stack, reactant gases must be distributed uniformly over all of the cells to maximize the power output.

Proton Exchange Membrane Fuel Cells (PEMFCs)

In the archetypical hydrogen–oxide proton exchange membrane fuel cell design, a proton-conducting polymer membrane (the electrolyte) separates the anode and cathode sides. This was called a "solid polymer electrolyte fuel cell" (SPEFC) in the early 1970s, before the proton exchange mechanism was well-understood. (Notice that the synonyms "polymer electrolyte membrane" and "proton exchange mechanism" result in the same acronym.)

On the anode side, hydrogen diffuses to the anode catalyst where it later dissociates into protons and electrons. These protons often react with oxidants causing them to become what are commonly referred to as multi-facilitated proton membranes. The protons are conducted through the membrane to the cathode, but the electrons are forced to travel in an external circuit (supplying power) because the membrane is electrically insulating. On the cathode catalyst, oxygen molecules react with the electrons (which have travelled through the external circuit) and protons to form water.

In addition to this pure hydrogen type, there are hydrocarbon fuels for fuel cells, including diesel, methanol and chemical hydrides. The waste products with these types of fuel are carbon dioxide and water.

Proton exchange membrane fuel cell

1 Hydrogen fuel is channeled through field flow plates to the anode on one side of the fuel cell, while oxidant (oxygen or air) is channeled to the cathode on the other side of the cell.

Backing layers
Hydrogen
gas
Oxidant
Oxidant flow field

Hydrogen
flow field

2 At the anode, a platinum catalyst causes the hydrogen to split into positive hydrogen ions (protons) and negatively charged electrons.

3 The polymer electrolyte membrane (PEM) allows only the positively charged ions to pass through it to the cathode. The negatively charged electrons must travel along an external circuit to the cathode, creating an electrical current.

Unused fuel
Water

Anode
(negative)
Cathode
(positive)
Polymer
electrolyte
membrane

4 At the cathode, the electrons and positively charged hydrogen ions combine with oxygen to form water, which flows out of the cell.

Fig. : Construction of a high-temperature PEMFC : Bipolar plate as electrode with in-milled gas channel structure, fabricated from conductive composites (enhanced with graphite, carbon black, carbon fiber, and/or carbon nanotubes for more conductivity); Porous carbon papers; reactive layer, usually on the polymer membrane applied; polymer membrane.

Fig. : Condensation of water produced by a PEMFC on the air channel wall. The gold wire around the cell ensures the collection of electric current.

The different components of a PEMFC are (i) bipolar plates, (ii) electrodes, (iii) catalyst, (iv) membrane, and (v) the necessary hardware. The materials used for different parts of the fuel cells differ by type. The bipolar plates may be made of different types of materials, such as, metal, coated metal, graphite, flexible graphite, C–C composite, carbon–polymer composites etc. The membrane electrode assembly (MEA) is referred as the heart of the PEMFC and is usually made of a proton exchange membrane sandwiched between two catalyst-coated carbon papers. Platinum and/or similar type of noble metals are usually used as the catalyst for PEMFC. The electrolyte could be a polymer membrane.

Proton Exchange Membrane Fuel Cell Design Issues

- Costs. In 2013, the Department of Energy estimated that 80-kW automotive fuel cell system costs of US$67 per kilowatt could be achieved, assuming volume production of 100,000 automotive units per year and US$55 per kilowatt could be achieved, assuming volume production of 500,000 units per year. In 2008, professor Jeremy P. Meyers estimated that cost reductions over a production ramp-up period will take about 20 years after fuel-cell cars are introduced before they will be able to compete commercially with current market technologies, including gasoline internal combustion engines. Many companies are working on techniques to reduce cost in a variety of ways including reducing the amount of platinum needed in each individual cell. Ballard Power Systems has experimented with a catalyst enhanced with carbon silk, which allows a 30% reduction (1 mg/cm^2 to 0.7 mg/cm^2) in platinum usage without reduction in performance. Monash University, Melbourne uses PEDOT as a cathode. A 2011 published study documented the first metal-free electrocatalyst using relatively inexpensive doped carbon nanotubes, which are less than 1% the cost of platinum and are of equal or superior performance.

- Water and air management (in PEMFCs). In this type of fuel cell, the membrane must be hydrated, requiring water to be evapourated at precisely the same rate that it is produced. If water is evapourated too quickly, the membrane dries, resistance across it increases, and eventually it will crack, creating a gas "short circuit" where hydrogen and oxygen combine directly, generating heat that will damage the fuel cell. If the water is evapourated too slowly, the electrodes will flood, preventing the reactants from reaching the catalyst and stopping the reaction. Methods to manage water in cells are being developed like electro-osmotic pumps focusing on flow control. Just as in a combustion engine, a steady ratio between the reactant and oxygen is necessary to keep the fuel cell operating efficiently.

- *Temperature management :* The same temperature must be maintained throughout the cell in order to prevent destruction of the cell through thermal loading. This is particularly challenging as the $2H_2 + O_2 \rightarrow 2H_2O$ reaction is highly exothermic, so a large quantity of heat is generated within the fuel cell.

- Durability, service life, and special requirements for some type of cells. Stationary fuel cell applications typically require more than 40,000 hours of reliable operation at a temperature of −35°C to 40°C (−31 °F to 104 °F), while automotive fuel cells require a 5,000-hour lifespan (the equivalent of 240,000 km (150,000 mi)) under extreme temperatures. Current service life is 7,300 hours under cycling conditions. Automotive engines must also be able to start reliably at −30°C (−22 °F) and have a high power-to-volume ratio (typically 2.5 kW per liter).

- Limited carbon monoxide tolerance of some (non-PEDOT) cathodes.

Phosphoric Acid Fuel Cell (PAFC)

Phosphoric acid fuel cells (PAFC) were first designed and introduced in 1961 by G. V. Elmore and H. A. Tanner. In these cells phosphoric acid is used as a non-conductive electrolyte to pass positive hydrogen ions from the anode to the cathode. These cells commonly work in temperatures of 150 to 200 degrees Celsius. This high temperature will cause heat and energy loss if the heat is not removed and used properly. This heat can be used to produce steam for air conditioning systems or any other thermal energy consuming system. Using this heat in co-generation can enhance the efficiency of phosphoric acid fuel cells from 40–50% to about 80%. Phosphoric acid, the electrolyte used in PAFCs, is a non-conductive liquid acid which forces electrons to travel from anode to cathode through an external electrical circuit. Since the hydrogen ion production rate on the anode is small, platinum is used as catalyst to increase this ionization rate. A key disadvantage of these cells is the use of an acidic electrolyte. This increases the corrosion or oxidation of components exposed to phosphoric acid.

High-temperature Fuel Cells

Solid Oxide Fuel Cell

A **solid oxide fuel cell (SOFC)** is an electro-chemical conversion device that produces electricity directly from *oxidizing* a fuel. *Fuel cells* are characterized by their electrolyte material; the SOFC has a solid oxide or *ceramic*, electrolyte. Advantages of this class of fuel cells include high efficiency, long-term stability, fuel flexibility, low emissions, and relatively low cost. The largest disadvantage is the high *operating temperature* which results in longer start-up times and mechanical and chemical compatibility issues.

Introduction

Solid oxide fuel cells are a class of fuel cells characterized by the use of a solid oxide material as the electrolyte. SOFCs use a solid oxide electrolyte to conduct negative oxygen ions from the cathode to the anode. The electro-chemical oxidation of the oxygen ions with hydrogen or carbon monoxide thus occurs on the anode side. More recently, proton-conducting SOFCs (PC-SOFC) are being developed which transport protons instead of oxygen ions through the electrolyte with the advantage of being able to be run at lower temperatures than traditional SOFCs.

They operate at very high temperatures, typically between 500 and 1,000°C. At these temperatures, SOFCs do not require expensive platinum catalyst material, as is currently necessary for lower-temperature fuel cells such as PEMFCs, and are not vulnerable to carbon monoxide catalyst poisoning. However, vulnerability to sulfur poisoning has been widely observed and the sulfur must be removed before entering the cell through the use of adsorbent beds or other means.

Solid oxide fuel cells have a wide variety of applications from use as auxiliary power units in vehicles to stationary power generation with outputs from 100 W to 2 MW. In 2009, Australian company, Ceramic Fuel Cells successfully achieved an efficiency of a SOFC device up to the previously theoretical mark of 60%. The higher operating temperature make SOFCs suitable candidates for application with heat engine energy recovery devices or combined heat and power, which further increases overall fuel efficiency.

Because of these high temperatures, light hydrocarbon fuels, such as methane, propane and butane can be internally reformed within the anode. SOFCs can also be fueled by externally reforming heavier hydrocarbons, such as gasoline, diesel, jet fuel (JP-8) or bio-fuels. Such reformates are mixtures of hydrogen, carbon monoxide, carbon dioxide, steam and methane, formed by reacting the hydrocarbon fuels with air or steam in a device upstream of the SOFC anode. SOFC power systems can increase efficiency by using the heat given off by the exothermic electro-chemical oxidation within the fuel cell for endothermic steam reforming process.

Thermal expansion demands a uniform and well-regulated heating process at startup. SOFC stacks with planar geometry require in the order of an hour to be heated to light-off temperature. Micro-tubular fuel cell design geometries promise much faster start up times, typically in the order of minutes.

Unlike most other types of fuel cells, SOFCs can have multiple geometries. The planar fuel cell design geometry is the typical sandwich type geometry employed by most types of fuel cells, where the electrolyte is sandwiched in between the electrodes. SOFCs can also be made in tubular geometries where either air or fuel is passed through the inside of the tube and the other gas is passed along the outside of the tube. The tubular design is advantageous because it is much easier to seal air from the fuel. The performance of the planar design is currently better than the performance of the tubular design however, because the planar design has a lower resistance comparatively. Other geometries of SOFCs include modified planar fuel cell designs (MPC or MPSOFC), where a wave-like structure replaces the traditional flat configuration of the planar cell. Such designs are highly promising, because they share the advantages of both planar cells (low resistance) and tubular cells.

Operation

A solid oxide fuel cell is made up of four layers, three of which are ceramics (hence the name). A single cell consisting of these four layers stacked together is typically only a few millimeters thick. Hundreds of these cells are then connected in series to form what most people refer to as an "SOFC stack". The ceramics used in SOFCs do not become electrically and ionically active until they reach very high temperature and as a consequence the stacks have to run at temperatures ranging from 500 to 1,000°C. Reduction of oxygen into oxygen ions occurs at the cathode. These ions can then diffuse through the solid oxide electrolyte to the anode where they can electro-chemically oxidize the fuel. In this reaction, a water by-product is given off as well as two electrons. These electrons then flow through an external circuit where they can do work. The cycle then repeats as those electrons enter the cathode material again.

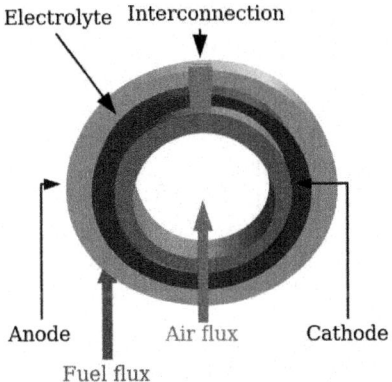

Fig.: Cross section of the three ceramic layers of an SOFC. From left to right: porous cathode, dense electrolyte, porous anode.

Balance of Plant

Most of the downtime of a SOFC stems from the mechanical balance of plant, the air preheater, prereformer, afterburner, water heat exchanger, anode tail gas oxidizer, and electrical balance of plant, power electronics, hydrogen sulfide sensor and fans. Internal reforming leads to a large decrease in the balance of plant costs in designing a full system.

Anode

The ceramic anode layer must be very porous to allow the fuel to flow towards the electrolyte. Consequently granular matter is often selected for anode fabrication procedures. Like the cathode, it must conduct electrons, with ionic conductivity a definite asset. The most common material used is a cermet made up of nickel mixed with the ceramic material that is used for the electrolyte in that particular cell, typically YSZ (yttria stabilized zirconia) nanomaterial-based catalysts, this YSZ part helps stop the grain growth of nickel. The anode is commonly the thickest and strongest layer in each individual cell, because it has the smallest polarization losses, and is often the layer that provides the mechanical support. Electrochemically speaking, the anode's job is to use the oxygen ions that diffuse through the electrolyte to oxidize the hydrogen fuel. The oxidation reaction between the oxygen ions and the hydrogen produces heat as well as water and electricity. If the fuel is a light hydrocarbon, for example methane, another function of the anode is to act as a catalyst for steam reforming the fuel into hydrogen. This provides another operational benefit to the fuel cell stack because the reforming reaction is endothermic, which cools the stack internally.

Electrolyte

The electrolyte is a dense layer of ceramic that conducts oxygen ions. Its electronic conductivity must be kept as low as possible to prevent losses from leakage currents. The high operating temperatures of SOFCs allow the kinetics of oxygen ion transport to be sufficient for good performance. However, as the operating temperature approaches the lower limit for SOFCs at around 600°C, the electrolyte begins to have large ionic transport resistances and affect the performance. Popular electrolyte materials include yttria stabilized zirconia (YSZ) (often the 8% form Y8SZ), scandia stabilized zirconia (ScSZ) (usually 9 mol%Sc2O3–9ScSZ) and Gadolinium doped ceria (GDC). The electrolyte material has crucial influence on the cell performances. Detrimental reactions between YSZ electrolytes and modern cathodes such as LSCF have been found, and can be prevented by thin (<100 nm) ceria diffusion barriers.

If the conductivity for oxygen ions in SOFC can remain high even at lower temperature (current target in research ~500°C), material choice for SOFC will broaden and many existing problems can potentially be solved. Certain processing technique such as thin film deposition can help solve this problem with existing material by

- Reducing the travelling distance of oxygen ions and electrolyte resistance as resistance is inversely proportional to conductor length;
- Producing grain structures that are less resistive such as columnar grain structure;
- Controlling the micro-structural nano-crystalline fine grains to achieve "fine-tuning" of electrical properties;
- Building composite with large interfacial areas as interfaces have shown to have extraordinary electrical properties.

Cathode

The cathode, or air electrode, is a thin porous layer on the electrolyte where oxygen reduction takes place. The overall reaction is written in Kröger-Vink Notation as follows :

$$\frac{1}{2}O_2(g) + 2e' + V_o^{\cdot\cdot} \rightarrow O_o^x$$

Cathode materials must be, at minimum, electronically conductive. Currently, lanthanum strontium manganite (LSM) is the cathode material of choice for commercial use because of its compatibility with doped zirconia electrolytes. Mechanically, it has similar coefficient of thermal expansion to YSZ and thus limits stresses buildup because of CTE mismatch. Also, LSM has low levels of chemical reactivity with YSZ which extends the lifetime of the material. Unfortunately, LSM is a poor ionic conductor, and so the electro-chemically active reaction is limited to the triple phase boundary (TPB) where the electrolyte, air and electrode meet. LSM works well as a cathode at high temperatures, but its performance quickly falls as the operating temperature is lowered below 800°C. In order to increase the reaction zone beyond the TPB, a potential cathode material must be able to conduct both electrons and oxygen ions. Composite cathodes consisting of LSM YSZ have been used to increase this triple phase boundary length. Mixed ionic/ electronic conducting (MIEC) ceramics, such as the perovskite LSCF, are also being researched for use in intermediate temperature SOFCs as they are more active and can makeup for the increase in the activation energy of reaction.

Interconnect

The interconnect can be either a metallic or ceramic layer that sits between each individual cell. Its purpose is to connect each cell in series, so that the electricity each cell generates can be combined. Because the interconnect is exposed to both the oxidizing and reducing side of the cell at high temperatures, it must be extremely stable. For this reason, ceramics have been more successful in the long term than metals as interconnect materials. However, these ceramic interconnect materials are very expensive as compared to metals. Nickel- and steel-based alloys are becoming more promising as lower temperature (600–800°C) SOFCs are developed. The material of choice for an interconnect in contact with Y8SZ is a metallic 95Cr-5Fe alloy. Ceramic-metal composites called 'cermet' are also under consideration, as they have demonstrated thermal stability at high temperatures and excellent electrical conductivity.

Polarizations

Polarizations, or over-potentials, are losses in voltage due to imperfections in materials, micro-structure, and design of the fuel cell. Polarizations result from ohmic resistance of oxygen ions conducting through the electrolyte (iRΩ), electrochemical activation barriers at the anode and cathode, and finally concentration polarizations due to inability of gases to diffuse at high rates through the porous anode and cathode (shown as ηA for the anode and ηC for cathode). The cell voltage can be calculated using the following equation :

$$V = E_0 - iR_\omega - \eta_{cathode} - \eta_{anode}$$

where E_0 is the Nernst potential of the reactants and R represents the Thévenin equivalent resistance value of the electrically conducting portions of the cell. $\eta_{cathode}$ and η_{anode} account for the remaining difference between the actual cell voltage and the Nernst potential. In SOFCs, it is often important to focus on the ohmic and concentration polarizations since high operating temperatures experience little activation polarization. However, as the lower limit of SOFC operating temperature is approached (~600°C), these polarizations do become important.

Above mentioned equation is used for determining the SOFC voltage (in fact for fuel cell voltage in general). This approach results in good agreement with particular experimental data (for which adequate factors were obtained) and poor agreement for other than original experimental working parameters. Moreover, most of the equations used require the addition of numerous factors which are difficult or impossible to determine. It makes very difficult any optimizing process of the SOFC working parameters as well as design architecture configuration selection. Because of those circumstances a few other equations were proposed :

$$E_{SOFC} = \frac{E_{max} - i_{max} \cdot \eta_f \cdot r_1}{\frac{r_1}{r_2} \cdot (1 - \eta_f) + 1}$$

where : E_{SOFC} – cell voltage, E_{max} – maximum voltage given by the Nernst equation, i_{max} – maximum current density (for given fuel flow), η_f – fuel utilization factor, r_1–ionic specific resistance of the electrolyte, and r_2– electric specific resistance of the electrolyte.

This method was validated and found to be suitable for optimization and sensitivity studies in plant-level modelling of various systems with solid oxide fuel cells. With this mathematical description it is possible to account for different properties of the SOFC. There are many parameters which impact cell working conditions, *e.g.* electrolyte material, electrolyte thickness, cell temperature, inlet and outlet gas compositions at anode and cathode, and electrode porosity, just to name some. The flow in these systems is often calculated using the Navier-stokes equation.

Ohmic Polarization

Ohmic losses in an SOFC result from ionic conductivity through the electrolyte. This is inherently a materials property of the crystal structure and atoms involved. However, to maximize the ionic conductivity, several methods can be done. Firstly, operating at higher temperatures can significantly decrease these ohmic losses. Substitutional doping methods to further refine the crystal structure and control defect concentrations can also play a significant role in increasing the conductivity. Another way to decrease ohmic resistance is to decrease the thickness of the electrolyte layer.

Ionic Conductivity

An ionic specific resistance of the electrolyte as a function of temperature can be described by the following relationship :

$$r_1 = \frac{\delta}{\sigma}$$

where : δ– electrolyte thickness, and σ–ionic conductivity.

The ionic conductivity of the solid oxide is defined as follows :

$$\sigma = \sigma_0 \cdot e^{\frac{-E}{R \cdot T}}$$

where : σ_0 and E – factors depended on electrolyte materials, T– electrolyte temperature, and R– ideal gas constant.

Concentration Polarization

The concentration polarization is the result of practical limitations on mass transport within the cell, and represents the voltage loss due to spatial variations in reactant concentration at the chemically active sites. This situation can be caused when the reactants are consumed by the electro-chemical reaction faster than they can diffuse into the porous electrode, and can also be caused by variation in bulk flow composition. The latter is due to the fact that the consumption of reacting species in the reactant flows causes a drop in reactant concentration as it travels along the cell, which causes a drop in the local potential near the tail end of the cell.

The concentration polarization occurs in both the anode and cathode. The anode can be particularly problematic, as the oxidation of the hydrogen produces steam, which further dilutes the fuel stream as it travels along the length of the cell. This polarization can be mitigated by reducing the reactant utilization fraction or increasing the electrode porosity, but these approaches each have significant design trade-offs.

Activation Polarization

The activation polarization is the result of the kinetics involved with the electro-chemical reactions. Each reaction has a certain activation barrier that must be overcome in order to proceed and this barrier leads to the polarization. The activation barrier is the result of many complex electro-chemical reaction steps where typically the rate limiting step is responsible for the polarization. The polarization equation shown below is found by solving the Butler–Volmer equation in the high current density regime (where the cell typically operates), and can be used to estimate the activation polarization :

$$\eta_{act} = \frac{RT}{\beta z F} \times \ln\left(\frac{i}{i_0}\right)$$

where :

- R = gas constant
- T_0 = operating temperature
- β = electron transfer coefficient
- z = electrons associated with the electro-chemical reaction
- F = Faraday's constant
- i = operating current
- i_0 = exchange current density.

The polarization can be modified by micro-structural optimization. The Triple Phase Boundary (TPB) length, which is the length where porous, ionic and electronically conducting pathways all meet, directly relates to the electro-chemically active length in the cell. The larger the length, the more reactions can occur and thus the less the activation polarization. Optimization of TPB length can be done by processing conditions to affect micro-structure or by materials selection to use a mixed ionic/electronic conductor to further increase TPB length.

Target

DOE target requirements are 40,000 hours of service for stationary fuel cell applications and greater than 5,000 hours for transportation systems (fuel cell vehicles) at a factory cost of $400/kW for a 10 kW coal-based system without additional requirements. Lifetime effects (phase stability, thermal expansion compatibility, element migration, conductivity and aging) must be addressed. The Solid State Energy Conversion Alliance 2008 (interim) target for overall degradation per 1,000 hours is 4.0%.

Research

Research is going now in the direction of lower-temperature SOFC (600°C) in order to decrease the materials cost, which will enable the use of metallic materials with better mechanical properties and thermal conductivity.

Research is currently under way to improve the fuel flexibility of SOFCs. While stable operation has been achieved on a variety of hydrocarbon fuels, these cells typically rely on external fuel processing. For the case of natural gas, the fuel is either externally or internally reformed and the sulfur compounds are removed. These processes add to the cost and complexity of SOFC systems. Work is under way at a number of institutions to improve the stability of anode materials for hydrocarbon oxidation and, therefore, relax the requirements for fuel processing and decrease SOFC balance of plant costs.

Research is also going on in reducing start-up time to be able to implement SOFCs in mobile applications. Due to their fuel flexibility they may run on partially reformed diesel, and this makes SOFCs interesting as auxiliary power units (APU) in refrigerated trucks.

Specifically, Delphi Automotive Systems are developing an SOFC that will power auxiliary units in automobiles and tractor-trailers, while BMW has recently stopped a similar project. A high-temperature SOFC will generate all of the needed electricity to allow the engine to be smaller and more efficient. The SOFC would run on the same gasoline or diesel as the engine and would keep the air conditioning unit and other necessary electrical systems running while the engine shuts off when not needed (*e.g.,* at a stop light or truck stop).

Rolls-Royce is developing solid-oxide fuel cells produced by screen printing onto inexpensive ceramic materials. Rolls-Royce Fuel Cell Systems Ltd is developing a SOFC gas turbine hybrid system fueled by natural gas for power generation applications in the order of a megawatt (*e.g.* Futuregen).

Ceres Power Ltd. has developed a low cost and low temperature (500–600 degrees) SOFC stack using cerium gadolinium oxide (CGO) in place of current industry standard ceramic, yttria stabilized zirconia (YSZ), which allows the use of stainless steel to support the ceramic.

Solid Cell Inc. has developed a unique, low cost cell architecture that combines properties of planar and tubular designs, along with a Cr-free cermet interconnect.

The high temperature electro-chemistry center (HITEC) at the University of Florida, Gainesville is focused on studying ionic transport, electrocatalytic phenomena and micro-structural characterization of ion conducting materials.

SiEnergy Systems, a Harvard spin-off company, has demonstrated the first macro-scale thin-film solid-oxide fuel cell that can operate at 500 degrees.

Acumentrics, a US company, manufactures and markets solid oxide fuel cell power generators for off-grid applications. Products range from a power rating of 250W to 1500W.

SOEC

A solid oxide electrolyser cell (SOEC) is a solid oxide fuel cell set in regenerative mode for the electrolysis of water with a solid oxide, or ceramic, electrolyte to produce oxygen and hydrogen gas.

ITSOFC

SOFCs that operate in an intermediate temperature (IT) range, meaning between 600 and 800°C, are named ITSOFCs. Because of the high degradation rates and materials costs incurred at temperatures in excess of 900°C, it is economically more favourable to operate SOFCs at lower temperatures. The push for high performance ITSOFCs is currently the topic of much research and development. One area of focus is the cathode material. It is thought that the oxygen reduction reaction is responsible for much of the loss in performance so the catalytic activity of the cathode is being studied and enhanced through various techniques, including catalyst impregnation.

LT-SOFC

LT stands for Low Temperature.

The University of Maryland and Redox Power Systems LLC demonstrated a 650°C operating temperature ceria/bismuth solid oxide fuel cell.

SOFC-GT

An SOFC-GT system is one which comprises a solid oxide fuel cell combined with a gas turbine. Such systems have been evaluated by Siemens Westinghouse and Rolls-Royce as a mean to achieve higher operating efficiencies by running the SOFC under pressure. SOFC-GT systems typically include anodic and/or cathodic atmosphere recirculation, thus increasing efficiency.

Theoretically, the combination of the SOFC and gas turbine can give result in high overall (electrical and thermal) efficiency. Further combination of the SOFC-GT in a combined cooling, heat and power (or trigeneration) configuration (*via* HVAC) also has the potential to yield even higher thermal efficiencies in some cases.

Molten Carbonate Fuel Cell

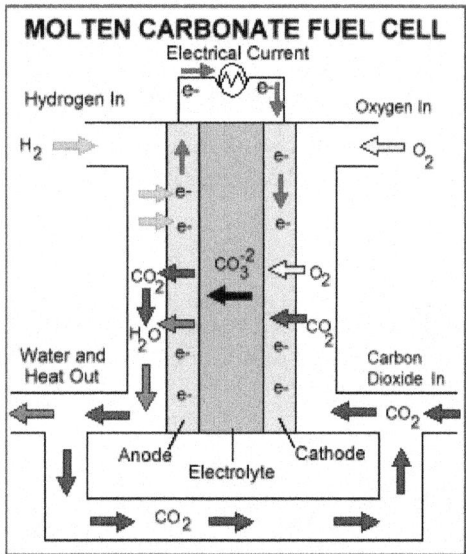

Fig. : Scheme of a molten-carbonate fuel cell.

Molten-carbonate fuel cells (MCFCs) are high-temperature *fuel cells* that *operate at temperatures* of 600°C and above.

Molten carbonate fuel cells (MCFCs) are currently being developed for *natural gas, biogas* (produced as a result of *anaerobic digestion* or *biomass gasification*), and *coal*-based power plants for *electrical utility*, industrial, and *military* applications. MCFCs are high-temperature fuel cells that use an *electrolyte* composed of a molten carbonate salt mixture suspended in a porous, chemically inert ceramic matrix of *beta-alumina solid electrolyte* (BASE). Since they operate at extremely high temperatures of 650°C (roughly 1,200 °F) and above, non-precious *metals* can be used as *catalysts* at the *anode* and *cathode*, reducing costs.

Improved efficiency is another reason MCFCs offer significant cost reductions over *phosphoric acid fuel cells* (PAFCs). Molten carbonate fuel cells can reach efficiencies approaching 60%, considerably higher than the 37–42% efficiencies of a phosphoric acid fuel cell plant. When the waste heat is *captured and used*, overall fuel efficiencies can be as high as 85%.

Unlike *alkaline*, phosphoric acid, and *polymer* electrolyte membrane fuel cells, MCFCs don't require an external reformer to convert more energy-dense fuels to *hydrogen*. Due to the high temperatures at which MCFCs operate, these fuels are converted to hydrogen within the fuel cell itself by a process called internal reforming, which also reduces cost.

Molten carbonate fuel cells are not prone to *poisoning* by *carbon monoxide* or *carbon dioxide*–they can even use carbon oxides as fuel–making them more attractive for fueling with gases made from coal. Because they are more resistant to impurities than other fuel cell types, scientists believe that they could even be capable of internal reforming of coal, assuming they can be made resistant to impurities such as sulfur and particulates that result from converting coal, a dirtier *fossil fuel* source than many others, into hydrogen. Alternatively, because MCFCs require CO_2 be delivered to the cathode along with the oxidizer, they can be used to electro-chemically separate carbon dioxide from the flue gas of other fossil fuel power plants for sequestration.

The primary disadvantage of current MCFC technology is durability. The high temperatures at which these cells operate and the corrosive electrolyte used accelerate component breakdown and corrosion, decreasing cell life. Scientists are currently exploring corrosion-resistant materials for components as well as fuel cell designs that increase cell life without decreasing performance.

MTU Fuel Cell

The German company MTU Friedrichshafen presented an MCFC at the Hannover Fair in 2006. The unit weighs 2 tonnes and can produce 240 kW of electric power from various gaseous fuels, including biogas. If fueled by fuels that contain carbon such as natural gas, the exhaust will contain CO_2 but will be reduced by up to 50% compared to diesel engines running on marine bunker fuel. The exhaust temperature is 400°C, hot enough to be used for many industrial processes. Another possibility is to make more electric power *via* a steam turbine. Depending on feed gas type, the electric efficiency is between 12% and 19%. A steam turbine can increase the efficiency by up to 24%. The unit can be used for co-generation.

Table : Efficiency of leading fuel cell types.

	Energy Efficiency & Renewable Energy	FUEL CELL TECHNOLOGIES PROGRAM					
Comparison of Fuel Cell Technologies							
Fuel Cell Type	Common Electrolyte	Operating Temperature	Typical Stack Size	Efficiency	Applications	Advantages	Disadvantages
Polymer Electrolyte Membrane (PEM)	Perfluoro sulfonic acid	50-100°C 122-212° typically 80°C	< 1kW-100kW	60% transportation 35% stationary	• Backup power • Portable power • Distributed generation • Transporation • Specialty vehicles	• Solid electrolyte reduces corrosion & electrolyte management problems • Low temperature • Quick start-up	• Expensive catalysts • Sensitive to fuel impurities • Low temperature waste heat
Alkaline (AFC)	Aqueous solution of potassium hydroxide soaked in a matrix	90-100°C 194-212°F	10-100 kW	60%	• Military • Space	• Cathode reaction faster in alkaline electrolyte, leads to high performance • Low cost components	• Sensitive to CO₂ in fuel and air • Electrolyte management
Phosphoric Acid (PAFC)	Phosphoric acid soaked in a matrix	150-200°C 302-392°F	400 kW 100 kW module	40%	• Distributed generation	• Higher temperature enables CHP • Increased tolerance to fuel impurities	• Pt catalyst • Long start up time • Low current and power
Molten Carbonate (MCFC)	Solution of lithium, sodium, and/ or potassium carbonates, soaked in a matrix	600-700°C 1112-1292°F	300 kW- 3 MW 300 kW module	45-50%	• Electric utility • Distributed generation	• High efficiency • Fuel flexibility • Can use a variety of catalysts • Suitable for CHP	• High temperature corrosion and breakdown of cell components • Long start up time • Low power density
Solid Oxide (SOFC)	Yttria stabilized zirconia	700-1000°C 1202-1832°F	1 kW- 2 MW	60%	• Auxiliary power • Electric utility • Distributed generation	• High efficiency • Fuel flexibility • Can use a variety of catalysts • Solid electrolyte • Suitable for CHP & CHHP • Hybrid/GT cycle	• High temperature corrosion and breakdown of cell components • High temperature operation requires long start up time and limits

For More Information
More information on the Fuel Cell Technologies Program is available at *http://www.hydrogenandfuelcells.energy.gov*.

ENERGY U.S. DEPARTMENT OF
Energy Efficiency & Renewable Energy

EERE Information Center
1-877-EERE-INFO (1-877-337-3463)
www.eere.energy.gov/informationcenter

Glossary of Terms in Table :

- *Anode :* The electrode at which oxidation (a loss of electrons) takes place. For fuel cells and other galvanic cells, the anode is the negative terminal; for electrolytic cells (where electrolysis occurs), the anode is the positive terminal.

- *Aqueous solution :* a : of, relating to, or resembling water b : made from, with, or by water.

- *Catalyst :* A chemical substance that increases the rate of a reaction without being consumed; after the reaction, it can potentially be recovered from the reaction mixture and is chemically unchanged. The catalyst lowers the activation energy required, allowing the reaction to proceed more quickly or at a lower temperature. In a fuel cell, the catalyst facilitates the reaction of oxygen and hydrogen. It is usually made of platinum powder very thinly coated onto carbon paper or cloth. The catalyst is rough and porous so the maximum surface area of the platinum can be exposed to the hydrogen or oxygen. The platinum-coated side of the catalyst faces the membrane in the fuel cell.

- *Cathode :* The electrode at which reduction (a gain of electrons) occurs. For fuel cells and other galvanic cells, the cathode is the positive terminal; for electrolytic cells (where electrolysis occurs), the cathode is the negative terminal.

- *Electrolyte :* A substance that conducts charged ions from one electrode to the other in a fuel cell, battery, or electrolyzer.

- *Fuel Cell Stack :* Individual fuel cells connected in a series. Fuel cells are stacked to increase voltage.

- *Matrix :* Something within or from which something else originates, develops, or takes form.

- *Membrane :* The separating layer in a fuel cell that acts as electrolyte (an ion-exchanger) as well as a barrier film separating the gases in the anode and cathode compartments of the fuel cell.

- *Molten Carbonate Fuel Cell (MCFC) :* A type of fuel cell that contains a molten carbonate electrolyte. Carbonate ions (CO_3^{2-}) are transported from the cathode to the anode. Operating temperatures are typically near 650°C.

- *Phosphoric acid fuel cell (PAFC) :* A type of fuel cell in which the electrolyte consists of concentrated phosphoric acid (H_3PO_4). Protons (H+) are transported from the anode to the cathode. The operating temperature range is generally 160–220°C.

- *Polymer Electrolyte Membrane (PEM) :* A fuel cell incorporating a solid polymer membrane used as its electrolyte. Protons (H+) are transported from the anode to the cathode. The operating temperature range is generally 60–100°C.

- *Solid Oxide Fuel Cell (SOFC) :* A type of fuel cell in which the electrolyte is a solid, non-porous metal oxide, typically zirconium oxide (ZrO_2) treated with Y_2O_3, and O^{2-} is transported from the cathode to the anode. Any CO in the reformate gas is oxidized to CO_2 at the anode. Temperatures of operation are typically 800–1,000°C.

- *Solution : a :* an act or the process by which a solid, liquid, or gaseous substance is homogeneously mixed with a liquid or sometimes a gas or solid, *b :* a homogeneous mixture formed by this process; especially : a single-phase liquid system, c : the condition of being dissolved

Theoretical Maximum Efficiency

The energy efficiency of a system or device that converts energy is measured by the ratio of the amount of useful energy put out by the system ("output energy") to the total amount of energy that is put in ("input energy") or by useful output energy as a percentage of the total input energy. In the case of fuel cells, useful output energy is measured in electrical energy produced by the system. Input energy is the energy stored in the fuel. According to the U.S. Department of Energy, fuel cells are generally between 40–60% energy efficient. This is higher than some other systems for energy generation. For example, the typical internal combustion engine of a car is about 25% energy efficient. In combined heat and power (CHP) systems, the heat produced by the fuel cell is captured and put to use, increasing the efficiency of the system to up to 85–90%.

The theoretical maximum efficiency of any type of power generation system is never reached in practice, and it does not consider other steps in power generation, such as production, transportation and storage of fuel and conversion of the electricity into mechanical power. However, this calculation allows the comparison of different types of power generation. The maximum theoretical energy efficiency of a fuel cell is 83%, operating at low power density and using pure hydrogen and oxygen as reactants (assuming no heat recapture). According to the World Energy Council, this compares with a maximum theoretical efficiency of 58% for internal combustion engines. While these efficiencies are not approached in most real world applications, high-temperature fuel cells (solid oxide fuel cells or molten carbonate fuel cells) can theoretically be combined with gas turbines to allow stationary fuel cells to come closer to the theoretical limit. A gas turbine would capture heat from the fuel cell and turn it into mechanical energy to increase the fuel cell's operational efficiency. This solution has been predicted to increase total efficiency to as much as 70%.

In Practice

The tank-to-wheel efficiency of a fuel-cell vehicle is greater than 45% at low loads and shows average values of about 36% when a driving cycle like the NEDC (New European Driving Cycle) is used as test procedure. The comparable NEDC value for a Diesel vehicle is 22%. In 2008 Honda released a demonstration fuel cell electric vehicle (the Honda FCX Clarity) with fuel stack claiming a 60% tank-to-wheel efficiency.

It is also important to take losses due to fuel production, transportation, and storage into account. Fuel cell vehicles running on compressed hydrogen may have a power-plant-to-wheel efficiency of 22% if the hydrogen is stored as high-pressure gas, and 17% if it is stored as liquid hydrogen. Fuel cells cannot store energy like a battery, except as hydrogen, but in some applications, such as stand-alone power plants based on discontinuous sources such as solar or wind power, they are combined with electrolyzers and storage systems to form an energy storage system. Most hydrogen, however, is produced by steam methane reforming, and so most hydrogen production emits carbon dioxide. The overall efficiency (electricity to hydrogen and back to electricity) of such plants (known as *round-trip efficiency*), using pure hydrogen and pure oxygen can be "from 35 up to 50 per cent", depending on gas density and other conditions. While a much cheaper lead–acid battery might return about 90%, the electrolyzer/fuel cell system can store indefinite quantities of hydrogen, and is therefore better suited for long-term storage.

Solid-oxide fuel cells produce exothermic heat from the recombination of the oxygen and hydrogen. The ceramic can run as hot as 800 degrees Celsius. This heat can be captured and used to heat water in a micro-combined heat and power (m-CHP) application. When the heat is captured, total efficiency can reach 80–90% at the unit, but does not consider production and distribution losses. CHP units are being developed today for the European home market.

Professor Jeremy P. Meyers, in the Electro-chemical Society journal *Interface* in 2008, wrote, "While fuel cells are efficient relative to combustion engines, they are not as efficient as batteries, due primarily to the inefficiency of the oxygen reduction reaction (and... the oxygen evolution reaction, should the hydrogen be formed by electrolysis of water).... [T]hey make the most sense for operation disconnected from the grid, or when fuel can be provided continuously. For applications that require frequent and relatively rapid start-ups... where zero emissions are a requirement, as in enclosed spaces such as warehouses, and where hydrogen is considered an acceptable reactant, a [PEM fuel cell] is becoming an increasingly attractive choice [if exchanging batteries is inconvenient]". In 2013 military organisations are evaluating fuel cells to significantly reduce the battery weight carried by soldiers.

Power

Stationary fuel cells are used for commercial, industrial and residential primary and backup power generation. Fuel cells are very useful as power sources in remote locations, such as spacecraft, remote weather stations, large parks, communications centers, rural locations including research stations, and in certain military applications. A fuel cell system running on hydrogen can be compact and lightweight, and have no major moving parts. Because fuel cells have no moving parts and do not involve combustion, in ideal conditions they can achieve up to 99.9999% reliability. This equates to less than one minute of downtime in a six-years period.

Since fuel cell electrolyzer systems do not store fuel in themselves, but rather rely on external storage units, they can be successfully applied in large-scale energy storage, rural areas being one example. There are many different types of stationary fuel cells so efficiencies vary, but most are between 40% and 60% energy efficient. However, when the fuel cell's waste heat is used to heat a building in a co-generation system this efficiency can increase to 85%. This is significantly more efficient than traditional coal power plants, which are only about one third energy efficient. Assuming production at scale, fuel cells could save 20–40% on energy costs when used in co-generation systems. Fuel cells are also much cleaner than traditional power generation; a fuel cell power plant using natural gas as a hydrogen source would create less than one ounce of pollution (other than CO_2) for every 1,000 kW · h produced, compared to 25 pounds of pollutants generated by conventional combustion systems. Fuel Cells also produce 97% less nitrogen oxide emissions than conventional coal-fired power plants.

Coca-Cola, Google, Walmart, Sysco, FedEx, UPS, Ikea, Staples, Whole Foods, Gills Onions, Nestle Waters, Pepperidge Farm, Sierra Nevada Brewery, Super Store Industries, Brigestone-Firestone, Nissan North America, Kimberly-Clark, Michelin and more have installed fuel cells to help meet their power needs. One such pilot program is operating on Stuart Island in Washington State. There the Stuart Island Energy Initiative has built a complete, closed-loop system : Solar panels power an electrolyzer, which makes hydrogen. The hydrogen is stored in a 500-U.S.-gallon (1,900 L) tank at 200 pounds per square inch (1,400 kPa), and runs a ReliOn fuel cell to provide full electric back-up to the off-the-grid residence. Another closed system loop was unveiled in late 2011 in Hempstead, NY.

Fuel cells can be used with low-quality gas from landfills or waste-water treatment plants to generate power and lower methane emissions. A 2.8 MW fuel cell plant in California is said to be the largest of the type.

Co-generation

Combined heat and power (CHP) fuel cell systems, including Micro combined heat and power (MicroCHP) systems are used to generate both electricity and heat for homes, office building and factories. The system generates constant electric power (selling excess power back to the grid when it is not consumed), and at the same time produces hot air and water from the waste heat. As the result CHP systems have the potential to save primary energy as they can make use of waste heat which is generally rejected by thermal energy conversion systems. A typical capacity range of home fuel cell is 1–3 kW_{el}/4–8 kW_{th}. CHP systems linked to absorption chillers use their waste heat for refrigeration.

The waste heat from fuel cells can be diverted during the summer directly into the ground providing further cooling while the waste heat during winter can be pumped directly into the building. The University of Minnesota owns the patent rights to this type of system

Co-generation systems can reach 85% efficiency (40–60% electric + remainder as thermal). Phosphoric-acid fuel cells (PAFC) comprise the largest segment of existing CHP products worldwide and can provide combined efficiencies close to 90%. Molten Carbonate (MCFC) and Solid Oxide Fuel Cells (SOFC) are also used for combined heat and power generation and have electrical energy efficiencies around 60%. Disadvantages of co-generation systems include slow ramping up and down rates, high cost and short lifetime. Also their need to have a hot water storage tank to smooth out the thermal heat production was a serious disadvantage in the domestic market place where space in domestic properties is at a great premium.

Fuel Cell Electric Vehicles (FCEVs)

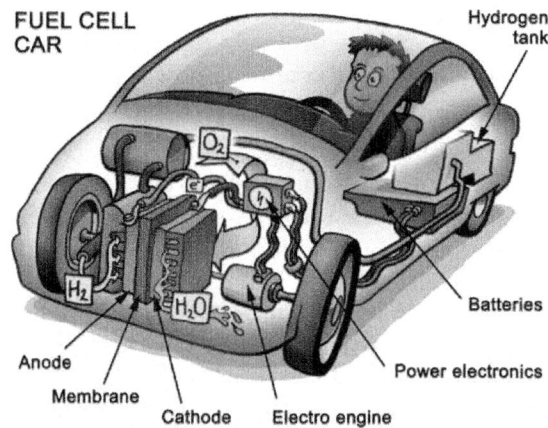

Fig. : Configuration of components in a fuel cell car.

Automobiles

Although there are currently no Fuel cell vehicles available for commercial sale, over 20 FCEVs prototypes and demonstration cars have been released since 2009. Demonstration models include the Honda FCX Clarity, Toyota FCHV-adv, and Mercedes-Benz F-Cell. As of June 2011 demonstration FCEVs had driven more than 4,800,000 km (3,000,000 mi), with more than 27,000 refuelings. Demonstration fuel cell vehicles have been produced with "a driving range of more than 400 km (250 mi) between refueling". They can be refueled in less than 5 minutes. The U.S. Department of Energy's Fuel Cell Technology Program claims that, as of 2011, fuel cells achieved 53–59% efficiency at one-quarter power and 42–53% vehicle efficiency at full power, and a durability of over 120,000 km (75,000 mi) with less than 10% degradation. In a Well-to-Wheels simulation analysis, that "did not address the economics and market constraints", General Motors and its partners estimated that per mile travelled, a fuel cell electric vehicle running on compressed gaseous hydrogen produced from natural gas could use about 40% less energy and emit 45% less greenhouse gasses than an internal combustion vehicle. A lead engineer from the Department of Energy whose team is testing fuel cell cars said in 2011 that the potential appeal is that "these are full-function vehicles with no limitations on range or refueling rate so they are a direct replacement for any vehicle. For instance, if you drive a full sized SUV and pull a boat up into the mountains, you can do that with this technology and you can't with current battery-only vehicles, which are more geared toward city driving."

Some experts believe, however, that fuel cell cars will never become economically competitive with other technologies or that it will take decades for them to become profitable. In July 2011, the chairman and CEO of General Motors, Daniel Akerson, stated that while the cost of hydrogen fuel cell cars is decreasing : "The car is still too expensive and probably won't be practical until the 2020-plus period, I don't know."

In 2012, Lux Research, Inc. issued a report that stated : "The dream of a hydrogen economy... is no nearer." It concluded that "Capital cost... will limit adoption to a mere 5.9 GW" by 2030, providing "a nearly insurmountable barrier to adoption, except in niche applications". The analysis concluded that, by 2030, PEM stationary market will reach $1 billion, while the vehicle market, including forklifts, will reach a total of $2 billion. Other analyses cite the lack of an extensive hydrogen infrastructure in the U.S. as an ongoing challenge to Fuel Cell Electric Vehicle commercialization. In 2006, a study for the IEEE showed that for hydrogen produced *via* electrolysis of water : "Only about 25% of the power generated from wind, water, or sun is converted to practical use." The study further noted that "Electricity obtained from hydrogen fuel cells appears to be four times as expensive as electricity drawn from the electrical transmission grid.... Because of the high energy losses [hydrogen] cannot compete with electricity." Furthermore, the study found : "Natural gas reforming is not a sustainable solution". "The large amount of energy required to isolate hydrogen from natural compounds (water, natural gas, biomass), package the light gas by compression or liquefaction, transfer the energy carrier to the user, plus the energy lost when it is converted to useful electricity with fuel cells, leaves around 25% for practical use."

Despite this, several major car manufacturers have announced plans to introduce a production model of a fuel cell car in 2015. In 2013, Toyota has stated that it plans to introduce such a vehicle at a price of less than US$100,000. Mercedes-Benz announced that they would move the scheduled production date of their fuel cell car from 2015 up to 2014, asserting that "The product is ready for the market technically.... The issue is infrastructure." At the Paris Auto Show in September 2012, Hyundai announced that it plans to begin producing a commercial production fuel cell model (based on the ix35) in December 2012 and hopes to deliver 1,000 of them by 2015. Other manufacturers planning to sell fuel cell electric vehicles commercially by 2016 or earlier include General Motors (2015), Honda, and Nissan (2016).

The Obama Administration sought to reduce funding for the development of fuel cell vehicles, concluding that other vehicle technologies will lead to quicker reduction in emissions in a shorter time. Steven Chu, the United States Secretary of Energy, stated in 2009 that hydrogen vehicles "will not be practical over the next 10 to 20 years". In 2012, however, Chu stated that he saw fuel cell cars as more economically feasible as natural gas prices have fallen and hydrogen reforming technologies have improved.

Buses

As of August 2011, there were a total of approximately 100 fuel cell buses deployed around the world. Most buses are produced by UTC Power, Toyota, Ballard, Hydrogenics, and Proton Motor. UTC Buses had accumulated over 970,000 km (600,000 mi) of driving by 2011. Fuel cell buses have a 39–141% higher fuel economy than diesel buses and natural gas buses. Fuel cell buses have been deployed around the world including in Whistler, Canada; San Francisco, United States; Hamburg, Germany; Shanghai, China; London, England; São Paulo, Brazil; as well as several others. The Fuel Cell Bus Club is a global cooperative effort in trial fuel cell buses. Notable Projects Include :

- 12 Fuel cell buses are being deployed in the Oakland and San Francisco Bay area of California.
- Daimler AG, with thirty-six experimental buses powered by Ballard Power Systems fuel cells completed a successful three-year trial, in eleven cities, in January 2007.
- A fleet of Thor buses with UTC Power fuel cells was deployed in California, operated by SunLine Transit Agency.

The first Brazilian hydrogen fuel cell bus prototype in Brazil was deployed in São Paulo. The bus was manufactured in Caxias do Sul and the hydrogen fuel will be produced in São Bernardo do Campo from water through electrolysis. The program, called "*Ônibus Brasileiro a Hidrogênio*" (Brazilian Hydrogen Autobus), includes three additional buses.

Forklifts

A fuel cell forklift (also called a fuel cell lift truck) is a fuel cell powered industrial forklift truck used to lift and transport materials. Most fuel cells used for material handling purposes are powered by PEM fuel cells.

In 2013 there were over 4,000 fuel cell forklifts used in material handling in the USA, of which only 500 received funding from DOE (2012). Fuel cell fleets are operated by a large number of companies, including Sysco Foods, FedEx Freight, GENCO (at Wegmans, Coca-Cola, Kimberly Clark, and Whole Foods), and H-E-B Grocers. Europe demonstrated 30 Fuel cell forklifts with Hylift and extended it with HyLIFT-EUROPE to 200 units, with other projects in France and Austria. Pike Research stated in 2011 that fuel-cell-powered forklifts will be the largest driver of hydrogen fuel demand by 2020.

PEM fuel-cell-powered forklifts provide significant benefits over both petroleum and battery powered forklifts as they produce no local emissions, can work for a full 8-hour shift on a single tank of hydrogen, can be refueled in 3 minutes and have a lifetime of 8–10 years. Fuel cell-powered forklifts are often used in refrigerated warehouses, as their performance is not degraded by lower temperatures. Many companies do not use petroleum powered forklifts, as these vehicles work indoors where emissions must be controlled and instead are turning to electric forklifts. In design the FC units are often made as drop-in replacements.

Motorcycles and Bicycles

In 2005 a British manufacturer of hydrogen-powered fuel cells, Intelligent Energy (IE), produced the first working hydrogen run motorcycle called the ENV (Emission Neutral Vehicle). The motorcycle holds enough fuel to run for four hours, and to travel 160 km (100 mi) in an urban area, at a top speed of 80 km/h (50 mph). In 2004 Honda developed a fuel-cell motorcycle that utilized the Honda FC Stack.

Other examples of motorbikes and bicycles that use hydrogen fuel cells include the Taiwanese company APFCT's scooter using the fueling system from Italy's Acta SpA and the Suzuki Burgman scooter with an IE fuel cell that received EU Whole Vehicle Type Approval in 2011. Suzuki Motor Corp. and IE have announced a joint venture to accelerate the commercialization of zero-emission vehicles.

Airplanes

Boeing researchers and industry partners throughout Europe conducted experimental flight tests in February 2008 of a manned airplane powered only by a fuel cell and lightweight batteries. The fuel cell demonstrator airplane, as it was called, used a proton exchange membrane (PEM) fuel cell/lithium-ion battery hybrid system to power an electric motor, which was coupled to a conventional propeller. In 2003, the world's first propeller-driven airplane to be powered entirely by a fuel cell was flown. The fuel cell was a unique FlatStack stack design, which allowed the fuel cell to be integrated with the aerodynamic surfaces of the plane.

There have been several fuel-cell-powered unmanned aerial vehicles (UAV). A Horizon fuel cell UAV set the record distance flown for a small UAV in 2007. The military is especially interested in this application because of the low noise, low thermal signature and ability to attain high altitude. In 2009 the Naval Research Laboratory's (NRL's) Ion Tiger utilized a hydrogen-powered fuel cell and flew for 23 hours and 17 minutes. Fuel cells are also being used to provide auxiliary power in aircraft, replacing fossil fuel generators that were previously used to start the engines and power on board electrical needs. Fuel cells can help airplanes reduce CO_2 and other pollutant emissions and noise.

Boats

The world's first fuel-cell boat HYDRA used an AFC system with 6.5 kW net output. Iceland has committed to converting its vast fishing fleet to use fuel cells to provide auxiliary power by 2015 and, eventually, to provide primary power in its boats. Amsterdam recently introduced its first fuel-cell-powered boat that ferries people around the city's famous and beautiful canals.

Submarines

The Type 212 submarines of the German and Italian navies use fuel cells to remain submerged for weeks without the need to surface.

The U212A is a non-nuclear submarine developed by German naval shipyard Howaldtswerke Deutsche Werft. The system consists of nine PEM fuel cells, providing between 30 kW and 50 kW each. The ship is silent giving it an advantage in the detection of other submarines.

Portable Power Systems

Portable power systems that use fuel cells can be used in the leisure sector (*i.e.* RV's, Cabins, Marine), the industrial sector (*i.e.* power for remote locations including gas/oil wellsites, communication towers, security, weather stations etc.), and in the military sector. SFC Energy is a German manufacturer of direct methanol fuel cells for a variety of portable power systems. Ensol Systems Inc. is an integrator of portable power systems, using the SFC Energy DMFC.

Other Applications

- Providing power for base stations or cell sites
- Distributed generation
- Emergency power systems are a type of fuel cell system, which may include lighting, generators and other apparatus, to provide backup resources in a crisis or when regular systems fail. They find uses in a wide variety of settings from residential homes to hospitals, scientific laboratories, data centers,
- Telecommunication equipment and modern naval ships.
- An uninterrupted power supply (**UPS**) provides emergency power and, depending on the topology, provide line regulation as well to connected equipment by supplying power from a separate source when utility power is not available. Unlike a standby generator, it can provide instant protection from a momentary power interruption.
- Base load power plants
- Solar Hydrogen Fuel Cell Water Heating
- Hybrid vehicles, pairing the fuel cell with either an ICE or a battery.
- Notebook computers for applications where AC charging may not be readily available.
- Portable charging docks for small electronics (*e.g.* a belt clip that charges your cell phone or PDA).
- Smartphones, laptops and tablets.
- Small heating appliances
- Food preservation, achieved by exhausting the oxygen and automatically maintaining oxygen exhaustion in a shipping container, containing, for example, fresh fish.
- Breathalyzers, where the amount of voltage generated by a fuel cell is used to determine the concentration of fuel (alcohol) in the sample.
- Carbon monoxide detector, electro-chemical sensor.

Fueling Stations

There were over 85 hydrogen refueling stations in the U.S. in 2010.

As of June 2012 California had 23 hydrogen refueling stations in operation. Honda announced plans in March 2011 to open the first station that would generate hydrogen through solar-powered renewable electrolysis. South Carolina also has two hydrogen fueling stations, in Aiken and Columbia, SC. The University of South Carolina, a founding member of the South Carolina Hydrogen & Fuel Cell Alliance, received 12.5 million dollars from the United States Department of Energy for its Future Fuels Program.

The first public hydrogen refueling station in Iceland was opened in Reykjavík in 2003. This station serves three buses built by Daimler Chrysler that are in service in the public transport net of Reykjavík. The station produces the hydrogen it needs by itself, with an electrolyzing unit (produced by Norsk Hydro), and does not need refilling : all that enters is electricity and water. Royal Dutch Shell is also a partner in the project. The station has no roof, in order to allow any leaked hydrogen to escape to the atmosphere.

The current 14 stations nationwide in Germany are planned to be expanded to 50 by 2015 through its public private partnership Now GMBH. Japan also has a hydrogen highway, as part of the Japan hydrogen fuel cell project. Twelve hydrogen fueling stations have been built in 11 cities in Japan, and additional hydrogen stations could potentially be operational by 2015. Canada, Sweden and Norway also have hydrogen highways being implemented.

Markets and Economics

In 2012, fuel cell industry revenues exceeded $1 billion market value worldwide, with Asian pacific countries shipping more than 3/4 of the fuel cell systems worldwide. However, as of October 2013, no public company in the industry had yet become profitable. There were 140,000 fuel cell stacks shipped globally in 2010, up from 11 thousand shipments in 2007, and from 2011 to 2012 worldwide fuel cell shipments had an annual growth rate of 85%. Tanaka Kikinzoku Kogyo K.K. expanded its production facilities for fuel cell catalysts in 2013 to meet anticipated demand as the Japanese ENE Farm scheme expects to install 50,000 units in 2013 and the company is experiencing rapid market growth.

Approximately 50% of fuel cell shipments in 2010 were stationary fuel cells, up from about a third in 2009, and the four dominant producers in the Fuel Cell Industry were the United States, Germany, Japan and South Korea. The Department of Energy Solid State Energy Conversion Alliance found that, as of January 2011, stationary fuel cells generated power at approximately $724 to $775 per kilowatt installed. In 2011, Bloom Energy, a major fuel cell supplier, said that its fuel cells generated power at 9–11 cents per kilowatt-hour, including the price of fuel, maintenance, and hardware.

Industry groups predict that there are sufficient platinum resources for future demand, and in 2007, research at Brookhaven National Laboratory suggested that platinum could be replaced by a gold-palladium coating, which may be less susceptible to poisoning and thereby improve fuel cell lifetime. Another method would use iron and sulphur instead of platinum. This would lower the cost of a fuel cell (as the platinum in a regular fuel cell costs around US$1,500, and the same amount of iron costs only around US$1.50). The concept was being developed by a coalition of the John Innes Centre and the University of Milan-Bicocca. PEDOT cathodes are immune to monoxide poisoning.

Research and Development

- *August 2005 :* Georgia Institute of Technology researchers use triazole to raise the operating temperature of PEM fuel cells from below 100°C to over 125°C, claiming this will require less carbon-monoxide purification of the hydrogen fuel.
- *2008 :* Monash University, Melbourne uses PEDOT as a cathode.
- *2009 :* Researchers at the University of Dayton, in Ohio, show that arrays of vertically grown carbon nanotubes could be used as the catalyst in fuel cells.
- *2009 :* Y-Carbon began to develop a carbide-derived-carbon-based ultra-capacitor, which they hoped would lead to fuel cells with higher energy density.
- *2009 :* A nickel bisdiphosphine-based catalyst for fuel cells is demonstrated.
- *2013 :* British firm ACAL Energy develops a fuel cell that it says runs for 10,000 hours in simulated driving conditions. It asserts that the cost of fuel cell construction can be reduced to $40/kW (roughly $9,000 for 300 HP).

FUEL CELL TYPES

Metal Hydride Fuel Cell

Metal hydride fuel cells are a subclass of *alkaline fuel cells* that are currently in the research and development phase. A notable feature is their ability to *chemically bond* and *store hydrogen* within the cell. This feature is shared with *direct borohydride fuel cells*, although the two differ in that MHFCs are refueled with pure hydrogen. Though the absorption characteristics of metal hydrides (around 2%) is far lower than sodium-borohydrides and other "light" metal hydrides (around 10,8%)[1], prototypes have been claimed to demonstrate a number of interesting characteristics :

- Ability to be recharged with electrical energy (similar to *NiMH batteries*);
- Low operating temperatures (down to -20°C);
- Fast kinetics;
- Extended shelf life;
- Fast "cold start" properties;
- Ability to operate for limited periods of time with no external hydrogen source, enabling "hot swapping" of fuel canisters.

Metal hydrides fuel cells are currently being researched by ECD Ovonics, as well as by the Japanese *National Institute of Advanced Industrial Science and Technology* (AIST). Though similar, the two MHFC concepts use different catalysts. Thus far, neither research project has produced a demonstratable model outside of a laboratory-only publications and patents-and significant efficiency hurdles have yet to be overcome. The Ovonics and AIST metal hydride fuel cells claim current densities of only 250mA/cm^2 and 20mA/cm^2, respectively, versus typical *PEMFC* performance at 1A/cm^2.

Electro-galvanic Fuel Cell

An **electro-galvanic fuel cell** is an electrical device, one form of which is commonly used to measure the concentration of oxygen gas in scuba diving and medical equipment.

A chemical reaction occurs in the fuel cell when the potassium hydroxide in the cell comes into contact with oxygen. This creates an electric current between the lead anode and the gold-plated cathode through a load resistance. The current produced is proportional to the concentration (partial pressure) of oxygen present.

They are used in oxygen analysers in technical diving to display the proportion of oxygen in a nitrox or trimix breathing gas before a dive. They are also used in electronic, closed-circuit rebreathers to monitor the oxygen partial pressure during the dive.

The partial pressure of oxygen in diving chambers and surface supplied breathing gas mixtures can also be monitored using these cells. This can either be done by placing the cell directly in the hyperbaric environment, wired through the hull to the monitor, or indirectly, by bleeding off gas from the hyperbaric environment or diver gas supply and analysing at atmospheric pressure, then calculating the partial pressure in the hyperbaric environment. This is frequently required in saturation diving and surface oriented surface supplied mixed gas commercial diving.

Electro-galvanic fuel cells have a limited lifetime which is reduced by exposure to high concentrations of oxygen. The reaction between oxygen and lead at the anode consumes lead, which eventually results in the cell failing to sense high concentrations of oxygen. Typically, a cell used for diving applications will function correctly for 3 years if stored in a sealed bag of air but only for four months if stored in pure oxygen.

Cell Limitations

Oxygen cells behave in a similar way to electrical batteries in that they have a finite lifespan which is dependent upon use. The chemical reaction described above causes the cell to create an electrical output that has a predicted voltage which is dependent on the materials used. In theory they should give that voltage from the day they are made until they are exhausted, except that one component of the planned chemical reaction has been left out of the assembly : oxygen.

Oxygen is one of the fuels of the cell so the more oxygen there is, the more electricity is generated. The chemistry sets the voltage and the fuel, the oxygen, sets how much electric current it can give. If you put an electric circuit on the cell that draws current you can draw up to this current but ask for more and the voltage from the cell fades.

Failures in cells can be life threatening for technical divers and in particular, rebreather divers. The failure modes common to these cells are : failing with a higher than expected output due to electrolyte leaks, current limitation due to exhausted cell life and non-linear output across its range. These failures are usually attributable to physical damage, contamination during manufacture or defects in manufacture.

Failing high is invariably a result of a manufacturing fault or mechanical damage. In rebreathers, failing high will result in the rebreather assuming that there is more oxygen in the loop than there actually is which results in hypoxia.

Current limited cells do give a high output in high concentrations of oxygen. The rebreather assumes there is insufficient oxygen in the loop and injects to reach a setpoint the cell will never achieve resulting in hyperoxia. Non-linear cells do not perform in an expected manner across its range of oxygen partial pressures. Calibration will not pick up this fault which results in inaccurate loop contents of a rebreather. This gives the potential for decompression illness.

Preventing accidents in rebreathers from cell failures is possible in most cases by accurately testing the cells before use. Some divers carry out in-water checks by pushing the oxygen content in the loop to a pressure that is above that of pure oxygen at sea level to indicate if the cell is capable of high outputs. This test is only a spot check and does not accurately assess the quality of prediction of failure of that cell. The only way to accurately test a cell is with a calibrated test chamber which can hold a static pressure without deviation and the ability to log the results and graph them.

Testing

The first certified cell checking device that was commercially available was launched in 2005 by Narked at 90 but did not achieve commercial success. A much revised model was released in 2007 and won the "Gordon Smith Award" for Innovation at the Diving Equipment Manufacturers Exhibition in Florida. Narked at 90 Ltd won the Award for Innovation for the Development of Advanced Diving products at Eurotek 2010 for the Cell Checker and its continuing Development. Now used throughout the world by organisations such as Teledyne/Vandegraph National Oceanic and Atmospheric Administration, NURC (NATO Underwater Research Centre) and Diving Diseases Research Centre.

Formic Acid Fuel Cell

Direct-**formic acid fuel cells** or DFAFCs are a sub-category of *proton exchange membrane fuel cells* where, the fuel, formic acid, is not reformed, but fed directly to the fuel cell. Their applications include small, portable electronics such as phones and laptop computers.

Advantages

Similar to methanol, formic acid is a small organic molecule fed directly into the fuel cell, removing the need for complicated catalytic reforming. Storage of formic acid is much easier and safer than that of hydrogen because it does not need to be done at high pressures and (or) low temperatures, as formic acid is a liquid at standard temperature and pressure. Formic acid does not cross over the polymer membrane, so its efficiency can be higher than that of methanol.

Reactions

DFAFCs convert formic acid and oxygen into carbon dioxide and water to produce energy. Formic acid oxidation occurs at the anode on a catalyst layer. Carbon dioxide is formed and protons (H^+) are passed through the polymer membrane to react with oxygen on a catalyst layer located at the cathode. Electrons are passed through an external circuit from anode to cathode to provide power to an external device.

Anode : $HCOOH \rightarrow CO_2 + 2\,H^+ + 2\,e^-$

Cathode : $1/2\,O_2 + 2\,H^+ + 2\,e^- \rightarrow H_2O$

Net reaction : $HCOOH + 1/2\,O_2 \rightarrow CO_2 + H_2O$.

History

During previous investigations, researchers dismissed formic acid as a practical fuel because of the high over-potential shown by experiments : this meant the reaction appeared to be too difficult to be practical. However, in 2005-2006, other researchers (in particular Richard Masel's group at the University of Illinois at Urbana-Champaign) found that the reason for the low performance was the usage of platinum as a catalyst, as it is common in most other types of fuel cells : using palladium instead, they claim to have obtained better performance than equivalent direct methanol fuel cells. As of April 2006, Tekion held the exclusive license to formic-acid fuel cell technology from the University of Illinois at Urbana-Champaign, and with an investment from Motorola, was partnering with BASF to design and manufacture power packs by late 2007, but development appears to have stalled, and almost all information was removed from Tekion's web site before April 24, 2010.

Zinc–air Battery

Zinc–air batteries (non-rechargeable), and **zinc–air fuel cells** (mechanically-rechargeable) are metal-air batteries powered by oxidizing zinc with oxygen from the air. These batteries have high energy densities and are relatively inexpensive to produce. Sizes range from very small button cells for hearing aids, larger batteries used in film cameras that previously used mercury batteries, to very large batteries used for electric vehicle propulsion.

During discharge, a mass of zinc particles forms a porous anode, which is saturated with an electrolyte. Oxygen from the air reacts at the cathode and forms hydroxyl ions which migrate into the zinc paste and form zincate ($Zn(OH)2-4$), releasing electrons to travel to the cathode. The zincate decays into zinc oxide and water returns to the electrolyte. The water and hydroxyl from the anode are recycled at the cathode, so the water is not consumed. The reactions produce a theoretical 1.65 volts, but this is reduced to 1.35–1.4 V in available cells.

Zinc–air batteries have some properties of fuel cells as well as batteries : the zinc is the fuel, the reaction rate can be controlled by varying the air flow, and oxidized zinc/electrolyte paste can be replaced with fresh paste.

Zinc-air batteries can be used to replace now discontinued 1.35 V mercury batteries (although with a significantly shorter operating life), which in the 1970s through 1980s were commonly used in photo cameras.

Possible future applications of this battery include its deployment as an electric vehicle battery and as a utility-scale energy storage system.

History

The effect of oxygen was known early in the 19th century when wet-cell Leclanche batteries absorbed atmospheric oxygen into the carbon cathode current collector. In 1878 a porous platinized carbon air electrode was found to work as well as the manganese dioxide (MnO_2) of the Leclanche cell. Commercial products began to be made on this principle in 1932 when George W. Heise and Erwin A. Schumacher of the National Carbon Company built cells, treating the carbon electrodes with wax to prevent flooding. This type is still used for large zinc–air cells for navigation aids and rail transportation. However, the current capacity is low and the cells are bulky.

Large primary zinc–air cells such as the Thomas A. Edison Industries *Carbonaire* type were used for railway signaling, remote communication sites, and navigation buoys.These were long-duration, low-rate applications. Development in the 1970s of thin electrodes based on fuel-cell research allowed application to small button and prismatic primary cells for hearing aids, pagers, and medical devices, especially cardiac telemetry.

First rechargeable zinc air batteries were manufactured in 1996 by a Slovenian innovator Miro Zoric. They were developed to power vehicles using the first AC based drive trains, also developed by Mr. Zoric. The first vehicles on roads to use zinc air batteries were small and mid sized buses in Singapore, where Mr. Zoric led the national electrification program at the Singapore Polytechnic university, during his technology transfer post. The Mass production assembly line for his zinc air batteries was put in place in 1997. The cells offered much higher energy density and specific energy (and weight) ratio, compared to then standard lead acid batteries.

Reaction Formulas

Here are the chemical equations for the zinc–air cell :

Anode : $Zn + 4OH^- \rightarrow Zn(OH)_4^{2-} + 2e^-$ ($E_0 = -1.25$ V)

Fluid : $Zn(OH)_4^{2-} \rightarrow ZnO + H_2O + 2OH^-$

Cathode : $1/2\ O_2 + H_2O + 2e^- \rightarrow 2OH^-$ ($E_0 = 0.34$ V pH=11)

Overall : $2Zn + O_2 \rightarrow 2ZnO$ ($E_0 = 1.59$ V)

Zinc–air batteries cannot be used in a sealed battery holder since some air must come in; the oxygen in 1 liter of air is required for every ampere-hour of capacity used.

Storage Density

Zinc-air batteries have higher energy density and specific energy (and weight) ratio than other types of battery because atmospheric air is one of the battery reactants. The air is not packaged with the battery, so that a cell can use more zinc in the anode than a cell that must also contain, for example, manganese dioxide. This increases capacity for a given weight or volume. As a specific example, a zinc–air battery of 11.6 mm diameter and height 5.4 mm from one manufacturer has a capacity of 620 mAh and weight 1.9 g; various silver oxide and alkaline cells of the same size supply 150–200 mAh and weigh 2.3–2.4 g.

Storage and Operating Life

Zinc-air cells have long shelf life if sealed to keep air out; even miniature button cells can be stored for up to 3 years at room temperature with little capacity loss if their seal is not removed. Industrial cells stored in a dry state have an indefinite storage life.

The operating life of a zinc–air cell is a critical function of its interaction with its environment. The electrolyte loses water more rapidly in conditions of high temperature and low humidity. Because the potassium hydroxide electrolyte is deliquescent, in very humid conditions excess water accumulates in the cell, flooding the cathode and destroying its active properties. Potassium hydroxide also reacts with atmospheric carbon dioxide; carbonate formation eventually reduces electrolyte conductivity. Miniature cells have high self-discharge once opened to air; the cell's capacity is intended to be used within a few weeks.

Discharge Properties

Because the cathode does not change properties during discharge, terminal voltage is quite stable until the cell approaches exhaustion.

Power capacity is a function of several variables : Cathode area, air availability, porosity, and the catalytic value of the cathode surface. Oxygen entry into the cell must be balanced against electrolyte water loss; cathode membranes are coated with (hydrophobic) Teflon material to limit water loss. Low humidity increases water loss; if enough water is lost the cell fails. Button cells have a limited current drain; for example an IEC PR44 cell has a capacity of 600 milliamp-hours (mAh) but a maximum current of only 22 milliamps (mA). Pulse load currents can be much higher since some oxygen remains in the cell between pulses.

Low temperature reduces primary cell capacity but the effect is small for low drains. A cell may deliver 80% of its capacity if discharged over 300 hours at 0°C (32°F), but only 20% of capacity if discharged at a 50 hour rate at that temperature. Lower temperature also reduces cell voltage.

Cell Types

Primary (Unrechargeable)

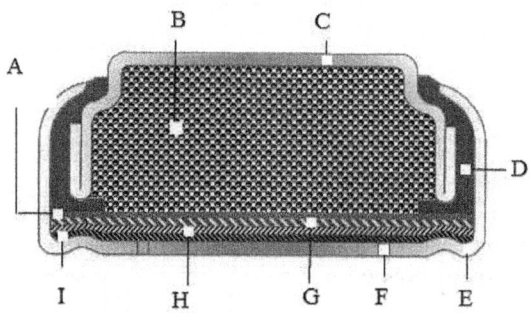

Fig. : Cross section through a zinc–air button cell. A : Separator, B : zinc powder anode and electrolyte, C : anode can, D : insulator gasket, E : cathode can, F : air hole, G : cathode catalyst and current collector, H : air distribution layer, I : Semi permeable membrane.

Large zinc–air batteries, with capacities up to 2,000 ampere–hours per cell, are used to power navigation instruments and marker lights, oceanographic experiments and railway signals.

Primary cells are made in button format to about 1 Ah. Prismatic shapes for portable devices are manufactured with capacities between 5 and 30 Ah. Hybrid cell cathodes include manganese dioxide to allow high peak currents.

Button cells are highly effective, but it is difficult to extend the same construction to larger sizes due to air diffusion performance, heat dissipation, and leakage problems. Prismatic and cylindrical cell designs address these problems. Stacking prismatic cells requires air channels in the battery and may require a fan to force air through the stack.

Secondary (Rechargeable)

Rechargeable zinc–air cells require zinc precipitation from the water-based electrolyte to be closely controlled. Challenges include dendrite formation, non-uniform zinc dissolution and limited solubility in electrolytes. Electrically reversing the reaction at a bi-functional air cathode, to liberate oxygen from discharge reaction products, is difficult; membranes tested to date have low overall efficiency. Charging voltage is much higher than discharge voltage, producing cycle energy efficiency as low as 50%. Providing charge and discharge functions by separate uni-functional cathodes, increases cell size, weight and complexity. A satisfactory electrically recharged system potentially offers low material cost and high specific energy. As of 2014, only one company has commercial units for sale, as described in a Dept. of Energy produced video at the ARPA-e Energy Innovation Summit in 2013. Fluidic Energy has apparently covered hundreds of thousands of outages in Asia at distributed critical load sites. And at least one firm claims to be in field tests for grid-scale backup applications.

Mechanical Recharge

Rechargeable systems may mechanically replace the anode and electrolyte, essentially operating as a refurbishable primary cell, or may use zinc powder or other methods to replenish the reactants. Mechanically-recharged systems were investigated for military electronics uses in the 1960s because of the high energy density and easy recharging. However, primary lithium batteries offered higher discharge rates and easier handling.

Mechanical recharging systems have been researched for decades for use in electric vehicles. Some approaches use a large zinc–air battery to maintain charge on a high discharge–rate battery used for peak loads during acceleration. Zinc granules serve as the reactant. Vehicles recharge *via* exchanging used electrolyte and depleted zinc for fresh reactants at a service station.

The term zinc–air fuel cell usually refers to a zinc–air battery in which zinc metal is added and zinc oxide is removed continuously. Zinc electrolyte paste or pellets are pushed into a chamber, and waste zinc oxide is pumped into a waste tank or bladder inside the fuel tank. Fresh zinc paste or pellets are taken from the fuel tank. The zinc oxide waste is pumped out at a refueling station for recycling. Alternatively, this term may refer to an electro-chemical system in which zinc is a co-reactant assisting the reformation of hydrocarbons at the anode of a fuel cell.

Materials

Catalysts

Cobalt oxide/carbon nanotube hybrid oxygen reduction catalyst and Nickel-iron layered double hydroxide oxygen evolution cathode catalysts exhibited higher catalytic activity and durability in concentrated alkaline electrolytes than precious metal Platinum and Iridium catalysts. The resulting primary zinc-air battery showed peak power density of ~265 mW cm−2, current density of ~200 mA cm−2 at 1 V and energy density >700 Wh kg−1.

Rechargeable Zn-air batteries in a tri-electrode configuration exhibited an unprecedented small charge–discharge voltage polarization of ~0.70 V at 20 mA cm−2, high reversibility and stability over long charge and discharge cycles.

Applications

Vehicle Propulsion

Metallic zinc could be used as an alternative fuel for vehicles, either in a zinc–air battery or to generate hydrogen near the point of use. Zinc's characteristics have motivated considerable interest as an energy source for electric vehicles. Gulf General Atomic demonstrated a 20 kW vehicle battery. General Motors conducted tests in the 1970s. Neither project led to a commercial product.

In addition to liquid, pellets that could be formed that are small enough to pump. Fuel cells using pellets would be able to quickly replace zinc-oxide with fresh zinc metal. The spent material can be recycled. The zinc–air cell is a primary cell (non-rechargeable); recycling is required to reclaim the zinc; much more energy is required to reclaim the zinc than is usable in a vehicle.

One advantage of utilizing zinc–air batteries for vehicle propulsion is that earth's supply of zinc metal is 100 times greater than that of lithium, per unit of battery energy. Current yearly global zinc production is sufficient to produce enough zinc-air batteries to power over one billion electric vehicles, whereas current lithium production is only sufficient to produce ten million lithium-ion powered vehicles. Approximately 35% of the world's supply, or 1.8 gigatons of zinc reserves are in the United States, whereas the U.S. holds only 0.38% of known lithium reserves.

Initial rechargeable zinc air batteries, developed for use in vehicles, were used for buses in Singapore. Their developer, Miro Zoric, chose zinc air chemistry specifically due to zinc air battery production requiring only abundant raw materials, which would, when used to power vehicular AC (induction) drive trains, which also do not require rare earth materials, allow global road transport electrification, without destabilizing global supply chains, or face and cause adverse raw material bottlenecks.

Grid Storage

The Eos Energy System battery is about half the size of a refrigerator and provides 1 MWh of storage. Con Edison, National Grid, Enel and GDF SUEZ began testing the battery for grid storage. Con Edison and City University of New York are testing a zinc-based battery from Urban Electric Power as part of a New York State Energy Research and Development Authority program. Eos projects that costs of $160 per kilowatt-hour and that it could provide electricity cheaper than a new natural gas power plant. Other battery technologies range from $400 to about $1,000 a kilowatt-hour.

Alternative Configurations

Attempts to address zinc–air's limitations include :

- Pumping zinc slurry through the battery in one direction for charging and reversing for discharge. Capacity is limited only by the slurry reservoir size.

- Alternate electrode shapes (*via* gelling and binding agents)
- Humidity management
- Careful catalyst dispersal to improve oxygen reduction and production
- Modularizing components for repair without complete replacement.

Safety and Environment

Zinc corrosion can produce potentially explosive hydrogen. Vent holes prevent pressure build-up within the cell. Manufacturers caution against hydrogen build-up in enclosed areas. A short-circuited cell gives relatively low current. Deep discharge below 0.5 V/cell may result in electrolyte leakage; little useful capacity exists below 0.9 V/cell.

Older designs used mercury amalgam amounting to about 1% of the weight of a button cell, to prevent zinc corrosion. Newer types have no added mercury. Zinc itself is relatively low in toxicity. Mercury-free designs require no special handling when discarded or recycled.

In United States waters, environmental regulations now require proper disposal of primary batteries removed from navigation aids. Formerly, discarded zinc–air primary batteries were dropped into the water around buoys, which allowed mercury to escape into the environment.

Microbial Fuel Cell

A **microbial fuel cell (MFC)** or **biological fuel cell** is a bio-*electro-chemical* system that drives a *current* by using *bacteria* and mimicking bacterial interactions found in *nature*. MFCs can be grouped into two general categories, those that use a mediator and those that are mediator-less. The first MFCs, demonstrated in the early 20th century, used a mediator : a chemical that transfers electrons from the bacteria in the cell to the anode. Mediator-less MFCs are a more recent development dating to the 1970s; in this type of MFC the bacteria typically have electro-chemically active *redox proteins* such as *cytochromes* on their outer membrane that can transfer electrons directly to the anode. Since the turn of the 21st century MFCs have started to find a commercial use in the treatment of wastewater.

History

The idea of using microbial cells in an attempt to produce electricity was first conceived in the early twentieth century. M. Potter was the first to perform work on the subject in 1911. A professor of botany at the University of Durham, Potter managed to generate electricity from *E. coli*, but the work was not to receive any major coverage. In 1931, however, Barnet Cohen drew more attention to the area when he created a number of microbial half fuel cells that, when connected in series, were capable of producing over 35 volts, though only with a current of 2 milliamps.

More work on the subject came with a study by DelDuca *et. al.* who used hydrogen produced by the fermentation of glucose by *Clostridium butyricum* as the reactant at the anode of a hydrogen and air fuel cell. Though the cell functioned, it was found to be unreliable owing to the unstable nature of hydrogen production by the micro-organisms. Although this issue was later resolved in work by Suzuki *et. al.* in 1976 the current design concept of an MFC came into existence a year later with work once again by Suzuki.

By the time of Suzuki's work in the late 1970s, little was understood about how microbial fuel cells functioned; however, the idea was picked up and studied later in more detail first by MJ Allen and then later by H. Peter Bennetto both from King's College London. People saw the fuel cell as a possible method for the generation of electricity for developing countries.

It is now known that electricity can be produced directly from the degradation of organic matter in a microbial fuel cell. Like a normal fuel cell, an MFC has both an anode and a cathode chamber. The anoxic anode chamber is connected internally to the cathode chamber *via* an ion exchange membrane with the circuit completed by an external wire.

In May 2007, the University of Queensland, Australia completed its prototype MFC as a cooperative effort with Foster's Brewing. The prototype, a 10 L design, converts brewery wastewater into carbon dioxide, clean water, and electricity. With the prototype proven successful, plans are in effect to produce a 660 gallon version for the brewery, which is estimated to produce 2 kilowatts of power. While this is a small amount of power, the production of clean water is of utmost importance to Australia, for which drought is a constant threat.

Types

Definition

A microbial fuel cell is a device that converts chemical energy to electrical energy by the catalytic reaction of micro-organisms.

A typical microbial fuel cell consists of anode and cathode compartments separated by a cation (positively charged ion) specific membrane. In the anode compartment, fuel is oxidized by micro-organisms, generating CO_2, electrons and protons. Electrons are transferred to the cathode compartment through an external electric circuit, while protons are transferred to the cathode compartment through the membrane. Electrons and protons are consumed in the cathode compartment, combining with oxygen to form water.

More broadly, there are two types of microbial fuel cell : mediator and mediator-less microbial fuel cells.

Mediator Microbial Fuel Cell

Most of the microbial cells are electro-chemically inactive. The electron transfer from microbial cells to the electrode is facilitated by mediators such as thionine, methyl viologen, methyl blue, humic acid, neutral red and so on. Most of the mediators available are expensive and toxic.

Mediator-free Microbial Fuel Cell

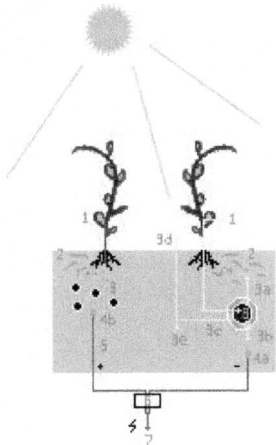

Fig. : A plant microbial fuel cell (PMFC).

Mediator-free microbial fuel cells do not require a mediator but use electro-chemically active bacteria to transfer electrons to the electrode (electrons are carried directly from the bacterial respiratory enzyme to the electrode). Among the electro-chemically active bacteria are, *Shewanella putrefaciens*, *Aeromonas hydrophila*, and others. Some bacteria, which have pili on their external membrane, are able to transfer their electron production *via* these pili. Mediator-less MFCs are a more recent area of research and, due to this, factors that affect optimum efficiency, such as the strain of bacteria used in the system, type of ion-exchange membrane, and system conditions (temperature, pH, etc.) are not particularly well understood.

Mediator-less microbial fuel cells can, besides running on wastewater, also derive energy directly from certain plants. This configuration is known as a plant microbial fuel cell. Possible plants include reed sweetgrass, cordgrass, rice, tomatoes, lupines, and algae. Given that the power is thus derived from living plants (in situ-energy production), this variant can provide additional ecological advantages.

Microbial Electrolysis Cell

A variation of the mediator-less MFC is the microbial electrolysis cells (MEC). Whilst MFC's produce electric current by the bacterial decomposition of organic compounds in water, MECs partially reverse the process to generate hydrogen or methane by applying a voltage to bacteria to supplement the voltage generated by the microbial decomposition of organics sufficiently lead to the electrolysis of water or the production of methane. A complete reversal of the MFC principle is found in microbial electrosynthesis, in which carbon dioxide is reduced by bacteria using an external electric current to form multi-carbon organic compounds.

Soil-based Microbial Fuel Cell

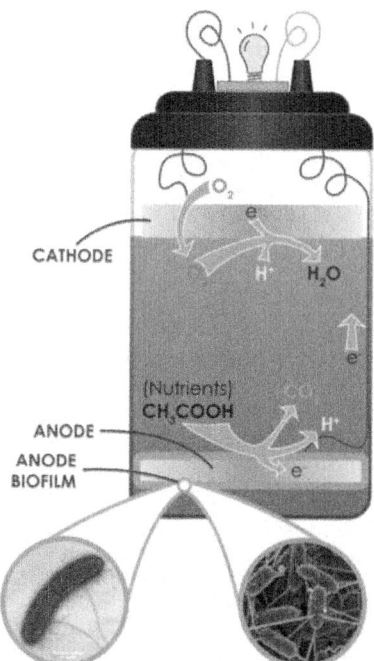

Fig. : A soil-based MFC.

Soil-based microbial fuel cells adhere to the same basic MFC principles as described above, whereby soil acts as the nutrient-rich anodic media, the inoculum, and the proton-exchange membrane (PEM). The anode is placed at a certain depth within the soil, while the cathode rests on top the soil and is exposed to the oxygen in the air above it.

Soils are naturally teeming with a diverse consortium of microbes, including the electrogenic microbes needed for MFCs, and are full of complex sugars and other nutrients that have accumulated over millions of years of plant and animal material decay. Moreover, the aerobic (oxygen consuming) microbes present in the soil act as an oxygen filter, much like the expensive PEM materials used in laboratory MFC systems, which cause the redox potential of the soil to decrease with greater depth. Soil-based MFCs are becoming popular educational tools for science class-rooms.

Phototrophic Bio-film Microbial Fuel Cell

Phototrophic bio-film MFCs (PBMFCs) are the ones that make use of anode with a phototrophic bio-film containing photo-synthetic micro-organism like chlorophyta, cyanophyta etc., since they could carry out photo-synthesis and thus they act as both producers of organic metabolites and also as electron donors.

A study conducted by Strik *et. al.* reveals that PBMFCs yield one of the highest power densities and, therefore, show promise in practical applications. Researchers face difficulties in increasing their power density and long-term performance so as to obtain a cost-effective MFC.

The sub-category of phototrophic microbial fuel cells that use purely oxygenic photo-synthetic material at the anode are sometimes called biological photo-voltaic systems.

Electrical Generation Process

When micro-organisms consume a substance such as sugar in aerobic conditions, they produce carbon dioxide and water. However, when oxygen is not present, they produce carbon dioxide, protons, and electrons, as described below :

$$C_{12}H_{22}O_{11} + 13H_2O \rightarrow 12CO_2 + 48H^+ + 48e^- \qquad (Eqt.\ 1)$$

Microbial fuel cells use inorganic mediators to tap into the electron transport chain of cells and channel electrons produced. The mediator crosses the outer cell lipid membranes and bacterial outer membrane; then, it begins to liberate electrons from the electron transport chain that normally would be taken up by oxygen or other intermediates.

The now-reduced mediator exits the cell laden with electrons that it transfers to an electrode where it deposits them; this electrode becomes the electro-generic anode (negatively charged electrode). The release of the electrons means that the mediator returns to its original oxidised state ready to repeat the process. It is important to note that this can happen only under anaerobic conditions; if oxygen is present, it will collect all the electrons, as it has a greater electronegativity than mediators.

In a microbial fuel cell operation, the anode is the terminal electron acceptor recognized by bacteria in the anodic chamber. Therefore, the microbial activity is strongly dependent on the redox potential of the anode. In fact, it was recently published that a Michaelis-Menten curve was obtained between the anodic potential and the power output of an acetate driven microbial fuel cell. A critical anodic potential seems to exist at which a maximum power output of a microbial fuel cell is achieved.

A number of mediators have been suggested for use in microbial fuel cells. These include natural red, methylene blue, thionine, or resorufin.

This is the principle behind generating a flow of electrons from most micro-organisms (the organisms capable of producing an electric current are termed exoelectrogens). In order to turn this into a usable supply of electricity, this process has to be accommodated in a fuel cell. In order to generate a useful current it is necessary to create a complete circuit, and not just transfer electrons to a single point.

The mediator and micro-organism, in this case yeast, are mixed together in a solution to which is added a suitable substrate such as glucose. This mixture is placed in a sealed chamber to stop oxygen entering, thus forcing the micro-organism to use anaerobic respiration. An electrode is placed in the solution that will act as the anode as described previously.

In the second chamber of the MFC is another solution and electrode. This electrode, called the cathode is positively charged and is the equivalent of the oxygen sink at the end of the electron transport chain, only now it is external to the biological cell. The solution is an oxidizing agent that picks up the electrons at the cathode. As with the electron chain in the yeast cell, this could be a number of molecules such as oxygen. However, this is not particularly practical as it would require large volumes of circulating gas. A more convenient option is to use a solution of a solid oxidizing agent.

Connecting the two electrodes is a wire (or other electrically conductive path, which may include some electrically powered device such as a light bulb) and completing the circuit and connecting the two chambers is a salt bridge or ion-exchange membrane. This last feature allows the protons produced, as described in *Eqt. 1* to pass from the anode chamber to the cathode chamber.

The reduced mediator carries electrons from the cell to the electrode. Here the mediator is oxidized as it deposits the electrons. These then flow across the wire to the second electrode, which acts as an electron sink. From here they pass to an oxidising material.

Applications

Power Generation

Microbial fuel cells have a number of potential uses. The most readily apparent is harvesting electricity produced for use as a power source. The use of MFCs is attractive for applications that require only low power but where replacing batteries may be time-consuming and expensive such as wireless sensor networks. Virtually any organic material could be used to feed the fuel cell, including coupling cells to wastewater treatment plants.

Bacteria would consume waste material from the water and produce supplementary power for the plant. The gains to be made from doing this are that MFCs are a very clean and efficient method of energy production. Chemical processing wastewater and designed synthetic wastewater have been used to produce bio-electricity in dual- and single-chamber mediatorless MFCs (non-coated graphite electrodes) apart from wastewater treatment.

Higher power production was observed with bio-film covered anode (graphite). A fuel cell's emissions are well below regulations. MFCs also use energy much more efficiently than standard combustion engines, which are limited by the Carnot Cycle. In theory, an MFC is capable of energy efficiency far beyond 50% (Yue & Lowther, 1986). According to new research conducted by René Rozendal, using the new microbial fuel cells, conversion of the energy to hydrogen is 8 times as high as conventional hydrogen production technologies.

However, MFCs do not have to be used on a large scale, as the electrodes in some cases need only be 7 μm thick by 2 cm long. The advantages to using an MFC in this situation as opposed to a normal battery is that it uses a renewable form of energy and would not need to be recharged like a standard battery would. In addition to this, they could operate well in mild conditions, 20°C to 40°C and also at pH of around 7. Although more powerful than metal catalysts, they are currently too unstable for long-term medical applications such as in pacemakers (Biotech/Life Sciences Portal).

Besides wastewater power plants, as mentioned before, energy can also be derived directly from crops. This allows the set-up of power stations based on algae platforms or other plants incorporating a large field of aquatic plants. According to Bert Hamelers, the fields are best set-up in synergy with existing renewable plants (*e.g.,* offshore wind turbines). This reduces costs as the microbial fuel cell plant can then make use of the same electricity lines as the wind turbines.

Education

Soil-based microbial fuel cells are popular educational tools, as they employ a range of scientific disciplines (microbiology, geochemistry, electrical engineering, etc.), and can be made using commonly available materials, such as soils and items from the refrigerator. There are also kits available for class-rooms and hobbyists, and research-grade kits for scientific laboratories and corporations.

Bio-sensor

Since the current generated from a microbial fuel cell is directly proportional to the energy content of wastewater used as the fuel, an MFC can be used to measure the solute concentration of wastewater (*i.e.,* as a bio-sensor system).

The strength of wastewater is commonly evaluated as bio-chemical oxygen demand (BOD) values. BOD values are determined incubating samples for 5 days with proper source of microbes, usually activate sludge collected from sewage works. When BOD values are used as a real-time control parameter, 5 days' incubation is too long.

An MFC-type BOD sensor can be used to measure real-time BOD values. Oxygen and nitrate are preferred electron acceptors over the electrode reducing current generation from an MFC. MFC-type BOD sensors underestimate BOD values in the presence of these electron acceptors. This can be avoided by inhibiting aerobic and nitrate respirations in the MFC using terminal oxidase inhibitors such as cyanide and azide. This type of BOD sensor is commercially available.

Bio-recovery

In 2010, A. ter Heijne *et. al.* constructed a device capable of producing electricity and reduce the ion Cu (II) to copper metal.

Current Research Practices

Some researchers point out some undesirable practices, such as recording the maximum current obtained by the cell when connecting it to a resistance as an indication of its performance, instead of the steady-state current that is often a degree of magnitude lower. Often the data about the values of the used resistance is minimal, or even non-existent, making much of the data non-comparable across all studies. This makes extrapolation from standardized procedures difficult if not impossible.

Commercial Applications

A number of companies have emerged to commercialize microbial fuel cells. These companies have attempted to tap into both the remediation and electricity generating aspects of the technologies. Some of these are companies are mentioned here.

Regenerative Fuel Cell

A **regenerative fuel cell** or **reverse fuel cell** (RFC) is a *fuel cell* run in reverse mode, which consumes electricity and chemical B to produce chemical A. By definition, the process of any fuel cell could be reversed. However, a given device is usually optimized for operating in one mode and may not be built in such a way that it can be operated backwards. Standard fuel cells operated backwards generally do not make very efficient systems unless they are purpose-built to do so as with *high-pressure electrolysers*, regenerative fuel cells, *solid-oxide electrolyser cells* and *unitized regenerative fuel cells*.

Process Description

A hydrogen fueled Proton exchange membrane fuel cell, for example, uses hydrogen gas (H_2) and oxygen (O_2) to produce electricity and water (H_2O); a regenerative hydrogen fuel cell uses electricity and water to produce hydrogen and oxygen.

When the fuel cell is operated in regenerative mode, the anode for the electricity production mode (fuel cell mode) becomes the cathode in the hydrogen generation mode (reverse fuel cell mode), and *vice versa*. When an external voltage is applied, water at the cathode side will undergo electrolysis to form hydrogen and oxide ions; oxide ions will be transported through the electrolyte to anode where it can be oxidized to form oxygen. In this reverse mode, the polarity of the cell is opposite to that for the fuel cell mode. The following reactions describe the chemical process in the hydrogen generation mode :

At cathode : $H_2O + 2e^- \rightarrow H_2 + O^{2-}$

At anode : $O^{2-} \rightarrow 1/2 O_2 + 2e^-$

Overall : $H_2O \rightarrow 1/2 O_2 + H_2$

Solid Oxide Regenerative Fuel Cell (SORFC)

One example of RFC is solid oxide regenerative fuel cell. Solid oxide fuel cell operates at high temperatures with high fuel to electricity conversion ratios and it is a good candidate for high temperature electrolysis. Less electricity is required for electrolysis process in SORFC due to high temperature.

The electrolyte can be O^{2-} conducting and/or proton(H^+) conducting. The state of the art for O^{2-} conducting yttria stabilized zirconia(YSZ) based SORFC using Ni–YSZ as the hydrogen electrode and LSM (or LSM–YSZ) as the oxygen electrode has been actively studied. Dönitz and Erdle reported on the operation of YSZ electrolyte cells with current densities of 0.3 A cm^{-2} and 100% Faraday efficiency at only 1.07 V. The recent study by researchers from Sweden shows that ceria-based composite electrolytes, where both proton and oxide ion conductions exist, produce high current output for fuel cell operation and high hydrogen output for electrolysis operation. Zirconia doped with scandia and ceria (10Sc1CeSZ) is also investigated as potential electrolyte in SORFC for hydrogen production at intermediate temperatures (500-750°C). It is reported that 10Sc1CeSZ shows good behaviour and produces high current densities, with suitable electrodes.

Current density–voltage (j–V) curves and impedance spectra are investigated and recorded. Impedance spectra are realized applying an ac current of 1–2A RMS (root-mean-square) in the frequency range from 30 kHz to 10^{-1} Hz. Impedance spectra shows that the resistance is high at low frequencies (<10 kHz) and near zero at high frequencies (>10 kHz). Since high frequency corresponds to electrolyte activities, while low frequencies corresponds to electrodes process, it can be deduced that only a small fraction of the overall resistance is from the electrolyte and most resistance comes from anode and cathode. Hence, developing high performance electrodes are essential for high efficiency SORFC. Area specific resistance (ASR) can be obtained from the slope of j–V curve. Commonly used/tested electrodes materials are nickel/zirconia cermet (Ni/YSZ) and lanthanum-substituted strontium titanate/ceria composite for SORFC cathode, and lanthanum strontium manganite (LSM) for SORFC anode. Other anode materials can be lanthanum strontium ferrite (LSF), lanthanum strontium copper ferrite (LSCuF) and lanthanum strontium cobalt ferrite (LSCoF). Studies show that Ni/YSZ electrode was less active in reverse fuel cell operation than in fuel cell operation, and this can be attributed to a diffusion-limited process in the electrolysis direction, or its susceptibility to aging in a high-steam environment, primarily due to coarsening of nickel particles. Therefore, alternative materials such as the titanate/ceria composite (La0.35Sr0.65TiO3–Ce0.5La0.5O2–δ) or (La0.75Sr0.25)0.95Mn0.5Cr0.5O3 (LSCM) have been proposed electrolysis cathodes. Both LSF and LSM/YSZ are reported as good anode candidates for electrolysis mode. Furthermore, higher operation temperature and higher absolute humidity ratio (AH) can result in lower ASR.

Direct Borohydride Fuel Cell

Direct borohydride fuel cells (DBFCs) are a sub-category of *alkaline fuel cells* which are directly fed by *sodium borohydride* or *potassium borohydride* as a fuel and either air/oxygen or hydrogen peroxide as the oxidant. DBFCs are relatively new types of fuel cells which are currently in the developmental stage and are attractive due to their high operating potential in relation to other type of fuel cells.

Chemistry

Sodium borohydride could potentially be used in more conventional hydrogen fuel cell systems as a means of storing hydrogen. The hydrogen can be regenerated for a fuel cell by catalytic decomposition of the borohydride :

$$NaBH_4 + 2H_2O \rightarrow NaBO_2 + 4H_2$$

Direct borohydride fuel cells decompose and oxidize the borohydride directly, side-stepping hydrogen production and even producing slightly higher energy yields :

Cathode : $2O_2 + 4H_2O + 8e^- \rightarrow 8OH^-$ ($E^0 = +0.4V$)

Anode : $NaBH_4 + 8OH^- \rightarrow NaBO_2 + 6H_2O + 8e^-$ ($E^0 = -1.24$ V)

Total $E^0 = +1.64V$

The simplified reaction is :

$$NaBH_4 + 2O_2 \rightarrow NaBO_2 + 2H_2O + Electricity$$

The working temperature of a direct sodium borohydride fuel cell is 70°C (158°F).

Advantages

DBFCs could be produced more cheaply than a traditional fuel cell because they do not need expensive platinum catalysts. In addition, they have a higher power density.

Disadvantages

Unfortunately, DBFCs do produce some hydrogen from a side reaction of $NaBH_4$ with water heated by the fuel cell. This hydrogen can either be piped out to the exhaust or piped to a conventional hydrogen fuel cell. Either fuel cell will produce water, and the water can be recycled to allow for higher concentrations of $NaBH_4$.

More importantly, the process of creating electricity *via* a DBFC is not easily reversible. For example, after sodium borohydride ($NaBH_4$) has released its hydrogen and has been oxidized, the product is $NaBO_2$ (sodium metaborate). Sodium metaborate might be hydrogenated back into sodium borohydride fuel by several different techniques, some of which might theoretically require nothing more than water and electricity or heat. However, these techniques are still in active development. As of June 30, 2010, many patents claiming to effectively achieve the conversion of sodium metaborate to sodium borohydride have been investigated but none have been confirmed — the current efficiency of "boron hydride recycling" seems to be well below 1% which is unsuitable for recharging a vehicle.

Cost

Sodium borohydride costs US$50 per kg, but with borax recycling and mass production projected prices for the fuel are as low as US$1/kg.

Alkaline Fuel Cell

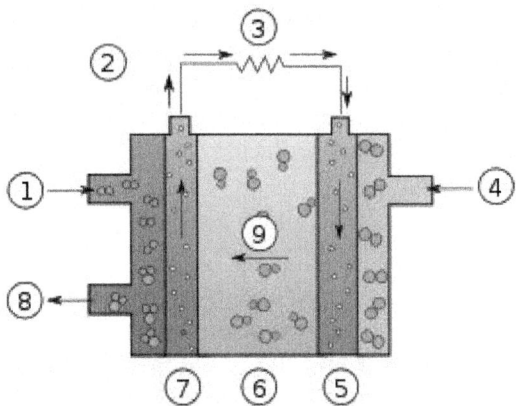

Fig. : Diagram of an Alkaline Fuel Cell. 1 : Hydrogen 2 : Electron flow 3 : Load 4 : Oxygen 5 : Cathode 6 : Electrolyte 7 : Anode 8 : Water 9 : Hydroxyl Ions.

The alkaline fuel cell (**AFC**), also known as the *Bacon* fuel cell after its British inventor, is one of the most developed *fuel cell* technologies. *NASA* has used alkaline fuel cells since the mid-1960s, in *Apollo*-series missions and on the *Space Shuttle*. AFCs consume hydrogen and pure oxygen producing potable water, heat, and electricity. They are among the most efficient fuel cells, having the potential to reach 70%.

Chemistry

The fuel cell produces power through a redox reaction between hydrogen and oxygen. At the anode, hydrogen is oxidized according to the reaction :

$$2H_2 + 4OH^- \rightarrow 4H_2O + 4e^-$$

producing water and releasing two electrons. The electrons flow through an external circuit and return to the cathode, reducing oxygen in the reaction :

$$O_2 + 2H_2O + 4e^- \rightarrow 4OH^-$$

producing hydroxide ions. The net reaction consumes one oxygen atom and two hydrogen atoms in the production of one water molecule. Electricity and heat are formed as by-products of this reaction.

Electrolyte

The two electrodes are separated by a porous matrix saturated with an aqueous alkaline solution, such as potassium hydroxide (KOH). Aqueous alkaline solutions do not reject carbon dioxide (CO_2) so the fuel cell can become "poisoned" through the conversion of KOH to potassium carbonate (K_2CO_3). Because of this, alkaline fuel cells typically operate on pure oxygen, or at least purified air and would incorporate a 'scrubber' into the design to clean out as much of the carbon dioxide as is possible. Because the generation and storage requirements of oxygen make pure-oxygen AFCs expensive, there are few companies engaged in active development of the technology. There is, however, some debate in the research community over whether the poisoning is permanent or reversible. The main mechanisms of poisoning are blocking of the pores in the cathode with K_2CO_3, which is not reversible, and reduction in the ionic conductivity of the electrolyte, which may be reversible by returning the KOH to its original concentration. An alternate method involves simply replacing the KOH which returns the cell back to its original output.

Basic Designs

Because of this poisoning effect, two main variants of AFCs exist : static electrolyte and flowing electrolyte. Static, or immobilized, electrolyte cells of the type used in the Apollo space craft and the Space Shuttle typically use an asbestos separator saturated in potassium hydroxide. Water production is managed by evapouration out the anode, as pictured above, which produces pure water that may be reclaimed for other uses. These fuel cells typically use platinum catalysts to achieve maximum volumetric and specific efficiencies.

Flowing electrolyte designs use a more open matrix that allows the electrolyte to flow either between the electrodes (parallel to the electrodes) or through the electrodes in a transverse direction (the ASK-type or EloFlux fuel cell). In parallel-flow electrolyte designs, the water produced is retained in the electrolyte, and old electrolyte may be exchanged for fresh, in a manner analogous to an oil change in a car. In the case of "parallel flow" designs, greater space is required between electrodes to enable this flow, and this translates into an increase in cell resistance, decreasing power output compared to immobilized electrolyte designs. A further challenge for the technology is that it is not clear how severe is the problem of permanent blocking of the cathode by K_2CO_3, however, some published reports indicate thousands of hours of operation on air. These designs have used both platinum and non-noble metal catalysts, resulting in increased volumetric and specific efficiencies and increased cost.

The EloFlux design, with its transverse flow of electrolyte, has the advantage of low-cost construction and replaceable electrolyte, but so far has only been demonstrated using oxygen.

Further variations on the alkaline fuel cell include the metal hydride fuel cell and the direct borohydride fuel cell.

Commercial Prospects

AFCs are the cheapest of fuel cells to manufacture. The catalyst required for the electrodes can be any of a number of different chemicals that are inexpensive compared to those required for other types of fuel cells.

The commercial prospects for AFCs lie largely with the recently developed bi-polar plate version of this technology, considerably superior in performance to earlier mono-plate versions.

The world's first Fuel Cell Ship HYDRA used an AFC system with 5 kW net output.

Another recent development is the solid-state alkaline fuel cell, utilizing alkali anion exchange membranes rather than a liquid. This resolves the problem of poisoning and allows the development of alkaline fuel cells capable of running on safer hydrogen-rich carriers such as liquid urea solutions or metal amine complexes.

Direct Methanol Fuel Cell

Direct-methanol fuel cells or **DMFCs** are a sub-category of proton-exchange fuel cells in which methanol is used as the fuel. Their main advantage is the ease of transport of methanol, an energy-dense yet reasonably stable liquid at all environmental conditions.

Efficiency is quite low for these cells, so they are targeted especially to portable applications, where energy and power density are more important than efficiency.

A more efficient version of a direct fuel cell would play a key role in the theoretical use of methanol as a general energy transport medium, in the hypothesized methanol economy.

The Cell

In contrast to indirect methanol fuel cells, where methanol is reacted to hydrogen by steam reforming, DMFCs use a methanol solution (usually around 1M, *i.e.* about 3% in mass) to carry the reactant into the cell; common operating temperatures are in the range 50–120°C, where high temperatures are usually pressurized. DMFCs themselves are more efficient at high temperatures and pressures, but these conditions end up causing so many losses in the complete system that the advantage is lost; therefore, atmospheric-pressure configurations are currently preferred.

Because of the methanol cross-over, a phenomenon by which methanol diffuses through the membrane without reacting, methanol is fed as a weak solution : this decreases efficiency significantly, since crossed-over methanol, after reaching the air side (the cathode), immediately reacts with air; though the exact kinetics are debated, the end result is a reduction of the cell voltage. Cross-over remains a major factor in inefficiencies, and often half of the methanol is lost to cross-over. Methanol cross-over and/or its effects can be alleviated by (a) developing alternative membranes (*e.g.*), (b) improving the electro-oxidation process in the catalyst layer and improving the structure of the catalyst and gas diffusion layers (*e.g.*), and (c) optimizing the design of the flow field and the membrane electrode assembly (MEA) which can be achieved by studying the current density distributions (*e.g.*).

Other issues include the management of carbon dioxide created at the anode, the sluggish dynamic behaviour, and the ability to maintain the solution water.

The only waste products with these types of fuel cells are carbon dioxide and water.

History

In 1990 super-acid specialist Dr. Surya Prakash, and Nobel laureate Dr. George A. Olah, both of the University of Southern California's Loker Hydrocarbon Research Institute, invented a fuel cell that would directly convert methanol to electricity. USC, in a collaborative effort with Jet Propulsion Laboratory (JPL) proceeded to invent the direct oxidation of liquid hydrocarbons subsequently coined as DMFC, Direct Methanol Fuel Cell Technology. Others were involved in the invention, including California Institute of Technology (Caltech) and DTI Energy, Inc. Some advances in DMFC have been : bringing down the fuel cross-over, miniaturizing the cell for consumer and military products, ceramic plates, carbon nanotubes, porous silicon layers and oxidation reduction.

Application

Current DMFCs are limited in the power they can produce, but can still store a high energy content in a small space. This means they can produce a small amount of power over a long period of time. This makes them ill-suited for powering large vehicles (at least directly), but ideal for smaller vehicles such as forklifts and tuggers and consumer goods such as mobile phones, digital cameras or laptops. Military applications of DMFCs are an emerging application since they have low noise and thermal signatures and no toxic effluent. These applications include power for man-portable tactical equipment, battery chargers, and autonomous power for test and training instrumentation. Units are available with power outputs between 25 watts and 5 kilowatts with durations up to 100 hours between refuelings.

Methanol

Methanol is a liquid from –97.0°C to 64.7°C at atmospheric pressure. The energy density of methanol is an order of magnitude greater than even highly compressed hydrogen, and 15 times higher than Lithium-ion batteries.

Methanol is toxic and flammable. However, the International Civil Aviation Organization's (ICAO) Dangerous Goods Panel (DGP) voted in November 2005 to allow passengers to carry and use micro fuel cells and methanol fuel cartridges when aboard airplanes to power laptop computers and other consumer electronic devices. On September 24, 2007, the US Department of Transportation issued a proposal to allow airline passengers to carry fuel cell cartridges on board.

The Department of Transportation issued a final ruling on April 30, 2008, permitting passengers and crew to carry an approved fuel cell with an installed methanol cartridge and up to two additional spare cartridges. It is worth noting that 200 ml maximum methanol cartridge volume allowed in the final ruling is double the 100 ml limit on liquids allowed by the Transportation Security Administration in carry-on bags.

Reaction

The DMFC relies upon the oxidation of methanol on a catalyst layer to form carbon dioxide. Water is consumed at the anode and is produced at the cathode. Protons (H^+) are transported across the proton exchange membrane-often made from Nafion-to the cathode where they react with oxygen to produce water. Electrons are transported through an external circuit from anode to cathode, providing power to connected devices.

The half-reactions are :

	Equation
Anode	$CH_3OH + H_2O \rightarrow 6\,H^+ + 6\,e^- + CO_2$ oxidation
Cathode	$\frac{3}{2}O_2 + 6\,H^+ + 6\,e^- \rightarrow 3\,H_2O$ reduction
Overall reaction	$CH_3OH + \frac{3}{2}\,O_2 \rightarrow 2\,H_2O + CO_2$ redox reaction

Methanol and water are adsorbed on a catalyst usually made of platinum and ruthenium particles, and lose protons until carbon dioxide is formed. As water is consumed at the anode in the reaction, pure methanol cannot be used without provision of water *via* either passive transport such as back diffusion (osmosis), or active transport such as pumping. The need for water limits the energy density of the fuel.

Platinum is used as a catalyst for both half-reactions. This contributes to the loss of cell voltage potential, as any methanol that is present in the cathode chamber will oxidize. If another catalyst could be found for the reduction of oxygen, the problem of methanol cross-over would likely be significantly lessened. Furthermore, platinum is very expensive and contributes to the high cost per kilowatt of these cells.

During the methanol oxidation reaction carbon monoxide (CO) is formed, which strongly adsorbs onto the platinum catalyst, reducing the number of available reaction sites and thus the performance of the cell. The addition of other metals, such as ruthenium or gold, to the platinum catalyst tends to ameliorate this problem. In the case of platinum-ruthenium catalysts, the oxophilic nature of ruthenium is believed to promote the formation of hydroxyl radicals on its surface, which can then react with carbon monoxide adsorbed on the platinum atoms. The water in the fuel cell is oxidized to a hydroxy radical *via* the following reaction : $H_2O \rightarrow OH\bullet + H^+ + e^-$. The hydroxy radical then oxidizes carbon monoxide to produce carbon dioxide, which is released from the surface as a gas : $CO + OH\bullet \rightarrow CO_2 + H^+ + e^-$.

Using these OH groups in the half reactions, they are also expressed as :

Equation

Anode $\qquad$ $CH_3OH + 6\ OH^- \rightarrow 5\ H_2O + 6\ e^- + CO_2$ oxidation

Cathode $\qquad$ $\dfrac{3}{2}\ O_2 + 3\ H_2O + 6\ e^- \rightarrow 6\ OH^-$ reduction

Overall reaction $\quad$ $CH_3OH + \dfrac{3}{2}\ O_2 \rightarrow 2\ H_2O + CO_2$ redox reaction

Cross-over Current

Methanol on the anodic side is usually in a weak solution (from 1M to 3M), because methanol in high concentrations has the tendency to diffuse through the membrane to the cathode, where its concentration is about zero because it is rapidly consumed by oxygen. Low concentrations help in reducing the cross-over, but also limit the maximum attainable current. The practical realization is usually that a solution loop enters the anode, exits, is refilled with methanol, and returns to the anode again. Alternatively, fuel cells with optimized structures can directly fed with high concentration methanol solutions or even pure methanol.

Water Drag

The water in the anodic loop is lost because of the anodic reaction, but mostly because of the associated water drag : every proton formed at the anode drags a number of water molecules to the cathode. Depending on temperature and membrane type, this number can be between 2 and 6.

Ancillary Units

A direct methanol fuel cell is usually part of a larger system including all the ancillary units that permit its operation. Compared to most other types of fuel cells, the ancillary system of DMFCs is relatively complex. The main reasons for its complexity are :

*   Providing water along with methanol would make the fuel supply more cumbersome, so water has to be recycled in a loop;
* CO_2 has to be removed from the solution flow exiting the fuel cell;
*   Water in the anodic loop is slowly consumed by reaction and drag; it is necessary to recover water from the cathodic side to maintain steady operation.

Reformed Methanol Fuel Cell

Reformed Methanol Fuel Cell (RMFC) or **Indirect Methanol Fuel Cell** (IMFC) systems are a sub-category of *proton-exchange fuel cells* where, the fuel, *methanol* (CH_3OH), is reformed, before being fed into the *fuel cell*. RMFC systems offer advantages over *direct methanol fuel cell* (DMFC) systems including higher efficiency, smaller cell stacks, no water management, better operation at low temperatures, and storage at sub-zero temperatures because methanol is a liquid from -97.0°C to 64.7°C (-142.6 °F to 148.5 °F). The tradeoff is that RMFC systems operate at hotter temperatures and therefore, need more advanced heat management and insulation. The waste products with these types of fuel cells are *carbon dioxide* and water.

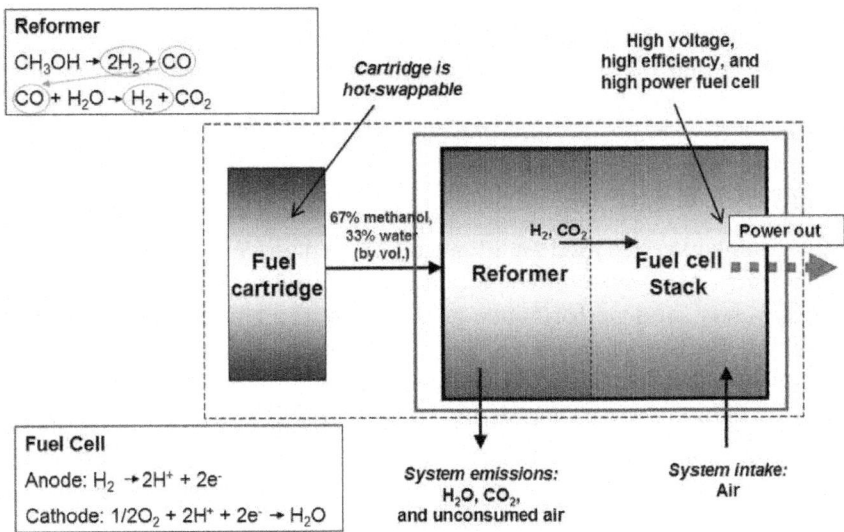

Fig. : Block diagram of a Reformed Methanol Fuel Cell.

Methanol is used as a fuel because it is naturally hydrogen dense (a hydrogen carrier) and can be *steam reformed* into hydrogen at low temperatures compared to other *hydrocarbon* fuels. Additionally, methanol is naturally occurring, *biodegradable*, and energy dense.

RMFC systems consist of a *fuel processing system* (FPS), a *fuel cell*, a fuel cartridge, and the BOP (the *balance of plant*).

Storage

The fuel cartridge stores the methanol fuel, which is often diluted with up to 40% (by volume) water.

Fuel Processing System (FPS) in

Methanol→Partial oxidation (POX)/Autothermal reforming (ATR)→Water gas shift reaction (WGS)→preferential oxidation (PROX) The reformer converts methanol to H_2 and CO_2, a reaction that occurs at temperatures of 250°C to 300°C.

Fuel Cell

→The membrane electrode assembly (MEA) fuel cell stack produces electricity in a reaction that combines H_2 (reformed from methanol in the fuel processor) and O_2 and produces water (H_2O) as a by-product.

Fuel Processing System (FPS) Out

→Tail gas combustor (TGC) catalytic combustion afterburner or (catalytic combustion) with a platinum-alumina (Pt–Al2O3) catalyst→condenser

Balance of Plant

The balance of plant (BOP) consists of any fuel pumps, air compressors, and fans required to circulate the gas and liquid in the system. A control system is also often needed to operate and monitor the RMFC.

State of Development

RMFC systems have reached an advanced stage of development. For instance, a small 25 watt RMFC system developed for the military has met environmental tolerance, safety, and performance goals set by the United States Army Communications-Electronics Research, Development and Engineering Center, and is commercially available.

Larger systems 350W to 8 MW] are also available for multiple applications, such as power plant generation, backup power generation and battery range extension. One application in the field is improving performance of a heavy duty smaller electric vehicle Video of solution

Direct-ethanol Fuel Cell

Direct-ethanol fuel cells or **DEFCs** are a sub-category of Proton-exchange fuel cells where the fuel, ethanol, is fed directly to the fuel cell.

Advantages

DEFC uses Ethanol in the fuel cell instead of the more toxic methanol. Ethanol is an attractive alternative to methanol because it comes with a supply chain that's already in place. Ethanol also remains the easier fuel to work with for widespread use by consumers.

Ethanol is a hydrogen-rich liquid and it has a higher energy density (8.0 kWh/kg) compared to methanol (6.1 kWh/kg). Ethanol can be obtained in great quantity from biomass through a fermentation process from renewable resources like from sugar cane, wheat, corn, or even straw. Bio-generated ethanol (or bio-ethanol) is thus attractive since growing crops for bio-fuels absorbs much of the carbon dioxide emitted into the atmosphere from fuel used to produce the bio-fuels, and from burning the bio-fuels themselves. This is in sharp contrast to the use of fossil fuels. The use of ethanol would also overcome both the storage and infrastructure challenge of hydrogen for fuel cell applications. In a fuel cell, the oxidation of any fuel requires the use of a catalyst in order to achieve the current densities required for commercially viable fuel cells, and platinum-based catalysts are some of the most efficient materials for the oxidation of small organic molecules.

Reaction

Direct Ethanol Fuel Cell

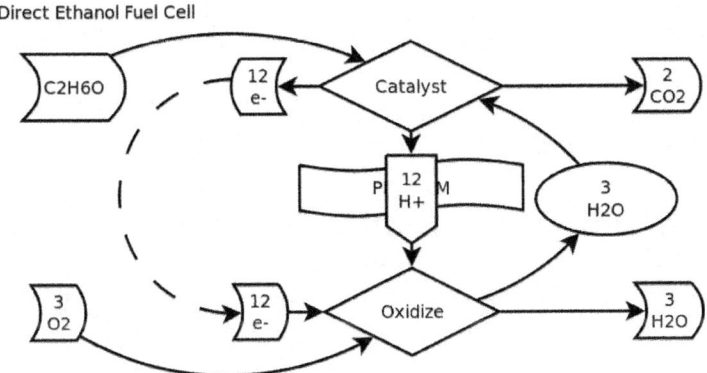

Fig. : Flow-chart of the reaction in a DEFC.

The DEFC, similar to the DMFC, relies upon the oxidation of ethanol on a catalyst layer to form carbon dioxide. Water is consumed at the anode and is produced at the cathode. Protons (H^+) are transported across the proton exchange membrane to the cathode where they react with oxygen to produce water. Electrons are transported through an external circuit from anode to cathode, providing power to connected devices.

The half-reactions are :

Equation

Anode $C_2H_5OH + 3\,H_2O \rightarrow 12\,H^+ + 12\,e^- + 2\,CO_2$ oxidation

Cathode $3\,O_2 + 12\,H^+ + 12\,e^- \rightarrow 6\,H_2O$ reduction

Overall reaction $C_2H_5OH + 3\,O_2 \rightarrow 3\,H_2O + 2\,CO_2$ redox reaction.

Issues

Platinum-based catalysts are expensive, so practical exploitation of ethanol as fuel for a PEM fuel cell requires a new catalyst. New nanostructured electrocatalysts (HYPERMEC by ACTA SpA for example) have been developed, which are based on non-noble metals, preferentially mixtures of Fe, Co, Ni at the anode, and Ni, Fe or Co alone at the cathode. With ethanol, power densities as high as 140 mW/ cm² at 0.5 V have been obtained at 25°C with self-breathing cells containing commercial anion-exchange membranes. This catalyst does not contain any precious metals. In practice tiny metal particles are fixed onto a substrate in such a way that they produce a very active catalyst.

A polymer acts as electrolyte. The charge is carried by the hydrogen ion (proton). The liquid ethanol (C_2H_5OH) is oxidized at the anode in the presence of water, generating CO_2, hydrogen ions and electrons. Hydrogen ions travel through the electrolyte. They react at the cathode with oxygen from the air and the electrons from the external circuit forming water.

Bio-Ethanol based fuel cells may improve the well-to-wheel balance of this bio-fuel because of the increased conversion rate of the fuel cell compared to the internal combustion engine. But real world figures may be only achieved in some years since the development of direct methanol and ethanol fuel cells is lagging behind hydrogen powered fuel cells.

Present Accomplishments

On 13 May, 2007 a team from the University of Applied Sciences in Offenburg did present world's first vehicle powered by a DEFC at Shell's Eco-marathon in France. The car "Schluckspecht" attended a successful test drive on Nogaro Circuit powered by a DEFC stack giving an output voltage of 20 to 45 V (depending on load).

Various prototypes of Direct Ethanol Fuel Cell Stack mobile phone chargers were built featuring voltages from 2V to 7V and powers from 800 mW to 2W were built and tested.

Proton Exchange Membrane Fuel Cell

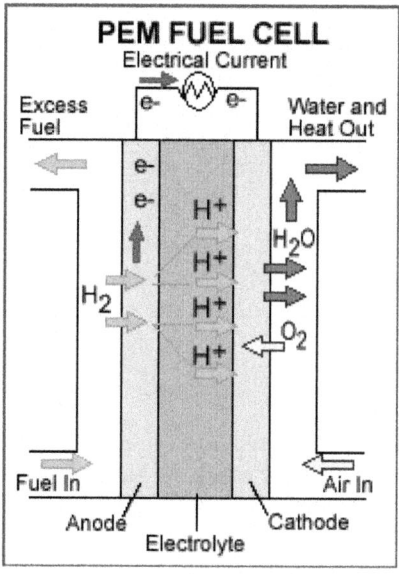

Fig. : Diagram of a PEM fuel cell.

Proton exchange membrane fuel cells, also known as polymer electrolyte membrane (PEM) fuel cells (**PEMFC**), are a type of fuel cell being developed for transport applications as well as for stationary fuel cell applications and portable fuel cell applications. Their distinguishing features include lower temperature/ pressure ranges (50 to 100°C) and a special polymer electrolyte membrane. PEMFCs operate on a similar principle to their younger sister technology PEM electrolysis. They are a leading candidate to replace the aging alkaline fuel cell technology, which was used in the Space Shuttle.

Science

Reactions

A Proton exchange membrane fuel cell transforms the chemical energy liberated during the electro-chemical reaction of hydrogen and oxygen to electrical energy, as opposed to the direct combustion of hydrogen and oxygen gases to produce thermal energy.

A stream of hydrogen is delivered to the anode side of the membrane electrode assembly (MEA). At the anode side it is catalytically split into protons and electrons. This oxidation half-cell reaction or Hydrogen Oxidation Reaction (HOR) is represented by :

At the Anode :

$$H_2 \rightarrow 2H^+ + 2e^- \quad E^o = 0V_{SHE} \tag{1}$$

The newly formed protons permeate through the polymer electrolyte membrane to the cathode side. The electrons travel along an external load circuit to the cathode side of the MEA, thus creating the current output of the fuel cell. Meanwhile, a stream of oxygen is delivered to the cathode side of the MEA. At the cathode side oxygen molecules react with the protons permeating through the polymer electrolyte membrane and the electrons arriving through the external circuit to form water molecules. This reduction half-cell reaction or oxygen reduction reaction (ORR) is represented by :

At the Cathode

$$1/2O_2 + 2H^+ + 2e^- \rightarrow H_2O \quad E^o = 1.229V_{SHE} \tag{2}$$

Overall Reaction

$$H_2 + 1/2O_2 \rightarrow H_2O \quad E^o = 1.229V_{SHE} \tag{3}$$

The reversible reaction is expressed in the equation and shows the reincorporation of the hydrogen protons and electrons together with the oxygen molecule and the formation of one water molecule.

Polymer Electrolyte Membrane

Splitting of the hydrogen molecule is relatively easy by using a platinum catalyst. Unfortunately however, splitting the oxygen molecule is more difficult, and this causes significant electric losses. An appropriate catalyst material for this process has not been discovered, and platinum is the best option. One promising catalyst that uses far less expensive materials — iron, nitrogen, and carbon — has long been known to promote the necessary reactions, but at rates that are far too slow to be practical. Recently, a Canadian research institute has dramatically increased the performance of this type of iron-based catalyst. Their material produces 99 A/cm³ at 0.8 volts, a key measurement of catalytic activity. That is 35 times better than the best non-precious metal catalyst so far, and close to the Department of Energy's goal for fuel-cell catalysts : 130 A/cm³. It also matches the performance of typical platinum catalysts. The only problem at the moment is its durability because after only 100 hours of testing the reaction rate dropped to half. Another significant source of losses is the resistance of the membrane to proton flow, which is minimized by making it as thin as possible, on the order of 50 µm.

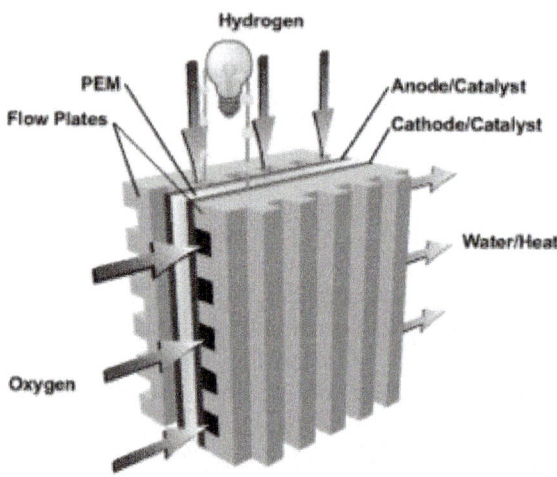

Fig. : To function, the membrane must conduct hydrogen ions (protons) but not electrons as this would in effect "short circuit" the fuel cell. The membrane must also not allow either gas to pass to the other side of the cell, a problem known as gas cross-over. Finally, the membrane must be resistant to the reducing environment at the cathode as well as the harsh oxidative environment at the anode.

A cheaper alternative to platinum is Cerium (IV) oxide catalysator used by professor Vladimír Matolín in the developement of PEMFC.

The PEMFC is a prime candidate for vehicle and other mobile applications of all sizes down to mobile phones, because of its compactness. However, the water management is crucial to performance : too much water will flood the membrane, too little will dry it; in both cases, power output will drop. Water management is a very difficult subject in PEM systems, primarily because water in the membrane is attracted toward the cathode of the cell through polarization. A wide variety of solutions for managing the water exist including integration of electro-osmotic pumps. Furthermore, the platinum catalyst on the membrane is easily poisoned by carbon monoxide (no more than one part per million is usually acceptable) and the membrane is sensitive to things like metal ions, which can be introduced by corrosion of metallic bipolar plates, metallic components in the fuel cell system or from contaminants in the fuel/oxidant.

PEM systems that use reformed methanol were proposed, as in Daimler Chrysler Necar 5; reforming methanol, *i.e.* making it react to obtain hydrogen, is however a very complicated process, that requires also purification from the carbon monoxide the reaction produces. A platinum-ruthenium catalyst is necessary as some carbon monoxide will unavoidably reach the membrane. The level should not exceed 10 parts per million. Furthermore, the start-up times of such a reformer reactor are of about half an hour. Alternatively, methanol, and some other bio-fuels can be fed to a PEM fuel cell directly without being reformed, thus making a direct methanol fuel cell (DMFC). These devices operate with limited success.

The most commonly used membrane is Nafion by DuPont, which relies on liquid water humidification of the membrane to transport protons. This implies that it is not feasible to use temperatures above 80 to 90°C, since the membrane would dry. Other, more recent membrane types, based on polybenzimidazole (PBI) or phosphoric acid, can reach up to 220°C without using any water management : higher temperature allow for better efficiencies, power densities, ease of cooling (because of larger allowable temperature differences), reduced sensitivity to carbon monoxide poisoning and better controllability (because of absence of water management issues in the membrane); however, these recent types are not as common.

Efficiency

The maximal theoretical efficiency applying the Gibbs free energy equation $\Delta G = -237.13$ kJ/mol and using the Lower Heating Value (LHV) of Hydrogen ($\Delta H = -285.84$ kJ/mol) is 83% at 298 K.

$$\eta = \frac{\Delta G}{\Delta H} = 1 - \frac{T \Delta S}{\Delta H}$$

The practical efficiency of a PEM's is in the range of 40–60% using the Higher Heating Value of hydrogen (HHV). Main factors that create losses are :

• Activation losses
• Ohmic losses
• Mass transport losses

Catalyst Research

Much of the current research on catalysts for PEM fuel cells can be classified as having one of two main objectives :

1. To obtain higher catalytic activity than the standard carbon-supported platinum particle catalysts used in current PEM fuel cells
2. To reduce the poisoning of PEM fuel cell catalysts by impurity gases.

Increasing Catalytic Activity

Platinum is by far the most effective element used for PEM fuel cell catalysts, and nearly all current PEM fuel cells use platinum particles on porous carbon supports to catalyze both hydrogen oxidation and oxygen reduction. However, due to their high cost, current Pt/C catalysts are not feasible for commercialization. The U.S.

Department of Energy estimates that platinum-based catalysts will need to use roughly four times less platinum than is used in current PEM fuel cell designs in order to represent a realistic alternative to internal combustion engines. Consequently, one main goal of catalyst design for PEM fuel cells is to increase the catalytic activity of platinum by a factor of four so that only one-fourth as much of the precious metal is necessary to achieve similar performance.

One method of increasing the performance of platinum catalysts is to optimize the size and shape of the platinum particles. Decreasing the particles' size alone increases the total surface area of catalyst available to participate in reactions per volume of platinum used, but recent studies have demonstrated additional ways to make further improvements to catalytic performance. For example, one study reports that high-index facets of platinum nanoparticles (that is Miller indexes with large integers, such as Pt (730)) provide a greater density of reactive sites for oxygen reduction than typical platinum nanoparticles.

A second method of increasing the catalytic activity of platinum is to alloy it with other metals. For example, it was recently shown that the $Pt_3Ni(111)$ surface has a higher oxygen reduction activity than pure Pt(111) by a factor of ten. The authors attribute this dramatic performance increase to modifications to the electronic structure of the surface, reducing its tendency to bond to oxygen-containing ionic species present in PEM fuel cells and hence increasing the number of available sites for oxygen adsorption and reduction.

Very recently, a new class of ORR electrocatalysts have been introduced in the case of Pt-M (M-Fe and Co) systems with an ordered intermetallic core encapsulated within a Pt-rich shell. These **intermetallic core-shell (IMCS) nanocatalysts** were found to exhibit an enhanced activity and most importantly, an extended durability compared to many previous designs. While the observed enhancement in the activities is ascribed to a strained lattice, the authors report that their findings on the degradation kinetics establish that the extended catalytic durability is attributable to a sustained atomic order.

Reducing Poisoning

The other popular approach to improving catalyst performance is to reduce its sensitivity to impurities in the fuel source, especially carbon monoxide (CO). Presently, pure hydrogen gas is not economical to mass-produce by electrolysis or any other means. Instead, hydrogen gas is produced by steam reforming light hydrocarbons, a process which produces a mixture of gases that also contains CO (1–3%), CO_2 (19–25%), and N_2 (25%). Even tens of parts per million of CO can poison a pure platinum catalyst, so increasing platinum's resistance to CO is an active area of research.

For example, one study reported that cube-shaped platinum nanoparticles with (100) faces displayed a fourfold increase in oxygen reduction activity compared to randomly faceted platinum nanoparticles of similar size. The authors concluded that the (111) facets of the randomly shaped nanoparticles bonded more strongly to sulfate ions than the (100) facets, reducing the number of catalytic sites open to oxygen molecules. The nanocubes they synthesized, in contrast, had almost exclusively (100) facets, which are known to interact with sulfate more weakly. As a result, a greater fraction of the surface area of those particles was available for the reduction of oxygen, boosting the catalyst's oxygen reduction activity.

In addition, researchers have been investigating ways of reducing the CO content of hydrogen fuel before it enters a fuel cell as a possible way to avoid poisoning the catalysts. One recent study revealed that ruthenium-platinum core–shell nanoparticles are particularly effective at oxidizing CO to form CO_2, a much less harmful fuel contaminant.

History

Before the invention of PEM fuel cells, existing fuel cell types such as solid-oxide fuel cells were only applied in extreme conditions. Such fuel cells also required very expensive materials and could only be used for stationary applications due to their size. These issues were addressed by the PEM fuel cell. The PEM fuel cell was invented in the early 1960s by Willard Thomas Grubb and Leonard Niedrach of General Electric. Initially, sulfonated polystyrene membranes were used for electrolytes, but they were replaced in 1966 by Nafion ionomer, which proved to be superior in performance and durability to sulfonated polystyrene.

PEM fuel cells were used in the NASA Gemini series of spacecraft, but they were replaced by Alkaline fuel cells in the Apollo program and in the Space shuttle.

Parallel with Pratt and Whitney Aircraft, General Electric developed the first proton exchange membrane fuel cells (PEMFCs) for the Gemini space missions in the early 1960s. The first mission to use PEMFCs was Gemini V. However, the Apollo space missions and subsequent Apollo-Soyuz, Skylab and Space Shuttle missions used fuel cells based on Bacon's design, developed by Pratt and Whitney Aircraft.

Extremely expensive materials were used and the fuel cells required very pure hydrogen and oxygen. Early fuel cells tended to require inconveniently high operating temperatures that were a problem in many applications. However, fuel cells were seen to be desirable due to the large amounts of fuel available (hydrogen and oxygen).

Despite their success in space programs, fuel cell systems were limited to space missions and other special applications, where high cost could be tolerated. It was not until the late 1980s and early 1990s that fuel cells became a real option for wider application base. Several pivotal innovations, such as low platinum catalyst loading and thin film electrodes, drove the cost of fuel cells down, making development of PEMFC systems more realistic. However, there is significant debate as to whether hydrogen fuel cells will be a realistic technology for use in automobiles or other vehicles.

Flow Battery

A **flow battery** is a type of rechargeable battery where rechargeability is provided by two chemical components dissolved in liquids contained within the system and separated by a membrane. Ion exchange (providing flow of electrical current) occurs through the membrane while both liquids circulate in their own respective space. Cell voltage is chemically determined by the Nernst equation and ranges, in practical applications, from 1.0 to 2.2 Volts.

A flow battery is technically akin both to a fuel cell and an electro-chemical accumulator cell (electro-chemical reversibility). While it has technical advantages such as potentially separable liquid tanks and near unlimited longevity over most conventional rechargeables, current implementations are comparatively less powerful and require more sophisticated electronics.

Construction Principle

A flow battery is a rechargeable fuel cell in which an electrolyte containing one or more dissolved electro-active elements flow through an electro-chemical cell that reversibly converts chemical energy directly to electricity (electro-active elements are "elements in solution that can take part in an electrode reaction or that can be adsorbed on the electrode"). Additional electrolyte is stored externally, generally in tanks, and is usually pumped through the cell (or cells) of the reactor, although gravity feed systems are also known. Flow batteries can be rapidly "recharged" by replacing the electrolyte liquid (in a similar way to refilling fuel tanks for internal combustion engines) while simultaneously recovering the spent material for re-energization.

In other words, a flow battery is just like an electro-chemical cell, with the exception that the ionic solution (electrolyte) is not stored in the cell around the electrodes. Rather, the ionic solution is stored outside of the cell, and can be fed into the cell in order to generate electricity. The total amount of electricity that can be generated depends on the size of the storage tanks. One benefit to this design is that the cell can be recharged simply by changing out the tanks.

Classes of Flow Batteries

Different classes of flow cells (batteries) have been developed, including redox, hybrid and membraneless. The fundamental difference between conventional batteries and flow cells is that energy is stored as the electrode material in conventional batteries but as the electrolyte in flow cells.

Redox

The redox (reduction-oxidation) cell is a reversible fuel cell in which all electro-chemical components are dissolved in the electrolyte. The energy capacity of the redox flow battery is fully independent of its power, because the energy available is related to the electrolyte volume (amount of liquid electrolyte) and the power to the surface area of the electrodes. Redox flow batteries are rechargeable (secondary cells). Because they employ heterogeneous electron transfer rather than solid-state diffusion or intercalation they are more appropriately called fuel cells than batteries. In industrial practice, fuel cells are usually, and unnecessarily, considered to be primary cells, such as the H_2/O_2 system. The unitized regenerative fuel cell on NASA's Helios Prototype is another reversible fuel cell. The European Patent Organisation classifies redox flow cells (H01M8/18C4) as a sub-class of regenerative fuel cells (H01M8/18). Examples of redox flow batteries are the vanadium redox flow battery, polysulfide bromide battery (Regenesys), and uranium redox flow battery. Redox fuel cells are less common commercially although many systems have been proposed.

Hybrid

The hybrid flow battery uses one or more electro-active components deposited as a solid layer. In this case, the electro-chemical cell contains one battery electrode and one fuel cell electrode. This type is limited in energy by the surface area of the electrode.

Hybrid flow batteries include the zinc-bromine, zinc-cerium and lead-acid flow batteries.

Membraneless

This battery employs a phenomenon called laminar flow in which two liquids are pumped through a channel. They undergo electro-chemical reactions to store or release energy. The solutions stream through in parallel, with little mixing. The flow naturally separates the liquids, eliminating the need for a membrane.

Membranes are often the most costly component and the most unreliable components of batteries, as they can corrode with repeated exposure to certain reactants. The absence of a membrane enabled the use of a liquid bromine solution and hydrogen. This combination is problematic when membranes are used, because they form hydrobromic acid that can destroy the membrane. Both materials are available at low cost.

The design uses a small channel between two electrodes. Liquid bromine flows through the channel over a graphite cathode and hydrobromic acid flows under a porous anode. At the same time, hydrogen gas flows across the anode. The chemical reaction can be reversed to recharge the battery — a first for any membraneless design.

One such membraneless flow battery published in August 2013 produced a maximum power density of $0.795w/cm^2$, three times as much power as other membraneless systems — and an order of magnitude higher than lithium-ion batteries.

Organic

In 2013 researchers announced the use of 9,10-anthraquinone-2,7-disulphonic acid (AQDS), a quinone, as a charge carrier in metal-free **flow batteries**. Each of the carbon-based molecules holds two units of electrical charge, compared with one unit in conventional batteries, implying that a battery could store twice as much energy in a given volume. AQDS undergoes rapid, reversible two-electron/two-proton reduction on a glassy carbon electrode in sulphuric acid. An aqueous flow battery with inexpensive carbon electrodes, combining the quinone/hydroquinone couple with the $Br2/Br-$redox couple, yields a peak galvanic power density exceeding $0.6\,W\,cm-2$ at $1.3\,A\,cm-2$. Cycling showed >99 per cent storage capacity retention per cycle. The organic anthraquinone species can be synthesized from inexpensive commodity chemicals. This organic approach permits tuning of the reduction potential and solubility by adding functional groups. Adding two hydroxy groups to AQDS increases the open circuit potential of the cell by 11%.

Metal Hydride

Proton flow batteries integrate a metal hydride storage electrode into a reversible proton exchange membrane (PEM) fuel cell. During charging, PFB combines hydrogen ions produced from splitting water with electrons and metal particles in one electrode of a fuel cell. The energy is stored in the form a solid-state metal hydride. Discharge produces electricity and water when the process is reversed and the protons are combined with ambient oxygen. Metals less expensive than lithium can be used and provide greater energy density than lithium cells.

Chemistries

Source

Couple	Max. cell voltage (V)	Average electrode power density (W/m^2)	Average fluid energy density $(W\,h/kg)$
Bromine-hydrogen		7,950	
Iron-tin	0.62	<200	
Iron-titanium	0.43	<200	
Iron-chrome	1.07	<200	
Vanadium-vanadium (sulphate)	1.4	~800	25
Vanadium-vanadium (bromide)			50
Sodium/bromine polysulfide	1.54	~800	
Zinc-bromine	1.85	~1,000	75
Lead-acid (methanesulfonate)	1.82	~1,000	
Zinc-cerium (methanesulfonate)	2.43	<1,200–2,500	

Advantages and Disadvantages

Redox flow batteries, and to a lesser extent hybrid flow batteries, have the advantages of flexible layout (due to separation of the power and energy components), long cycle life (because there are no solid-solid phase transitions), quick response times, no need for "equalisation" charging (the over charging of a battery to ensure all cells have an equal charge) and no harmful emissions. Some types also offer easy state-of-charge determination (through voltage dependence on charge), low maintenance and tolerance to overcharge/over-discharge.

On the negative side, flow batteries are rather complicated in comparison with standard batteries as they may require pumps, sensors, control units and secondary containment vessels. The energy densities vary considerably but are, in general, rather low compared to portable batteries, such as the Li-ion.

Applications

Flow batteries are normally considered for relatively large (1 kWh–10 MWh) stationary applications. These are for :

- Load balancing-where the battery is connected to an electrical grid to store excess electrical power during off-peak hours and release electrical power during peak demand periods.
- Storing energy from renewable sources such as wind or solar for discharge during periods of peak demand.
- Peak shaving, where spikes of demand are met by the battery.
- UPS, where the battery is used if the main power fails to provide an uninterrupted supply.
- Power conversion — because all cells share the same electrolyte/s. Therefore, the electrolyte/s may be charged using a given number of cells and discharged with a different number. Because the voltage of the battery is proportional to the number of cells used the battery can therefore act as a very powerful DC/DC converter. In addition, if the number of cells is continuously changed (on the input and/or output side) power conversion can also be AC/DC, AC/AC, or DC/AC with the frequency limited by that of the switching gear.
- Electric vehicles — Because flow batteries can be rapidly "recharged" by replacing the electrolyte, they can be used for applications where the vehicle needs to take on energy as fast as a combustion engined vehicle.
- Stand-alone power system — An example of this is the telecom industry for use in cellphone base stations where there is no grid power available. The battery can be used alongside solar or wind power sources to compensate for their fluctuating power levels and alongside a generator to make the most efficient use of it to save fuel.

Phosphoric Acid Fuel Cell

Phosphoric acid fuel cells (PAFC) are a type of fuel cell that uses liquid phosphoric acid as an electrolyte. They were the first fuel cells to be commercialized. Developed in the mid-1960s and field-tested since the 1970s, they have improved significantly in stability, performance, and cost. Such characteristics have made the PAFC a good candidate for early stationary applications.

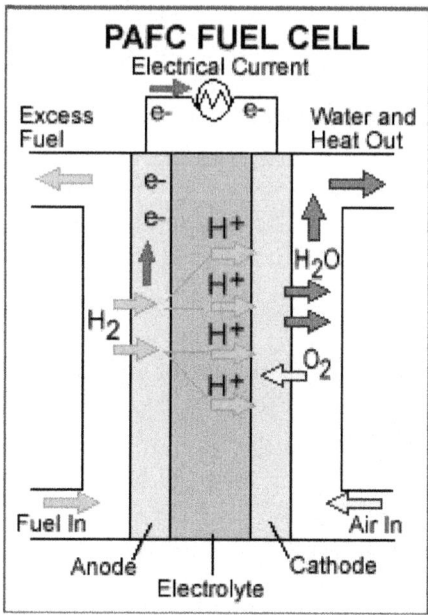

Fig. : Scheme of a phosphoric-acid fuel cell.

Design

Electrolyte is highly concentrated or pure liquid phosphoric acid (H_3PO_4) saturated in a silicon carbide matrix (SiC). Operating range is about 150 to 210°C. The electrodes are made of carbon paper coated with a finely dispersed platinum catalyst.

Electrode Reactions

Anode reaction : $2H_2 \rightarrow 4H^+ + 4e^-$

Cathode reaction : $O_2(g) + 4H^+ + 4e^- \rightarrow 2H_2O$

Overall cell reaction : $2\,H_2 + O_2 \rightarrow 2H_2O$

Advantages and Disadvantages

At an operating range of 150 to 200°C, the expelled water can be converted to steam for air and water heating (combined heat and power). This potentially allows efficiency increases of up to 70%. PAFCs are CO_2-tolerant and even can tolerate a CO concentration of about 1.5 per cent, which broadens the choice of fuels they can use. If gasoline is used, the sulfur must be removed. At lower temperatures phosphoric acid is a poor ionic conductor, and CO poisoning of the platinum electro-catalyst in the anode becomes severe. However, they are much less sensitive to CO than PEFCs and AFCs.

Disadvantages include rather low power density and aggressive electrolyte.

Applications

PAFC have been used for stationary power generators with output in the 100 kW to 400 kW range and they are also finding application in large vehicles such as buses.

Major manufacturers of PAFC technology include Clear Edge Power (formerly UTC Power) and Fuji Electric. India's DRDO are developing PAFC for air independent propulsion in their Scorpène class submarines, and the Indian Navy have requested a fully engineered system by 2014.

Molten Carbonate Fuel Cell

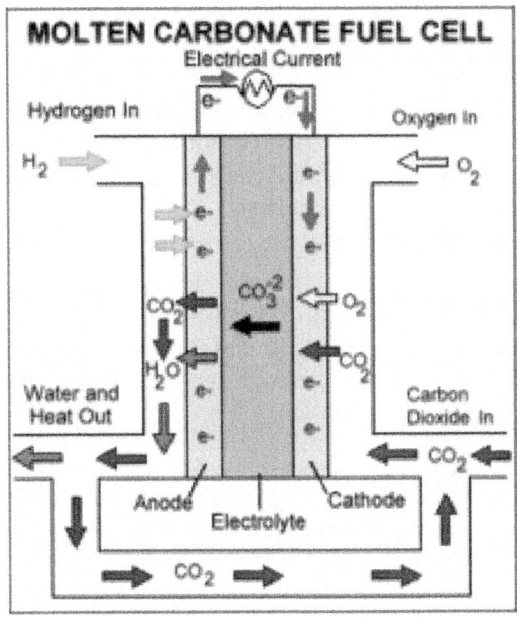

Fig. : Scheme of a molten-carbonate fuel cell.

Molten-carbonate fuel cells (MCFCs) are high-temperature fuel cells that operate at temperatures of 600°C and above.

Molten carbonate fuel cells (MCFCs) are currently being developed for natural gas, biogas (produced as a result of anaerobic digestion or biomass gasification), and coal-based power plants for electrical utility, industrial, and military applications. MCFCs are high-temperature fuel cells that use an electrolyte composed of a molten carbonate salt mixture suspended in a porous, chemically inert ceramic matrix of beta-alumina solid electrolyte (BASE). Since they operate at extremely high temperatures of 650°C (roughly 1,200°F) and above, non-precious metals can be used as catalysts at the anode and cathode, reducing costs.

Improved efficiency is another reason MCFCs offer significant cost reductions over phosphoric acid fuel cells (PAFCs). Molten carbonate fuel cells can reach efficiencies approaching 60%, considerably higher than the 37–42% efficiencies of a phosphoric acid fuel cell plant. When the waste heat is captured and used, overall fuel efficiencies can be as high as 85%.

Unlike alkaline, phosphoric acid, and polymer electrolyte membrane fuel cells, MCFCs don't require an external reformer to convert more energy-dense fuels to hydrogen. Due to the high temperatures at which MCFCs operate, these fuels are converted to hydrogen within the fuel cell itself by a process called internal reforming, which also reduces cost.

Molten carbonate fuel cells are not prone to poisoning by carbon monoxide or carbon dioxide — they can even use carbon oxides as fuel — making them more attractive for fueling with gases made from coal. Because they are more resistant to impurities than other fuel cell types, scientists believe that they could even be capable of internal reforming of coal, assuming they can be made resistant to impurities such as sulfur and particulates that result from converting coal, a dirtier fossil fuel source than many others, into hydrogen. Alternatively, because MCFCs require CO_2 be delivered to the cathode along with the oxidizer, they can be used to electro-chemically separate carbon dioxide from the flue gas of other fossil fuel power plants for sequestration.

The primary disadvantage of current MCFC technology is durability. The high temperatures at which these cells operate and the corrosive electrolyte used accelerate component breakdown and corrosion, decreasing cell life. Scientists are currently exploring corrosion-resistant materials for components as well as fuel cell designs that increase cell life without decreasing performance.

MTU Fuel Cell

The German company MTU Friedrichshafen presented an MCFC at the Hannover Fair in 2006. The unit weighs 2 tonnes and can produce 240 kW of electric power from various gaseous fuels, including biogas. If fueled by fuels that contain carbon such as natural gas, the exhaust will contain CO_2 but will be reduced by up to 50% compared to diesel engines running on marine bunker fuel. The exhaust temperature is 400°C, hot enough to be used for many industrial processes. Another possibility is to make more electric power *via* a steam turbine. Depending on feed gas type, the electric efficiency is between 12% and 19%. A steam turbine can increase the efficiency by up to 24%. The unit can be used for co-generation.

Protonic Ceramic Fuel Cell

A **protonic ceramic fuel cell** or **PCFC** is a fuel cell based on a ceramic electrolyte material that exhibits high protonic conductivity at elevated temperatures.

PCFCs share the thermal and kinetic advantages of high temperature operation at 700 degrees Celsius with molten carbonate and solid oxide fuel cells, while exhibiting all of the intrinsic benefits of proton conduction in proton exchange membrane fuel cells (PEMFC) and phosphoric acid fuel cells (PAFC). The high operating temperature is necessary to achieve very high electrical fuel efficiency with hydrocarbon fuels. PCFCs can operate at high temperatures and electro-chemically oxidize fossil fuels directly to the anode. This eliminates the intermediate step of producing hydrogen through the costly reforming process. Gaseous molecules of the hydrocarbon fuel are absorbed on the surface of the anode in the presence of water vapour, and hydrogen atoms are efficiently stripped off to be absorbed into the electrolyte, with carbon dioxide as the primary reaction product. PCFCs have a solid electrolyte, so that the membrane cannot dry out as with PEM fuel cells, and liquid cannot leak out as with PAFCs.

CoorsTek is primarily researching this type of fuel cell.

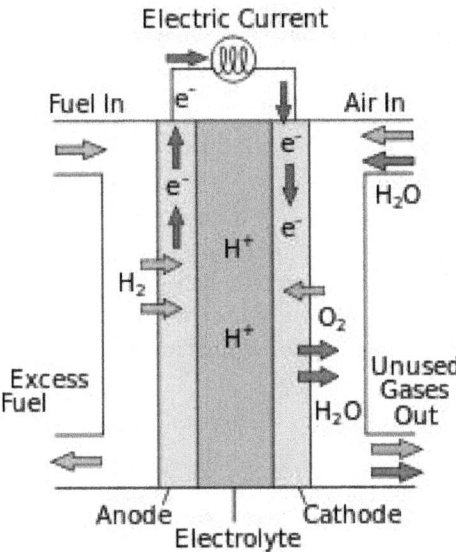

Fig. : Scheme of a proton conducting fuel cell.

Direct Carbon Fuel Cell

A **Direct Carbon Fuel Cell** (DCFC) is a fuel cell that uses a carbon rich material as a fuel. The cell produces energy by combining carbon and oxygen, which releases carbon dioxide as a by-product.

The total reaction of the cell is $C + O_2 \rightarrow CO_2$. The process in half cell notation :

- *Anode* : $C + 2O^{2-} \rightarrow CO_2 + 4e^-$
- *Cathode* : $O_2 + 4e^- \rightarrow 2O^{2-}$

Despite this release of carbon dioxide, the direct carbon fuel cell is more environmentally friendly than traditional carbon burning techniques. Due to its higher efficiency, it requires less carbon to produce the same amount of energy. Also, because pure carbon dioxide is emitted, carbon capture techniques are much cheaper than for conventional power stations. Utilized carbon can be in the form of coal, coke, char, or a non-fossilized source of carbon.

At least four types of DCFC exist :

- The first one is based on the Solid oxide fuel cell (SOFC) concept.
- The second one is molten hydroxides fuel cell. William W. Jacques obtained an US Patent 555,511 in this type of fuel cell in 1896. Prototypes have been demonstrated by the research group, SARA, Inc.
- The third one is based on the Molten Carbonate Fuel Cell (MCFC) concept. William W. Jacques obtained a Canadian patent in this type of fuel cell in 1897. It has been developed further at the Lawrence Livermore Laboratory.
- The fourth is a molten tin anode solid oxide fuel cell design, which utilizes molten tin and tin oxide as an inter stage reaction between oxidation of the carbon dissolving in the anode and reduction of oxygen at the solid oxide cathode.

Enzymatic Bio-fuel Cell

An **Enzymatic bio-fuel cell** is a specific type of fuel cell that uses enzymes as a catalyst to oxidize its fuel, rather than precious metals. Enzymatic bio-fuel cells, while currently confined to research facilities, are widely prized for the promise they hold in terms of their relatively inexpensive components and fuels, as well as a potential power source for bionic implants.

Operation

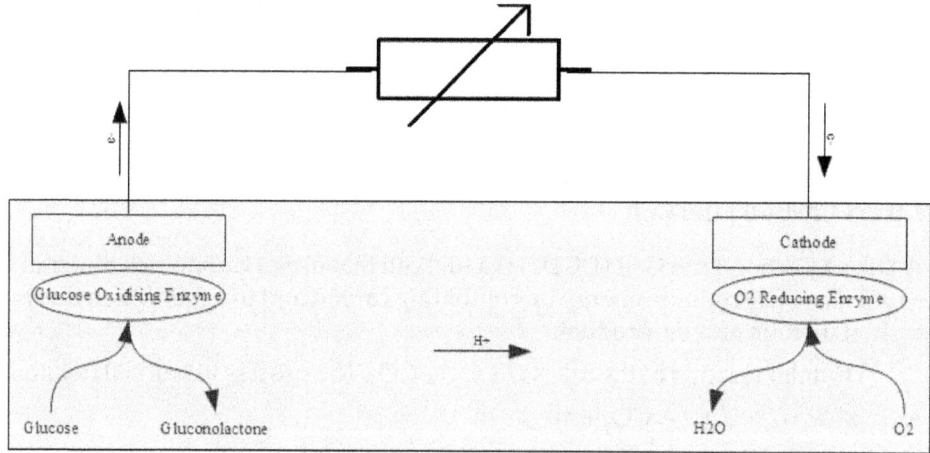

Fig. : A general diagram for an enzymatic bio-fuel cell using Glucose and Oxygen. The blue area indicates the electrolyte.

Enzymatic bio-fuel cells work on the same general principles as all fuel cells : use a catalyst to separate electrons from a parent molecule and force it to go around an electrolyte barrier through a wire to generate an electric current. What makes the enzymatic bio-fuel cell distinct from more conventional fuel cells are the catalysts they use and the fuels that they accept. Whereas most fuel cells use metals like platinum and nickel as catalysts, the enzymatic bio-fuel cell uses enzymes derived from living cells (although not within living cells; fuel cells that use whole cells to catalyze fuel are called microbial fuel cells). This offers a couple of advantages for enzymatic bio-fuel cells : Enzymes are relatively easy to mass-produce and so benefit from economies of scale, whereas precious metals must be mined and so have an inelastic supply. Enzymes are also specifically designed to process organic compounds such as sugars and alcohols, which are extremely common in nature. Most organic compounds cannot be used as fuel by fuel cells with metal catalysts because the carbon monoxide formed by the interaction of the carbon molecules with oxygen during the fuel cell's functioning will quickly "poison" the precious metals that the cell relies on, rendering it useless. Because sugars and other bio-fuels can be grown and harvested on a massive scale, the fuel for enzymatic bio-fuel cells is extremely cheap and can be found in nearly any part of the world, thus making it an extraordinarily attractive option from a logistics standpoint, and even more so for those concerned with the adoption of renewable energy sources.

Enzymatic bio-fuel cells also have operating requirements not shared by traditional fuel cells. What is most significant is that the enzymes that allow the fuel cell to operate must be "immobilized" near the anode and cathode in order to work properly; if not immobilized, the enzymes will diffuse into the cell's fuel and most of the liberated electrons will not reach the electrodes, compromising its effectiveness. Even with immobilization, a means must also be provided for electrons to be transferred to and from the electrodes. This can be done either directly from the enzyme to the electrode ("direct electron transfer") or with the aid of other chemicals that transfer electrons from the enzyme to the electrode ("mediated electron transfer"). The former technique is possible only with certain types of enzymes whose activation sites are close to the enzyme's surface, but doing so presents fewer toxicity risks for fuel cells intended to be used inside the human body. Finally, completely processing the complex fuels used in enzymatic bio-fuel cells requires a series of different enzymes for each step of the 'metabolism' process; producing some of the required enzymes and maintaining them at the required levels can pose problems.

History

Early work with bio-fuel cells, which began in the early 20th century, was purely of the microbial variety. Research on using enzymes directly for oxidation in bio-fuel cells began in the early 1960s, with the first enzymatic bio-fuel cell being produced in 1964. This research began as a product of NASA's interest in finding ways to recycle human waste into usable energy on board spacecraft, as well as a component of the quest for an artificial heart, specifically as a power source that could be put directly into the human body. These two applications–use of animal or vegetable products as fuel and development of a power source that can be directly implanted into the human body without external refueling–remain the primary goals for developing these bio-fuel cells. Initial results, however, were disappointing. While the early cells did successfully produce electricity, there was difficulty in transporting the electrons liberated from the glucose fuel to the fuel cell's electrode and further difficulties in keeping the system stable enough to produce electricity at all due to the enzymes' tendency to move away from where they needed to be in order for the fuel cell to function. These difficulties led to an abandonment by bio-fuel cell researchers of the enzyme-catalyst model for nearly three decades in favor of the more conventional metal catalysts (principally platinum), which are used in most fuel cells. Research on the subject did not begin again until the 1980s after it was realized that the metallic-catalyst method was not going to be able to deliver the qualities desired in a bio-fuel cell, and since then work on enzymatic bio-fuel cells has revolved around the resolution of the various problems that plagued earlier efforts at producing a successful enzymatic bio-fuel cell.

However, many of these problems were resolved in 1998. In that year, it was announced that researchers had managed to completely oxidize methanol using a series (or "cascade") of enzymes in a bio-fuel cell. Previous to this time, the enzyme catalysts had failed to completely oxidize the cell's fuel, delivering far lower amounts of energy than what was expected given what was known about the energy capacity of the fuel. While methanol is now far less relevant in this field as a fuel, the demonstrated method of using a series of enzymes to completely oxidize the cell's fuel gave researchers a way forward, and much work is now devoted to using similar methods to achieve complete oxidation of more complicated compounds, such as glucose. In addition, and perhaps what is more important, 1998 was the year in which enzyme "immobilization" was successfully demonstrated, which increased the usable life of the methanol fuel cell from just eight hours to over a week. Immobilization also provided researchers with the ability to put earlier discoveries into practice, in particular the discovery of enzymes that can be used to directly transfer electrons from the enzyme to the electrode. This process had been understood since the 1980s but depended heavily on placing the enzyme as close to the electrode as possible, which meant that it was unusable until after immobilization techniques were devised. In addition, developers of enzymatic bio-fuel cells have applied some of the advances in nanotechnology to their designs, including the use of carbon nanotubes to immobilize enzymes directly. Other research has gone into exploiting some of the strengths of the enzymatic design to dramatically miniaturize the fuel cells, a process that must occur if these cells are ever to be used with implantable devices. One research team took advantage of the extreme selectivity of the enzymes to completely remove the barrier between anode and cathode, which is an absolute requirement in fuel cells not of the enzymatic type. This allowed the team to produce a fuel cell that produces 1.1 microwatts operating at over half a volt in a space of just 10 cubic micrometers.

While enzymatic bio-fuel cells are not currently in use outside of the laboratory, as the technology has advanced over the past decade non-academic organizations have shown an increasing amount of interest in practical applications for the devices. In 2007, Sony announced that it had developed an enzymatic bio-fuel cell that can be linked in sequence and used to power an mp3 player, and in 2010 an engineer employed by the US Army announced that the Defense Department was planning to conduct field trials of its own "bio-batteries" in the following year. In explaining their pursuit of the technology, both organizations emphasized the extraordinary abundance (and extraordinarily low expense) of fuel for these cells, a key advantage of the technology that is likely to become even more attractive if the price of portable energy sources goes up, or if they can be successfully integrated into electronic human implants.

Magnesium-Air Fuel Cell

A Magnesium-Air Fuel Cell (MAFC) is a type of battery often called a type of fuel cell. It is made up of a magnesium anode which is consumed, oxygen from air as a cathode and a salt water electrolyte.

Research and commercialization of the technology by Mag Power Systems has shown an efficiency of 90% and an operating range of-20 to 55 degrees Celsius.

Solid Oxide Electrolyser Cell

A solid oxide electrolyzer cell (SOEC) is a solid oxide fuel cell that runs in regenerative mode to achieve the electrolysis of water and which uses a solid oxide, or ceramic, electrolyte to produce oxygen and hydrogen gas Principle.

Solid oxide electrolyzer cells operate at temperatures which allow high-temperature electrolysis to occur, typically between 500 and 850°C. These operating temperatures are similar to those conditions for an SOFC. The net cell reaction yields hydrogen and oxygen gases. The reactions for one mole of water are shown below, with oxidation of water occurring at the anode and reduction of water occurring at the cathode.

Anode : $H_2O -> 1/2O_2 + 2H^+ + 2e^-$

Cathode : $2H_2O + 2e^- -> H_2 + 2OH^-$

Net Reaction : $H_2O -> H_2 + 1/2O_2$

Electrolysis of water at 298 K (25°C) requires 285.83 kJ of energy in order to occur, and the reaction is increasingly endothermic with increasing temperature. However, the energy demand may be reduced due to the Joule heating of an electrolysis cell, which may be utilized in the water splitting process at high temperatures. Research is ongoing to add heat from external heat sources such as concentrating solar thermal collectors and geothermal sources.

Considerations

Advantages of solid oxide-based regenerative fuel cells include high efficiencies, as they are not limited by Carnot efficiency. Additional advantages include long-term stability, fuel flexibility, low emissions, and low operating costs. However, the greatest disadvantage is the high operating temperature, which results in long start-up times and break-in times. The high operating temperature also leads to mechanical compatibility issues such as thermal expansion mismatch and chemical stability issues such as diffusion between layers of material in the cell.

In principle, the process of any fuel cell could be reversed, due to the inherent reversibility of chemical reactions. However, a given fuel cell is usually optimized for operating in one mode and may not be built in such a way that it can be operated in reverse. Fuel cells operated backwards may not make very efficient systems unless they are constructed to do so such as in the case of solid oxide electrolyzer cells, high pressure electrolyzers, unitized regenerative fuel cells and regenerative fuel cells. However, current research is being conducted to investigate systems in which a solid oxide cell may be run in either direction efficiently.

Applications

SOECs have possible application in fuel production, carbon dioxide recycling, and chemicals synthesis. In addition to the production of hydrogen and oxygen, an SOEC could be used to create syngas by electrolyzing water vapour and carbon dioxide. This conversion could be useful for energy generation and energy storage applications.

Chapter 2

ELECTRO-CHEMICAL ENERGY STORAGE

FLYWHEEL ENERGY STORAGE

Flywheel energy storage (FES) works by accelerating a rotor (flywheel) to a very high speed and maintaining the energy in the system as rotational energy. When energy is extracted from the system, the flywheel's rotational speed is reduced as a consequence of the principle of conservation of energy; adding energy to the system correspondingly results in an increase in the speed of the flywheel.

Most FES systems use electricity to accelerate and decelerate the flywheel, but devices that directly use mechanical energy are being developed.

Advanced FES systems have rotors made of high strength carbon-fiber composites, suspended by magnetic bearings, and spinning at speeds from 20,000 to over 50,000 rpm in a vacuum enclosure. Such flywheels can come up to speed in a matter of minutes–reaching their energy capacity much more quickly than some other forms of storage.

Main Components

A typical system consists of a rotor suspended by bearings inside a vacuum chamber to reduce friction, connected to a combination electric motor and electric generator.

First generation flywheel energy storage systems use a large steel flywheel rotating on mechanical bearings. Newer systems use carbon-fiber composite rotors that have a higher tensile strength than steel and are an order of magnitude less heavy.

Magnetic bearings are sometimes used instead of mechanical bearings, to reduce friction.

The expense of refrigeration led to the early dismissal of low temperature super-conductors for use in magnetic bearings. However, high-temperature super-conductor (HTSC) bearings may be economical and could possibly extend the time energy could be stored economically. Hybrid bearing systems are most likely to see use first. High-temperature super-conductor bearings have historically had problems providing the lifting forces necessary for the larger designs, but can easily provide a stabilizing force. Therefore, in hybrid bearings, permanent magnets support the load and high-temperature super-conductors are used to stabilize it. The reason super-conductors can work well stabilizing the load is because they are perfect diamagnets. If the rotor tries to drift off center, a restoring force due to flux pinning restores it. This is known as the magnetic stiffness of the bearing. Rotational axis vibration can occur due to low stiffness and damping, which are inherent problems of super-conducting magnets, preventing the use of completely super-conducting magnetic bearings for flywheel applications.

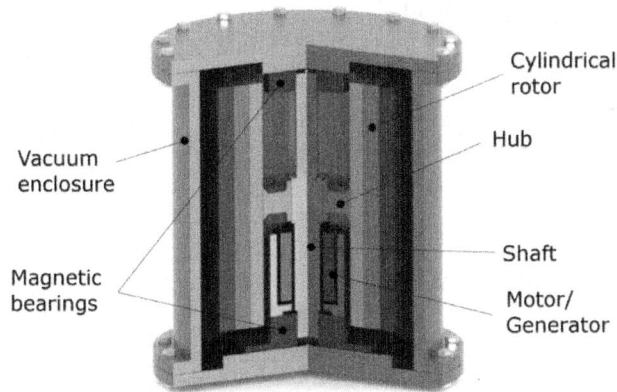

Fig. : The main components of a typical flywheel.

Since flux pinning is the important factor for providing the stabilizing and lifting force, the HTSC can be made much more easily for FES than for other uses. HTSC powders can be formed into arbitrary shapes so long as flux pinning is strong. An ongoing challenge that has to be overcome before super-conductors can provide the full lifting force for an FES system is finding a way to suppress the decrease of levitation force and the gradual fall of rotor during operation caused by the flux creep of SC material.

PHYSICAL CHARACTERISTICS

General

Compared with other ways to store electricity, FES systems have long lifetimes (lasting decades with little or no maintenance; full-cycle lifetimes quoted for fly-wheels range from in excess of 10^5, up to 10^7, cycles of use), high energy density (100–130 W · h/kg, or 360–500 kJ/kg), and large maximum power output. The energy efficiency (*ratio of energy out per energy in*) of flywheels can be as high as 90%. Typical capacities range from 3 kWh to 133 kWh. Rapid charging of a system occurs in less than 15 minutes. The high energy densities often cited with flywheels can be a little misleading as commercial systems built have much lower energy density, for example 11 W · h/kg, or 40 kJ/kg.

Energy Density

The maximum energy density of a flywheel rotor is mainly dependent on two factors, the first being the rotor's geometry, and the second being the properties of the material being used. For single-material, isotropic rotors this relationship can be expressed as :

$$\frac{E}{m} = K\left(\frac{\sigma}{\rho}\right),$$

where the variables are defined as follows :

E-kinetic energy of the rotor [J]

m-the rotor's mass [kg]

K-the rotor's geometric shape factor [dimensionless]

σ-the tensile strength of the material [Pa]

ρ-the material's density [kg/m³].

Geometry (Shape Factor)

The highest possible value for the shape factor of a flywheel rotor, is $K = 1$, which can only be achieved by the theoretical *constant-stress disc* geometry. A constant-thickness disc geometry has a shape factor of $K = 0.606$, while for a rod of constant thickness the value is $K = 0.333$. A thin cylinder has a shape factor of $K = 0.5$.

Material Properties

For energy storage purposes, materials with high strength, and low density are desirable. For this reason, composite materials are frequently being used in advanced flywheels. The strength-to-density ratio of a material can be expressed in the units [Wh/kg], and values greater than 400 Wh/kg can be achieved by certain composite materials.

Composite Rotors

Several modern flywheel rotors are made from composite materials. Examples include the *Smart Energy 25* flywheel from Beacon Power Corporation, and the *PowerThru* flywheel from Phillips Service Industries.

For these rotors, the relationship between material properties, geometry and energy density can be expressed by using a weighed-average approach.

Tensile Strength and Failure Modes

One of the primary limits to flywheel design is the tensile strength of the material used for the rotor. Generally speaking, the stronger the disc, the faster it may be spun, and the more energy the system can store.

When the tensile strength of a composite flywheel's outer binding cover is exceeded, the binding cover will fracture, followed by the wheel shattering as the outer wheel compression is lost around the entire circumference, releasing all of its stored energy at once; this is commonly referred to as "flywheel explosion" since wheel fragments can reach kinetic energy comparable to that of a bullet. Composite materials that are wound and glued in layers tend to disintegrate quickly, first into small-diameter filaments that entangle and slow each other, and then into red-hot powder, instead of large chunks of high-velocity shrapnel as can occur with a cast metal flywheel.

For a cast metal flywheel, the failure limit is the binding strength of the grain boundaries of the polycrystalline molded metal. Aluminum in particular suffers from fatigue and can develop micro-fractures due to repeated low-energy stretching. Angular forces may cause portions of a metal flywheel to bend outward and begin dragging on the outer containment vessel, or to separate completely and bounce randomly around the interior. The rest of the flywheel is now severely unbalanced, which may lead to rapid bearing failure from vibration, and sudden shock fracturing of large segments of the flywheel.

Traditional flywheel systems require strong containment vessels as a safety precaution, which increases the total mass of the device. The energy release from failure can be dampened with a gelatinous or encapsulated liquid inner housing lining, which will boil and absorb the energy of destruction. Still, many customers of large-scale flywheel energy-storage systems prefer to have them embedded in the ground to halt any material that might escape the containment vessel.

Energy Storage Efficiency

An additional limitation for some flywheel types is energy storage time. Flywheel energy storage systems using mechanical bearings can lose 20% to 50% of their energy in two hours. Much of the friction responsible for this energy loss results from the flywheel changing orientation due to the rotation of the earth (a concept similar to a Foucault pendulum). This change in orientation is resisted by the gyroscopic forces exerted by the flywheel's angular momentum, thus exerting a force against the mechanical bearings. This force increases friction. This can be avoided by aligning the flywheel's axis of rotation parallel to that of the earth's axis of rotation.

Conversely, flywheels with magnetic bearings and high vacuum can maintain 97% mechanical efficiency, and 85% round trip efficiency.

Effects of Angular Momentum in Vehicles

When used in vehicles, flywheels also act as gyroscopes, since their angular momentum is typically of a similar order of magnitude as the forces acting on the moving vehicle. This property may be detrimental to the vehicle's handling characteristics while turning or driving on rough ground; driving onto the side of a sloped embankment may cause wheels to partially lift off the ground as the flywheel opposes sideways tilting forces. On the other hand, this property could be utilized to keep the car balanced so as to keep it from rolling over during sharp turns.

The resistance of angular tilting can be almost completely removed by mounting the flywheel within an appropriately applied set of gimbals, allowing the flywheel to retain its original orientation without affecting the vehicle. This doesn't avoid the complication of gimbal lock, and so a compromise between the number of gimbals and the angular freedom is needed.

The center axle of the flywheel acts as a single gimbal, and if aligned vertically, allows for the 360 degrees of yaw in a horizontal plane. However, for instance driving up-hill requires a second pitch gimbal, and driving on the side of a sloped embankment requires a third roll gimbal.

Full-motion Gimbals

Although the flywheel itself may be of a flat ring shape, a free-movement gimbal mounting inside a vehicle requires a spherical volume for the flywheel to freely rotate within. Left to its own, a spinning flywheel in a vehicle would slowly precess following the Earth's rotation, and precess further yet in vehicles that travel long distances over the Earth's curved spherical surface.

A full-motion gimbal has additional problems of how to communicate power into and out of the flywheel, since the flywheel could potentially flip completely over once a day, precessing as the Earth rotates. Full free rotation would require slip rings around each gimbal axis for power conductors, further adding to the design complexity.

Limited-motion Gimbals

To reduce space usage, the gimbal system may be of a limited-movement design, using shock absorbers to cushion sudden rapid motions within a certain number of degrees of out-of-plane angular rotation, and then gradually forcing the flywheel to adopt the vehicle's current orientation. This reduces the gimbal movement space around a ring-shaped flywheel from a full sphere, to a short thickened cylinder, encompassing for example +/-30 degrees of pitch and +/-30 degrees of roll in all directions around the flywheel.

Counterbalancing of Angular Momentum

An alternative solution to the problem is to have two joined flywheels spinning synchronously in opposite directions. They would have a total angular momentum of zero and no gyroscopic effect. A problem with this solution is that when the difference between the momentum of each flywheel is anything other than zero the housing of the two flywheels would exhibit torque. Both wheels must be maintained at the same speed to keep the angular velocity at zero. Strictly speaking, the two flywheels would exert a huge torqueing moment at the central point, trying to bend the axle. However, if the axle were sufficiently strong, no gyroscopic forces would have a net effect on the sealed container, so no torque would be noticed.

To further balance the forces and spread out strain, a single large flywheel can be balanced by two half-size flywheels on each side, or the flywheels can be reduced in size to be a series of alternating layers spinning in opposite directions. However this increases housing and bearing complexity.

APPLICATIONS

Transportation

Automotive

In the 1950s, flywheel-powered buses, known as gyrobuses, were used in Yverdon, Switzerland and there is ongoing research to make flywheel systems that are smaller, lighter, cheaper and have a greater capacity. It is hoped that flywheel systems can replace conventional chemical batteries for mobile applications, such as for electric vehicles. Proposed flywheel systems would eliminate many of the disadvantages of existing battery power systems, such as low capacity, long charge times, heavy weight and short usable lifetimes. Flywheels may have been used in the experimental Chrysler Patriot, though that has been disputed.

Flywheels have also been proposed for use in continuously variable transmissions. Punch Powertrain is currently working on such a device.

During the 1990s, Rosen Motors developed a gas turbine powered series hybrid automotive powertrain using a 55,000 rpm flywheel to provide bursts of acceleration which the small gas turbine engine could not provide. The flywheel also stored energy through regenerative braking. The flywheel was composed of a titanium hub with a carbon fiber cylinder and was gimbal-mounted to minimize adverse gyroscopic effects on vehicle handling. The prototype vehicle was successfully road tested in 1997 but was never mass-produced.

In 2013, Volvo announced a flywheel system fitted to the rear axle of its S60 sedan. Braking action spins the flywheel at up to 60,000 rpm and stops the front-mounted engine. Flywheel energy is applied *via* a special transmission to partially or completely power the vehicle. The 20 centimetres (7.9 in), 6 kilograms (13 lb) carbon fiber flywheel spins in a vacuum to eliminate friction. When partnered with a four-cylinder engine, it offers up to a 25 per cent reduction in fuel consumption *versus* a comparably performing turbo six-cylinder, providing an 80 hp boost and allowing it to reach 100 kilometres per hour (62 mph) in 5.5 seconds. The company did not announce specific plans to include the technology in its product line.

Rail Vehicles

Flywheel systems have also been used experimentally in small electric locomotives for shunting or switching, *e.g.* the Sentinel-Oerlikon Gyro Locomotive. Larger electric locomotives, *e.g.* British Rail Class 70, have sometimes been fitted with flywheel boosters to carry them over gaps in the third rail. Advanced flywheels, such as the 133 kWh pack of the University of Texas at Austin, can take a train from a standing start up to cruising speed.

The Parry People Mover is a railcar which is powered by a flywheel. It was trialled on Sundays for 12 months on the Stourbridge Town Branch Line in the West Midlands, England during 2006 and 2007 and was intended to be introduced as a full service by the train operator London Midland in December 2008 once two units had been ordered. In January 2010, both units are in operation.

Rail Electrification

FES can be used at the lineside of electrified railways to help regulate the line voltage thus improving the acceleration of unmodified electric trains and the amount of energy recovered back to the line during regenerative braking, thus lowering energy bills. Trials have taken place in London, New York, Lyon and Tokyo, and New York MTA's Long Island Rail Road is now investing $5.2m in a pilot project on LIRR's West Hempstead Branch line.

Uninterruptible Power Supplies

Flywheel power storage systems in production as of 2001 have storage capacities comparable to batteries and faster discharge rates. They are mainly used to provide load levelling for large battery systems, such as an uninterruptible power supply for data centers as they save a considerable amount of space compared to battery systems.

Flywheel maintenance in general runs about one-half the cost of traditional battery UPS systems. The only maintenance is a basic annual preventive maintenance routine and replacing the bearings every five to ten years, which takes about four hours. Newer flywheel systems completely levitate the spinning mass using maintenance-free magnetic bearings, thus eliminating mechanical bearing maintenance and failures.

Costs of a fully installed flywheel UPS are about $330 per kilowatt in combination with a diesel generator set or integrated design, it supplies continuous power as long as there is fuel.

Laboratories

A long-standing niche market for flywheel power systems are facilities where circuit-breakers and similar devices are tested : even a small household circuit-breaker may be rated to interrupt a current of 10,000 or more amperes, and larger units may have interrupting ratings of 100,000 or 1,000,000 amperes. The enormous transient loads produced by deliberately forcing such devices to demonstrate their ability to interrupt simulated short circuits would have unacceptable effects on the local grid if these tests were done directly from building power. Typically such a laboratory will have several large motor-generator sets, which can be spun up to speed over some minutes; then the motor is disconnected before a circuit breaker is tested.

Other similar high power applications are in tokamak fusion (like the Joint European Torus) and laser experiments, where very high currents are also used for very brief intervals. JET has two 775 ton flywheels that spin up to 225 rpm. Each flywheel stores 3 GJ.

Amusement Rides

The Incredible Hulk roller coaster at Universal's Islands of Adventure features a rapidly accelerating uphill launch as opposed to the typical gravity drop. This is achieved through powerful traction motors that throw the car up the track. To achieve the brief very high current required to accelerate a full coaster train to full speed uphill, the park utilizes several motor generator sets with large flywheels. Without these stored energy units, the park would have to invest in a new sub-station or risk browning-out the local energy grid every time the ride launches.

Pulse Power

Since FES can store and release energy quickly, they have found a niche providing pulsed power.

Motor Sports

The FIA re-allowed the use of KERS (see kinetic energy recovery system) as part of its Formula One 2009 Sporting Regulations. KERS were used in the 2011 Formula 1 season. Using a continuously variable transmission (CVT), energy is recovered from the drive train during braking and stored in a flywheel. This stored energy is then used during acceleration by altering the ratio of the CVT. In motor sports applications this energy is used to improve acceleration rather than reduce carbon dioxide emissions–although the same technology can be applied to road cars to improve fuel efficiency.

Automobile Club de l'Ouest, the organizer behind the annual 24 Hours of Le Mans event and the Le Mans Series, is currently "studying specific rules for LMP1 which will be equipped with a kinetic energy recovery system."

Williams Hybrid Power, a subsidiary of Williams F1 Racing team, have supplied Porsche and Audi with flywheel based hybrid system for Porsche's 911 GT3 R Hybrid and Audi's R18 e-Tron Quattro. Audi's victory in 2012 24 Hours of Le Mans is the first for a hybrid(diesel-electric) vehicle.

Grid Energy Storage

Beacon Power opened a 5MWh, (20 MW over 15 mins) flywheel energy storage plant in Stephentown, New York in 2011. Lower carbon emissions, faster response times and ability to buy power at off-peak hours are among some advantages of using flywheels instead of traditional sources of energy for peaking power plants.

Wind Turbines

Flywheels may be used to store energy generated by wind turbines during off-peak periods or during high wind speeds.

Beacon Power began testing of their Smart Energy 25 (Gen 4) flywheel energy storage system at a wind farm in Tehachapi, California. The system is part of a wind power/flywheel demonstration project being carried out for the California Energy Commission (Beacon Power Press Release March 2010).

Toys

Friction motors used to power many toy cars, trucks, trains, action toys and such, are simple flywheel motors.

Toggle Action Presses

In industry, toggle action presses are still popular. The usual arrangement involves a very strong crankshaft and a heavy duty connecting rod which drives the tup. Large and heavy flywheels are driven by electric motors but the flywheels only turn the crankshaft when clutches are activated.

Comparison to Batteries

Flywheels are not as adversely affected by temperature changes, can operate at a much wider temperature range, and are not subject to many of the common failures of chemical rechargeable batteries. They are also less potentially damaging to the environment, being largely made of inert or benign materials. Another advantage of flywheels is that by a simple measurement of the rotation speed it is possible to know the exact amount of energy stored.

Unlike lithium ion polymer batteries which operate for a finite period of roughly 36 months, a flywheel can potentially have an indefinite working lifespan. Flywheels built as part of James Watt steam engines have been continuously working for more than two hundred years. Working examples of ancient flywheels used mainly in milling and pottery can be found in many locations in Africa, Asia, and Europe.

However, this is a somewhat unfair comparison because batteries are typically a complex sealed device that is minimally maintained throughout its service life. Flywheels in a sealed device would have a similar lifespan because eventually components such as bearings wear out and need replacement. Open flywheels are subject to airborne dust collecting in the bearing grease, which will lead to loss of efficiency and eventually seizure if the grease is not periodically replaced or replenished. Closed flywheels are not subject to such deterioration.

Bearing replacement can be quite difficult due to the high mass of the flywheel, and may require a large crane to lift and support it while the bearings are serviced. Severe injury or death from pinching and crushing can occur during servicing due to the very high mass of the flywheel, which also has the potential to roll away at high speed on gently sloped surfaces, if not properly restrained or supported during servicing.

HOW TO MEASURE CAPACITY

The traditional charge/discharge/charge cycle still offers a dependable way to measure battery capacity. Alternative methods have been tried but none deliver reliable readings. Inaccuracies have led users to adhere to the proven discharge methods even if the process is time-consuming and removes the battery from service for the duration of the test.

While portable batteries can be discharged and recharged relatively quickly, a full discharge and recharge on large lead acid batteries gets quite involved, and service personnel continue to seek faster methods even if the readings are less accurate.

Discharge Method

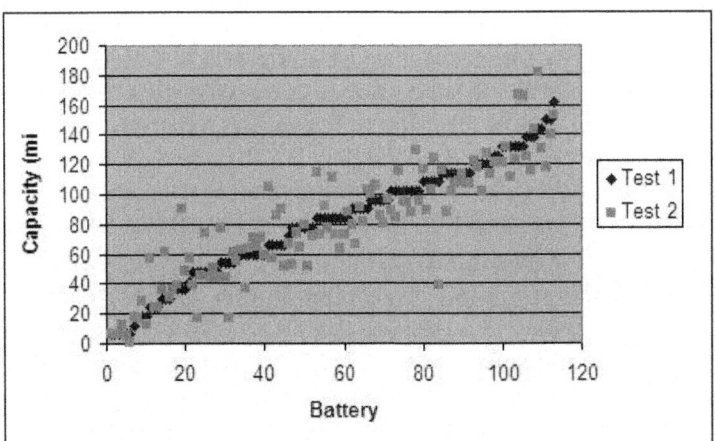

Fig. : Capacity fluctuations on two identical charge/discharge tests of 91 starter batteries. The capacities differ +/–15% between Test 1 and Test 2.
Courtesy of Cadex (2005).

One would assume that capacity measurement with discharge is accurate but this is not always the case, especially with lead acid batteries. In fact, there are large variations between identical tests, even when using highly accurate equipment and following established charge and discharge standards, with temperature control and mandated rest periods. This behaviour is not fully understood except to consider that batteries exhibit human-like qualities. Our IQ levels also vary depending on the time of day and other conditions. Nickel-and lithium-based chemistries provide more consistent results than lead acid on discharge/charge tests.

The horizontal x-axis shows the batteries from weak to strong, and the vertical y-axis reflects capacity. The batteries were prepared in the Cadex laboratories according to SAE J537 standards by giving them a full charge and a 24-hour rest. The capacity was then measured by applying a regulated 25A discharge to 10.50V (1.75V/cell) and the results plotted in diamonds. The test was repeated under identical conditions and the resulting capacities added in squares. The second reading exhibits differences in capacity of +/−15 per cent across the battery population. Other laboratories that test lead acid batteries experience similar discrepancies.

Capacity *vs.* CCA

Starter batteries have two distinct values, CCA and capacity. These two readings are close to each other like lips and teeth, but the characteristics are uniquely different; one cannot predict the other. [BU-806, Changes in Capacity and Resistance]

Measuring the internal battery resistance, which relates to CCA on a starter battery, is relatively simple but the reading provides only a snapshot of the battery at time of measurement. Resistance alone cannot predict the end of life of a battery. For example, at a CCA of 560A and a capacity of 25 per cent, a starter battery will still crank well but it can surprise the motorist with a sudden failure of not turning the engine (as I have experienced).

The leading health indicator of a battery is capacity, but this estimation is difficult to read. A capacity test by discharge is not practical with starter batteries; this would cause undue stress and take a day to complete. Most battery testers do not measure capacity but look at the internal resistance, which is an approximation of CCA. The term approximation is correct—laboratory tests at Cadex and at a German luxury car manufacturer reveal that the readings are only about 70 per cent accurate. A full CCA test is seldom done; one battery can take a week to measure.

The SAE J537 CCA test mandates to cool a fully charged battery to-18°C (0°F) for 24 hours, and while at sub-freezing temperature apply a high-current discharge that simulates the cranking of an engine. A 500 CCA battery would need to supply 500A for 30 seconds and stay above 7.2V (1.2V/cell) to pass. If it fails the test, the battery has a CCA rating of less than 500A. To find the CCA rating, the test must be repeated several times with different current settings to find the triggering point when the battery passes through 7.2V line. Between each test, the battery must be brought to ambient temperature for recharging and cooled again for testing. (For CCA DIN and IEC norms, please refer to "Test Method" on this essay.)

To examine the relationship between CCA and capacity, Cadex measured CCA and capacity of 175 starter batteries at various performance levels. Figure shows the CCA on the vertical y-axis and reserve capacity* readings on the horizontal x-axis. The batteries are arranged from low to high, and the values are given as a percentage of the original ratings.

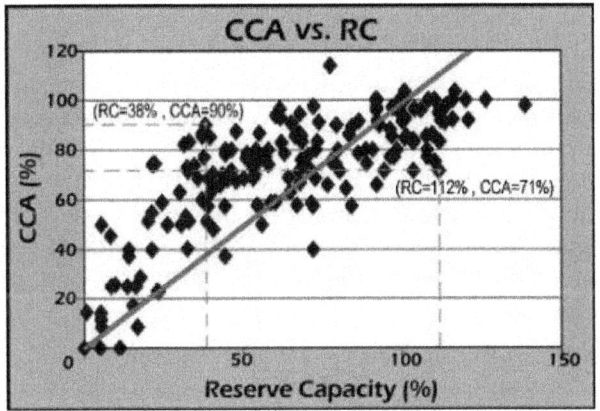

Fig. : CCA and reserve capacity (RC) of 175 aging starter batteries.

The CCA of aging starter batteries gravitates above the diagonal reference line. (Few batteries have low CCA and high capacity.)

Test method : The CCA and RC readings were obtained according to SAE J537 standards (BCI). CCA (BCI) loads a fully charged battery at–18°C (0°F) for 30s at the CCA-rated current of the battery. The voltage must stay above 7.2V to pass. CCA DIN and IEC are similar with these differences : DIN discharges for 30s to 9V, and 150s to 6V; IEC discharges for 60s to 8.4V. RC applies a 25A discharge to 1.75V/cell and measures the elapsed time in minutes.

The table shows noticeable discrepancies between CCA and capacity, and there is little correlation between these readings. Rather than converging along the diagonal reference line, CCA and RC wander off in both directions and resemble the stars in a clear sky. A closer look reveals that CCA gravitates above the reference line, leaving the lower right vacant. High CCA with low capacity is common, however, low CCA with high capacity is rare. In our table, one battery has 90 per cent CCA and produces a low 38 per cent capacity; another delivers 71 per cent CCA and delivers a whopping 112 per cent capacity (these are indicated by the dotted lines).

As discussed earlier, a battery check must include several test points. An analogy can be made with a medical doctor who examines a patient with several instruments to find the diagnosis. A serious illness could escape the doctor's watchful eyes if only blood pressure or temperature was taken. While medical staff are well trained to evaluate multiple data points, most battery personnel do not have the knowledge to read a Nyquist plot and other data on a battery scan. Nor are test devices available that give reliable diagnosis of all battery ills.

TESTING LEAD ACID BATTERIES

Many manufacturers of battery testers claim to measure battery health on the fly. These instruments work well in finding battery defects that involve voltage anomalies and elevated internal resistance, but other performance criteria remain unknown. Stating that a battery tester based on internal resistance can also measure capacity is misleading. Advertising features that are outside the equipment's capabilities confuses the industry into believing that multifaceted results are attainable with basic methods. Manufacturers of these instruments are aware of the complexity involved, but some like to add a flair of mystery in their marketing scheme, similar to a maker of a shampoo product promising to grow lush hair on a man's bald head. Here is a brief history of battery testers for lead acid and what they can do.

The *carbon pile*, introduced in the 1980s, applies a DC load of short duration to a starter battery, simulating cranking. The voltage drop and recovery time provide a rough indication of battery health. The test works reasonably well and offers evidence that power is present. A major advantage is the ability to detect batteries that have failed due to a shorted cell (low specific gravity in a cell due to high self-discharge). Capacity estimation, however, is not possible, and a battery that has a low state-of-charge appears as *weak*. A skilled mechanic can, however, detect a faulty battery based on the voltage signature and loading behaviour. To do a CCA pass/fail test, load a fully charged starter battery with half the rated CCA value for 15 seconds. To pass, the voltage must stay above 9.6V at 10°C (50°F) and higher. Colder temperatures cause a large voltage drop.

The *AC conductance* meters appeared in 1992 and were hailed as a breakthrough. The non-invasive method injects an AC signal into the battery to measure the internal resistance. Today, these testers are commonly used to check the CCA of starter batteries and verify resistance change in stationary batteries. While small and easier to use, AC conductance cannot read capacity, and the resistive value gives only an approximation of the real CCA of a starter battery. A shorted cell could pass as good because in such a battery the overall conductivity and terminal voltage are close to normal, even though the battery cannot crank the motor. AC conductance testers are common in North America; Europe prefers the DC load method.

Critical progress has been made towards *electro-chemical impedance spectroscopy* (EIS). Cadex took the EIS technology a step further and developed battery specific models that are able to estimate the health of a lead acid battery.

The Spectro CA-12 handheld device, in which the Spectro™ technology is embedded, excites the battery with signals from 20–2000Hz. A DSP deciphers the 40 million transactions churned out during the 15-second test into readable results. To check a battery, the user simply selects the battery voltage, Ah and designated matrix. Tests can be done under a steady load of up to 30A and a partial charge, however, if the state-of-charge is less than 40 per cent, the instrument advises the user to charge and retest.

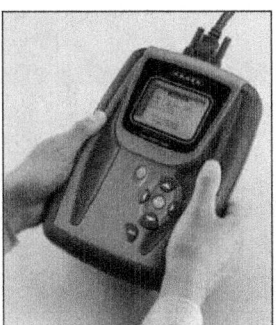

Fig. : Spectro CA-12 battery tester.

Compact battery rapid tester displays capacity, CCA and state-of-charge in 15 seconds.
Courtesy Cadex

The Spectro method is a further development of EIS, a technology that had been around for several decades. What's new is the use of multi-models and faster process times. Cost and size have also shrunk. Earlier models cost tens of thousands of dollars and travelled on wheels. The heart of Spectro is not so much the mechanics but the algorithm. No longer do modern EIS devices accompany a team of scientist to decipher tons of data. Experts predict that the battery industry is moving towards the multi-model EIS technology to estimate batter performance.

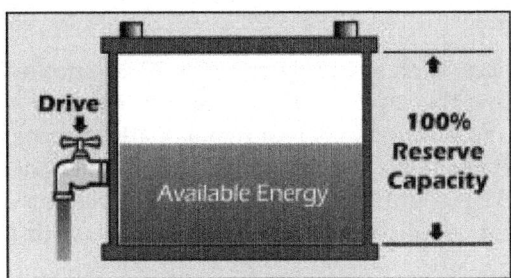

Fig. : Low charge.

Drive is sluggish; Spectroäreads low SoC. Capacity estimation is correct in spite of low charge.

Fig. : Low capacity.

Battery has good drive but short run-times. Spectroäreads good impedance but low capacity.

Fig. : Faulty set.

Spectroäfinds low performing and shorted blocks in a string. Good batteries can be regrouped and reused.
All figures Courtesy of Cadex

Nowhere is the ability to read capacity more meaningful than with deep-cycle batteries in golf cars, aerial work platforms and wheelchairs, as well as military and naval carriers. Getting a readout in seconds without putting the vehicles out of commission allows for a quick performance check on a suspect battery before deployment in the field. Figures show typical battery problems and how modern test technologies can detect them.

Matrices

Measurement devices, such as the Spectro CA-12, are not universal instruments capable of estimating the capacity of any battery that may come along; they require battery specific matrices, also known as pattern recognition algorithm. A matrix is a multi-dimensional lookup table against which the measured readings are compared. Text recognition, fingerprint identification and visual imaging operate on a similar principle in that a model exists, with which to equate the derived readings.

This book identifies three commonly used measuring methods. The principle in all is to take one or several sets of readings and compare them against known reference settings or images to disclose the characteristics of a battery. The three methods are as follows.

Scalar : The *single value scalar test* takes a reading and compares the result with a stored reference value. In battery testing this could be measuring a voltage, interrogating the battery by applying discharge pulses or injecting a frequency and then comparing the derived result against a single reference point. This is the simplest test, and most DC load and single-frequency AC conductance testers use this method.

Vector : The *vector method* applies pulses of different currents, or excites the battery with several frequencies, and evaluates the results against preset vector points to study the battery under various stress conditions. Typical applications for this one-dimensional scalar model are battery testers that apply multi-tier DC loads or inject several test frequencies. Because of added complexity in evaluating the different data points and limited benefits, the vector method is seldom used.

Matrix : The *matrix method* scans a battery with a frequency spectrum as if to capture the image of a landscape and compare the imprint with a stored model of known characteristics. This multi-dimensional set of scalars, which form the foundation of Spectroä, provides the most in-depth information but is complex in terms of evaluating the data generated. With a proprietary algorithm, the Spectroätechnology is able to estimate battery capacity, CCA and SoC.

Matrices are primarily used to estimate battery capacity, however, CCA and state-of-charge also require matrices. These are easier to assemble and serve a broad range of starter batteries. While the Spectroämethod offers an accuracy of 80 to 90 per cent on capacity, CCA is 95 per cent exact. This compares to 60 to 70 per cent with battery testers using the scalar method. Service personnel are often unaware of the low accuracy; verifications are seldom done, as this would involve several days of laboratory testing.

A further drawback of scalar battery testers is obtaining a reading that is neither resistance nor CCA. While there are similarities between the two, no standard exists and each instrument gives different values. In terms of assessing a dying battery, however, this method is adequate as it reflects conductivity. The larger disadvantage is not being able to read capacity. Table 5 illustrates test accuracies using scalar, vector and matrix methods.

Table : Accuracy in battery readings with different measuring methods.

Measuring units	Scalar Single value	Vector One-dimensional set of scalars	Matrix Multi-dimensional set of scalars
CCA	60–70% accurate		90–95% accurate
Capacity	N/A		80–90% accurate
SoC	Voltage-based; requires rest after charge and discharge		90–95% accurate (with new battery)

Scalar and vector provide resistance with references to CCA on starter batteries. The matrix method is more accurate and provides capacity estimations but needs reference matrices.

To generate a matrix, batteries with different state-of-health are scanned. The more batteries of the same model but diverse capacity mix are included in the mix, the stronger the matrix will become. If, for example, the matrix consists only of two batteries, one showing a capacity of 60 per cent and the other 100 per cent, then the accuracy would be low for the batteries in between. Adding a third battery with an 80 per cent capacity will solidify the matrix, similar to placing a pillar at the center of a bridge. To cover the full spectrum, a well-developed matrix should include battery samples capturing capacities of 50, 60, 70, 80, 90 and 100 per cent. Batteries much below 50 per cent are less important because they constitute a fail.

It is difficult to obtain aged batteries, especially with newer models. Forced aging by cycling in an environmental chamber is of some help; however, age-related stresses from the field are not represented accurately. It also helps to include batteries from different regions to represent unique environmental user patterns. A starter battery in a Las Vegas taxi has different strains than that of a car driven by a grandmother in northern Germany.

Different state-of-charge levels increase the complexity to estimate battery health. The SoC on a new battery can be determined relatively easily with imped-ance spectroscopy, however, the formula changes as the battery ages. A battery tester should therefore be capable of examining new and old batteries with a charge level of 40 to 100 per cent. With ample data, this is possible because natural aging of a battery is predictable and the scanned information can be massaged to calculate age. This is similar to face recognition that correctly identifies a person even if he or she has developed a few wrinkles and has grown gray hair.

Simplifications in matrix development are possible by grouping batteries that share the same chemistry, voltage and a similar capacity range into a generic matrix. This simplifies logistics; however, the readout is classified into categories rather than numbers. Figure illustrates the classification scheme of Good, Low and Poor. Good passes as a good battery; Low is suspect and predicts the end of life; and Poor is a fail that mandates replacement.

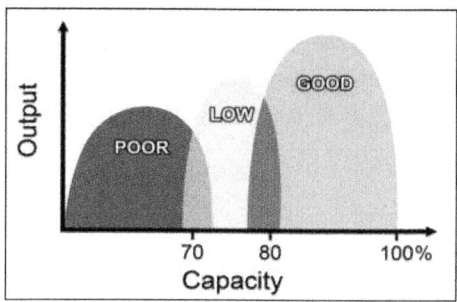

Fig. : 6 : Classifying batteries into categories.

The classification method provides an intelligent assessment of what constitutes a usable battery for a given application. Some classifications have pass/fail; others provide GOOD, LOW and POOR.
Courtesy of Cadex

Service personnel appreciate the classification method because it gives them an intelligent assessment of what constitutes a usable battery for a given application and eliminates customer interference. If numeric capacity readings are mandatory for a given battery type, a designated matrix can be developed and downloaded into the tester from the Internet.

TESTING NICKEL-BASED BATTERIES

Nickel-based batteries have unique properties, and Cadex developed a rapid-test method for these battery systems called *QuickTest*™. The process takes three minutes and uses an inference algorithm. Figure illustrates the general structure of the algorithm applied.

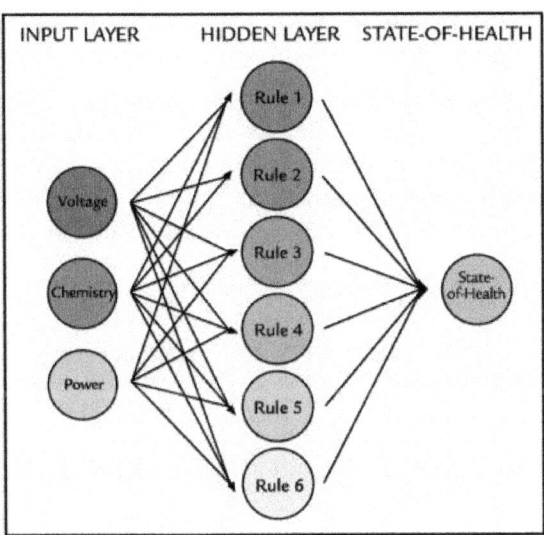

Fig. : QuickTest™ structure.

Multiple variables are fed to the micro controller, "'fuzzified" and processed by parallel logic. The data is averaged and weighted according to battery application.
Courtesy of Cadex

QuickTest™ fuses data from six variables, which are capacity, internal resistance, self-discharge, charge acceptance, discharge capabilities and mobility of electrolyte. A trend-learning algorithm combines the data to provide a dependable state-of-health (SoH) reading in percentage. The system uses battery-specific matrices stored in battery adapters of a designated battery analyzer (Cadex). The user can create a matrix in the field by scanning two or more batteries on the analyzer's Learn program. The battery must be at least 20 per cent charged.

Among other parameters, QuickTest™ relies on the internal resistance of a battery pack, and the welding joints between the cells might cause a problem, especially on packs with 10 cells or more. Although seemingly insignificant in terms of added resistance, mechanical linkages behave differently to a chemical cell and this causes an unwanted error. The linkage error is not seen on a conventional discharge test or when doing a resistance check but interferes with rapid-test methods on voltages above 20V. It is also possible that each cell of a multi-cell pack behaves on its own when excited with a common signal and the result gets muddled.

TESTING LITHIUM-BASED BATTERIES

With the large number of lithium-ion batteries in use and the population growing rapidly, developing an effective testing method has become an urgent task. *QuickSort*™ (Cadex) is a further development of QuickTest™ using a generic matrix. The simplification was made possible by limiting the battery population to single-cell Li-ion from 500 to 1,500mAh. (Larger cells and higher voltages will need a different generic matrix.) Rather than capacity readout in percentage, QuickSort™ classifies the battery health as Good, Low or Poor.

Electro-chemical dynamic response, the method used for QuickSort™, measures the mobility of ion flow between the electrodes on a digital load. The response can be compared with a mechanical arm under load. A strong arm resembling a good battery remains firm, and a weak arm synonymous to a faded battery bends and becomes sluggish under load. Figure illustrates the concept of the technology.

DISCHARGE PULSE

Fig. : Electro-chemical dynamic response.

The electro-chemical dynamic response measures the ion flow between the positive and negative plates. This process can be compared to a mechanical arm under load.
Courtesy of Cadex

The test takes 30 seconds, is 90 per cent accurate regardless of battery cathode material and can be performed with a state-of-charge range of between 40 and 100 per cent. QuickSort™ requires the correct mAh, and setting a wrong value does not shift the reading on a linear scale from good to poor, as one would expect, but makes the sorting less accurate. The system does not rely on internal resistance *per se*. This would produce unreliable readings because modern lithium-ion maintains a low resistance with use and time. Read more about How to Measure Internal Resistance. At the conclusion of the test, however, an overall resistance check is performed.

Lithium-ion batteries have different diffusion rates, and in terms of electrochemical dynamic response, Li-ion polymer with gelled electrolyte appears to be faster than Li-ion containing liquefied electrolyte. Li-polymer may need a different matrix to produce accurate readings.

Scientists explore new ways to evaluate the health of a battery with scanning frequencies ranging from several kilohertz to milihertz. High frequencies reveal the resistive qualities of a battery, which presents a bird-eye's view in landscape form. By lowering the frequency, diffusion begins to provide insight into unique battery characteristics that allow capacity estimation, sulfation detection and revealing of dry-out condition.

Evaluating batteries at sub one-hertz frequency needs long test times. At one milihertz, for example, a cycle takes 1,000 seconds and several data points are required to assess a battery with certainty. Low-frequency tests can take several minutes for one measurement, however, with clever software simulation, the duration can be shortened to just a few seconds.

Research engineers at Cadex are working on a technique called *Low Frequency Pulse Train* (LFPT), also known as *diffusion technology*. Diffusion works with most chemistries and the information retrieved provides vital information relating to battery capacity and underlying deficiencies. This knowledge enables the all-important *state-of-life estimation*, the ultimate goal for advanced battery management systems (BMS).

There is a critical need for practical battery testers that can examine the state-of-health of batteries in medical equipment, military instruments, computing devices, power tools and UPS systems. There are currently no instruments that can reliably predict battery state-of-life on the fly, although many device manufacturers may claim their instruments will do so.

HOW TO MONITOR A BATTERY

One of the most urgent requirements for battery-powered devices is the development of a reliable and economical way to monitor battery state-of-function (SoF). This is a demanding task when considering that there is still no dependable method to read state-of-charge, the most basic characteristic of a battery. Even if SoC were displayed accurately, charge information alone has limited benefits without knowing the capacity. The objective is to identify *battery readiness,* which describes what the battery can deliver at a given moment. SoF includes capacity (the amount of energy the battery can hold), internal resistance (the delivery of power), and state-of-charge (the amount of energy the battery holds at that moment).

Stationary batteries were among the first to include monitoring systems, and the most common form of supervision is voltage measurement of individual cells. Some systems also include cell temperature and current measurement. Knowing the voltage drop of each cell at a given load reveals cell resistance. Cell failure caused by rising resistance through plate separation, corrosion and other malfunctions can thus be identified. Battery monitoring also serves in medical, defense and communication devices, as well as wheeled mobility and electric vehicle applications.

In many ways, present battery monitoring falls short of meeting the basic requirements. Besides assuring *readiness*, battery monitoring should also keep track of aging and offer end-of-life predictions so that the user knows when to replace a fading battery. This is currently not being done in a satisfactory manner. Most monitoring systems are tailored for new batteries and adjust poorly to aging ones. As a result, battery management systems (BMS) tend to lose accuracy gradually until the information obtained gets so far off that it becomes a nuisance. This is not an oversight by the manufacturers; engineers know about this shortcoming. The problem lies in technology, or lack thereof.

Another limitation of current monitoring systems is the bandwidth in which battery conditions can be read. Most systems only reveal anomalies once the battery performance has dropped below 70 per cent and the performance is being affected. Assessment in the all-important 80–100 per cent operating range is currently impossible, and systems give the batteries a good bill of health. This complicates end-of-life predictions, and the user needs to wait until the battery has sufficiently deteriorated to make an assessment. Measuring a battery once the performance has dropped or the battery has died is ineffective, and this complicates battery exchange systems proposed for the electric vehicle market. One maker of a battery tester proudly states in a brochure that their instrument "Detects any faulty battery." So, eventually, does the user.

Some medical devices use date stamp or cycle count to determine the end of service life of a battery. This does not work well either, because batteries that are used little are not exposed to the same stresses as those in daily operation. To reduce the risk of failure, authorities may mandate an earlier replacement of all batteries. This causes the replacement of many packs that are still in good working condition. Old habits are hard to break, and it is often easier to leave the procedure as written rather than to revolt. This satisfies the battery vendor but increases operating costs and creates environmental burdens.

Portable devices such as laptops use coulomb counting that keeps track of the in-and out flowing currents. Such a monitoring device should be flawless, but as mentioned earlier, the method is not ideal either. Internal losses and inaccuracies in capturing current flow add to an unwanted error that must be corrected with periodic calibrations.

Over-expectation with monitoring methods is common, and the user is stunned when suddenly stranded without battery power. Let's look at how current systems work and examine up-and-coming technologies that may change the way batteries are monitored.

Voltage-Current-Temperature Method

The Volkswagen Beetle in simpler days had minimal battery problems. The only management system was ensuring that the battery was being charged while driving. Onboard electronics for safety, convenience, comfort and pleasure have greatly added to the demands on the battery in modern cars since then. For the accessories to function reliably, the state-of-charge of the battery must be known at all times. This is especially critical with start-stop technologies, a mandated requirement on new European cars to improve fuel economy.

When the engine stops at a red light, the battery draws 25–50 amperes of current to feed the lights, ventilators, windshield wipers and other accessories. When the light changes, the battery must have enough charge to crank the engine, which requires an additional 350A. With the engine started again and accelerating to the posted speed limit, the battery begins charging after a 10-second delay.

Realizing the importance of battery monitoring, car manufacturers have added battery sensors that measure voltage, current and temperature. Packaged in a small housing that forms part of the positive clamp, the *electronic battery monitor* (EBM)provides useful information about the battery and provides an accuracy of about +/–15 per cent when the battery is new. As the battery ages, the EBM begins drifting and the accuracy drops to 20–30 per cent. The model used for monitoring the battery is simply not able to adjust. To solve this problem, EBM would need to know the state-of-health of the battery, and that includes the all-important capacity. No method exists today that is fully satisfactory, and some mechanics disconnect the battery management system to stop the false warning messages.

A typical start-stop vehicle goes through about 2,000 micro cycles per year. Test data obtained from automakers and the Cadex laboratories indicate that with normal usage in a start-stop configuration, the battery capacity drops to approximately 60 per cent in two years. Field use reveals that the standard flooded lead acid lacks robustness, and carmakers are reverting to a modified version lead acid battery.

Automakers want to ensure that no driver gets stuck in traffic with a dead battery. To conserve energy, modern cars automatically turn off unnecessary accessories when the battery is low and the motor stays running at a stoplight. Even with this measure, state-of-charge can remain low if commuting in gridlock conditions because motor idling does not provide much charge to the battery, and with essential accessories like lights and windshield wipers on, the net effect could be a small discharge.

Battery monitoring is also important on hybrid vehicles to optimize charge levels. Intelligent charge management prevents stressful over-charge and avoids deep discharges. When the charge level is low, the internal combustion (IC) engine engages earlier than normal and is left running longer for additional charge. On a fully charged battery, the IC engine turns off and the car moves on the electrical motor in slow traffic.

Improved battery management is of special interest to the manufacturers of the electric vehicle. In terms of state-of-charge, a discerning driver expects similar accuracies in energy reserve as are possible with a fuel-powered vehicle, and current technologies do not yet allow this. Furthermore, the driver of an EV anticipates a fully charged battery will power the vehicle for the same distance as the car ages. This is not the case and the drivable distance will get shorter with each passing year. Distances will also be shorter when driving in cold temperatures because of reduced battery performance.

BATTERY TEST EQUIPMENT

Batteries are commonly tested by measuring the capacity through a full discharge. While voltage and internal resistance provide a rough indication of the battery condition, these readings do not disclose the capacity, the leading health indicator of a battery. Voltage and resistance tend to reveal anomalies only when the battery is in a fault mode. Most batteries keep a normal voltage and low resistance while the capacity gradually fades with age.

There is a move towards rapid testing, however, current methods only provide an estimation of the battery performance and the results can be in accurate. Rapid-test methods work best with single-cell Li-ion packs; series and parallel connection of cells can distort the readings. Public safety, medical and defense organizations still apply a periodic full discharge/charge cycles, and this is normally done with a battery analyzers.

Battery Analyzer

Battery analyzers became popular in the 1980s and 1990s to restore nickel-cadmium batteries affected by "memory." Today, battery analyzers serve in identifying packs that no longer meet requirements; they form a vital part in maintaining fleet batteries. Read about How to Maintain Fleet Batteries. Typical battery analyzers are the Cadex C7000 Series, workhorses that serve a broad range of batteries. These devices accommodate lead-, nickel-and lithium-based batteries, feature automated service programs and operate in stand-alone mode or with PC software.

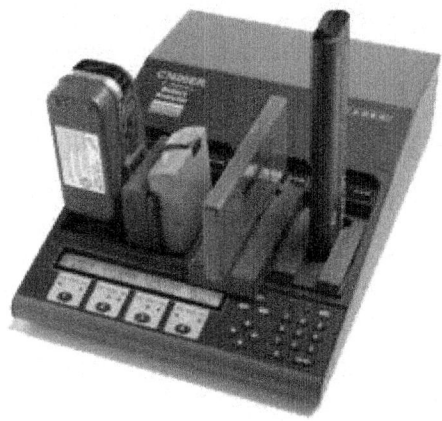

Fig. : Cadex C7400ER battery analyzer.

Four-station battery analyzer services batteries of up to 36V and 6A per station. Custom and universal battery adapters accommodate lead-, nickel-and lithium-based batteries. Courtesy of Cadex

The Cadex analyzers include *Custom* programs in which the user sets a unique sequence of charge, discharge, recondition, wait and repeat. The *Lifecycle* program cycles battery until the capacity drops to the preset target capacity while counting the delivered cycles. *OhmTest* measures the internal battery resistance, and *Runtime* discharges at three different current levels to test battery run-times within a simulated user pattern. *QuickSort*™ sorts lithium-ion batteries in 30 seconds into Good, Low and Poor; *Boost* reactivates packs that fell asleep due to over-discharge. Further programs include *Self-Discharge* to measure losses in 24 hours, and *Prime* to prepare new and stored batteries for field use.

Connecting the batteries for service has always been a challenge. Cadex solved the battery interface with the *SnapLock™* adapter system consisting of *custom adapters* for common batteries and *universal adapters* for specialty packs. The custom adapters are easiest to use as they are designed for a battery type and the pack can go in only one way. The adapters include configuration codes that store up to 10 unique battery types and feature a thermistor to monitor temperature. Installing the adapter configures the analyzer to the correct setting. Editing is possible with analyzer's menu function or *via* the PC-Battery Shop software.

With the proliferation of cellular batteries and the need for a quick and simple battery interchange, Cadex developed the *Rigid Arm™*. This universal battery adapter features spring-loaded arms that meet the battery contacts from the top down. Read about How to Service Mobile Phone Batteries. A third option is the *Smart Cables* featuring alligator clips and a temperature sensor to monitor battery temperature.

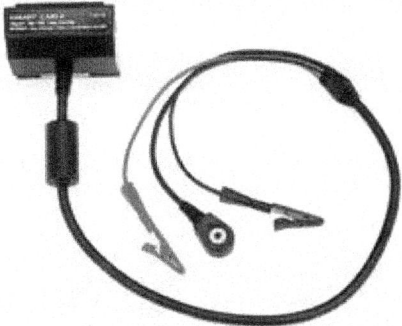

Fig. : Programmable Smart Cable.

The cable stores 10 different battery types; programming is by menu or *via* PC-Battery Shop™ software; a thermistor monitors temperature.
Courtesy of Cadex.

With PC-Battery Shop™ software (Cadex), the PC becomes the master and the analyzers the slave. Clicking the mouse on any of the 2,000 batteries listed in the database or swiping the bar code on the battery label configures the analyzer to the correct setting. You can extend the library by adding new battery models or downloading the latest listing from the Cadex website. PC-Battery Shop™ software is optional; it displays the readings and real-time graphic and is designed to operate 32 analyzers for simultaneous service of 128 batteries (with most PCs).

While battery analyzers are primarily used as a service tool, *battery test systems* provide multi-purpose test functions for research laboratories. Typical applications are life-cycle and stress-testing to verify batteries for field use. Much of this testing can be automated.

The Cadex C8000 is such an automated battery test system. You can measure the battery run-times by capturing and storing load signatures from mobile phones, laptops, power tools and electric drivetrain and then replicate the load condition in the lab. A further test involves checking the longevity of a battery under the discharge conditions reminiscent in the field. SMBus capability displays the register settings of a smart battery to read flags and to check for correct function. If higher discharge currents than 10A are needed, the C8000 connects to designated external load banks. The C8000 forms a laboratory system that controls environmental chambers, monitors analog signals and triggers user-set alarm conditions. PC-BatteryLab™ software provides interface to a PC for the control and monitoring of up to 8 units to service 32 batteries independently.

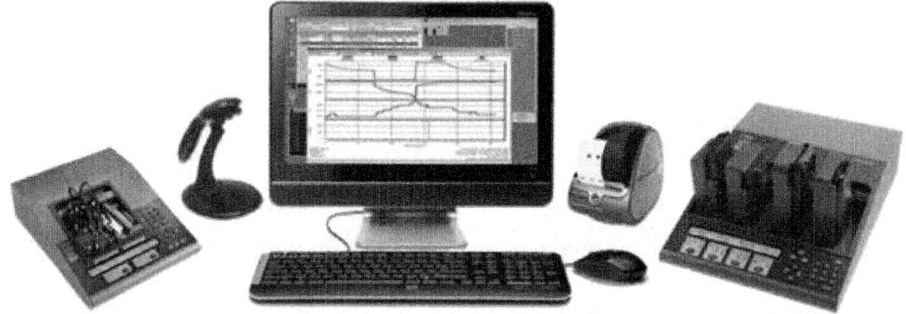

Fig. : PC-Battery Shop™ software provides practical PC-interface to control and monitor Cadex C7000 Series battery analyzers. The monitor provides real-time graphic; the system stores vital data.
Courtesy of Cadex.

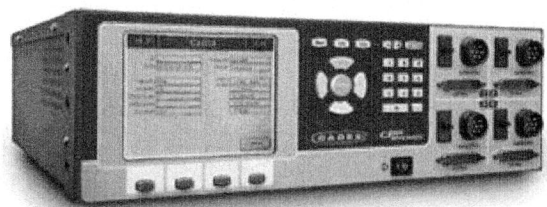

Fig. : Cadex C8000 Battery Test System.

Four independent channels provide up to 10A each and 36V. Maximum charge power is 400W, discharge is 320W. The discharge power can be enhanced with external load banks. Courtesy of Cadex.

The alternate to a battery test system is a programmable power supply controlled by a computer. Such a platform offers high flexibility but requires careful programming to prevent stress to the battery and avoid damage or fire, should an anomaly occur. A battery test system, such as the Cadex C8000, offers protected charge and discharge programs that will identify a faulty battery and terminate a service safely. The system can be overridden to perform destructive tests, however.

RECHARGEABLE BATTERY

A **rechargeable battery, storage battery,** or **accumulator** is a type of electrical battery. It comprises one or more electro-chemical cells, and is a type of energy accumulator. It is known as a **secondary cell** because its electro-chemical reactions are electrically reversible. Rechargeable batteries come in many different shapes and sizes, ranging from button cells to megawatt systems connected to stabilize an electrical distribution network. Several different combinations of chemicals are commonly used, including : lead–acid, nickel cadmium (NiCd), nickel metal hydride (NiMH), lithium ion (Li-ion), and lithium ion polymer (Li-ion polymer).

Rechargeable batteries have lower total cost of use and environmental impact than disposable batteries. Some rechargeable battery types are available in the same sizes as disposable types. Rechargeable batteries have higher initial cost but can be recharged very cheaply and used many times.

Usage and Applications

Rechargeable batteries are used for automobile starters, portable consumer devices, light vehicles (such as motorized wheelchairs, golf carts, electric bicycles, and electric forklifts), tools, and uninterruptible power supplies. Emerging applications in hybrid electric vehicles and electric vehicles are driving the technology to reduce cost and weight and increase lifetime.

Traditional rechargeable batteries have to be charged before their first use; newer low self-discharge NiMH batteries hold their charge for many months, and are typically charged at the factory to about 70% of their rated capacity before shipping.

Grid energy storage applications use rechargeable batteries for load levelling, where they store electric energy for use during peak load periods, and for renewable energy uses, such as storing power generated from photovoltaic arrays during the day to be used at night. By charging batteries during periods of low demand and returning energy to the grid during periods of high electrical demand, load-levelling helps eliminate the need for expensive peaking power plants and helps amortize the cost of generators over more hours of operation.

The US National Electrical Manufacturers Association has estimated that US demand for rechargeable batteries is growing twice as fast as demand for non-rechargeables.

Rechargeable batteries are used for mobile phones, laptops, mobile power tools like cordless screwdrivers. They are used as electric vehicle battery for example in electric cars, electric motorcycles and scooters, electric buses, electric trucks. In most submarines they are used to drive under water. In diesel-electric transmission they are used in ships, in locomotives and huge trucks. They are also used in distributed electricity generation and stand-alone power systems.

Charging and Discharging

During charging, the positive active material is oxidized, producing electrons, and the negative material is reduced, consuming electrons. These electrons constitute the current flow in the external circuit. The electrolyte may serve as a simple buffer for internal ion flow between the electrodes, as in lithium-ion and nickel-cadmium cells, or it may be an active participant in the electro-chemical reaction, as in lead–acid cells.

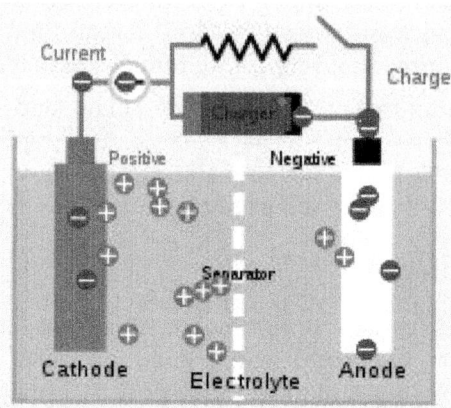

Fig. : Diagram of the charging of a secondary cell battery.

Fig. : Battery charger.

Fig. : A solar-powered charger for rechargeable AA batteries.

The energy used to charge rechargeable batteries usually comes from a battery charger using AC mains electricity, although some are equipped to use a vehicle's 12-volt DC power outlet. Regardless, to store energy in a secondary cell, it has to be connected to a DC voltage source. The negative terminal of the cell has to be connected to the negative terminal of the voltage source and the positive terminal of the voltage source with the positive terminal of the battery. Further, the voltage output of the source must be higher than that of the battery, but not *much* higher : the greater the difference between the power source and the battery's voltage capacity, the faster the charging process, but also the greater the risk of over-charging and damaging the battery.

Chargers take from a few minutes to several hours to charge a battery. Slow "dumb" chargers without voltage-or temperature-sensing capabilities will charge at a low rate, typically taking 14 hours or more to reach a full charge. Rapid chargers can typically charge cells in two to five hours, depending on the model, with the fastest taking as little as fifteen minutes. Fast chargers must have multiple ways of detecting when a cell reaches full charge (change in terminal voltage, temperature, etc.) to stop charging before harmful over-charging or overheating occurs. The fastest chargers often incorporate cooling fans to keep the cells from overheating.

Battery charging and discharging rates are often discussed by referencing a "C" rate of current. The C rate is that which would theoretically fully charge or discharge the battery in one hour. For example, trickle charging might be performed at C/20 (or a "20 hour" rate), while typical charging and discharging may occur at C/2 (two hours for full capacity). The available capacity of electro-chemical cells varies depending on the discharge rate. Some energy is lost in the internal resistance of cell components (plates, electrolyte, interconnections), and the rate of discharge is limited by the speed at which chemicals in the cell can move about. For lead-acid cells, the relationship between time and discharge rate is described by Peukert's law; a lead-acid cell that can no longer sustain a usable terminal voltage at a high current may still have usable capacity, if discharged at a much lower rate. Data sheets for rechargeable cells often list the discharge capacity on 8-hour or 20-hour or other stated time; cells for uninterruptible power supply systems may be rated at 15 minute discharge.

Battery manufacturers' technical notes often refer to VPC; this is volts per cell, and refers to the individual secondary cells that make up the battery. (This is typically in reference to 12-volt lead-acid batteries.) For example, to charge a 12 V battery (containing 6 cells of 2 V each) at 2.3 VPC requires a voltage of 13.8 V across the battery's terminals.

Non-rechargeable alkaline and zinc–carbon cells output 1.5V when new, but this voltage drops with use. Most NiMH AA and AAA cells are rated at 1.2 V, but have a flatter discharge curve than alkalines and can usually be used in equipment designed to use alkaline batteries.

Damage from Cell Reversal

Subjecting a discharged cell to a current in the direction which tends to discharge it further, rather than charge it, is called reverse charging. Generally, pushing current through a discharged cell in this way causes undesirable and irreversible chemical reactions to occur, resulting in permanent damage to the cell. Reverse charging can occur under a number of circumstances, the two most common being:

• When a battery or cell is connected to a charging circuit the wrong way around.

• When a battery made of several cells connected in series is deeply discharged.

In the latter case, the problem occurs due to the different cells in a battery having slightly different capacities. When one cell reaches discharge level ahead of the rest, the remaining cells will force the current through the discharged cell. This is known as "cell reversal". Many battery-operated devices have a low-voltage cutoff that prevents deep discharges from occurring that might cause cell reversal.

Cell reversal can occur to a weakly charged cell even before it is fully discharged. If the battery drain current is high enough, the cell's internal resistance can create a resistive voltage drop that is greater than the cell's forward emf. This results in the reversal of the cell's polarity while the current is flowing. The higher the required discharge rate of a battery, the better matched the cells should be, both in kind of cell and state of charge, in order to reduce the chances of cell reversal.

In some situations (such as when correcting Ni-Cad batteries that have been previously over-charged), it may be desirable to fully discharge a battery. To avoid damage from the cell reversal effect, it is necessary to access each cell separately: each cell is individually discharged by connecting a load clip across the terminals of each cell, thereby avoiding cell reversal.

Damage During Storage in Fully Discharged State

If a multi-cell battery is fully discharged, it will often be damaged due to the cell reversal effect mentioned above. It is possible however to fully discharge a battery without causing cell reversal — either by discharging each cell separately, or by allowing each cell's internal leakage to dissipate its charge over time.

Even if a cell is brought to a fully discharged state without reversal, however, damage may occur over time simply due to remaining in the discharged state. An example of this is the sulfation that occurs in lead-acid batteries that are left sitting on a shelf for long periods. For this reason it is often recommended to charge a battery that is intended to remain in storage, and to maintain its charge level by periodically recharging it. Since damage may also occur if the battery is over-charged, the optimal level of charge during storage is typically around 30% to 70%.

Depth of Discharge

Depth of discharge (DOD) is normally stated as a percentage of the nominal ampere-hour capacity; 0% DOD means no discharge. Seeing as the usable capacity of a battery system depends on the rate of discharge and the allowable voltage at the end of discharge, the depth of discharge must be qualified to show the way it is to be measured. Due to variations during manufacture and aging, the DOD for complete discharge can change over time or number of charge cycles. Generally a rechargeable battery system will tolerate more charge/discharge cycles if the DOD is lower on each cycle.

Active Components

The active components in a secondary cell are the chemicals that make up the positive and negative active materials, and the electrolyte. The positive and negative are made up of different materials, with the positive exhibiting a reduction potential and the negative having an oxidation potential. The sum of these potentials is the standard cell potential or voltage.

In primary cells the positive and negative electrodes are known as the cathode and anode, respectively. Although this convention is sometimes carried through to rechargeable systems — especially with lithium-ion cells, because of their origins in primary lithium cells — this practice can lead to confusion. In rechargeable cells the positive electrode is the cathode on discharge and the anode on charge, and vice versa for the negative electrode.

Table of Rechargeable Battery Types

Type	Voltage[a]	Energy density[b]			Power[c]	E/$[e]	Disch.[f]	Cycles[g]	Life[h]
	(V)	(MJ/kg)	(Wh/kg)	(Wh/L)	(W/kg)	(Wh/$)	(%/month)	(#)	(years)
Lead–acid	2.1	0.11–0.14	30-40	60-75	180	5-8	3-4%	500-800	5-8 (automotive battery), 20 (stationary)
Alkaline	1.5	0.31	85	250	50	7.7	<0.3	100-1000	<5
Nickel–iron	1.2	0.18	50		100	5-7.3	20-40%		50+
Nickel–cadmium	1.2	0.14-0.22	40-60	50-150	150	1.25-2.5	20%	1500	
Nickel–hydrogen	1.5	0.27	75	60	220			20,000+	15+ (satellite application with frequent charge-discharge cycles)
Nickel–metal hydride	1.2	0.11-0.29	30-80	140-300	250-1000	2.75	30%	500-1000	
Nickel–zinc	1.7	0.22	60	170	900	2-3.3		100-500	
Lithium-air (organic)	2.7	7.2	2000	2000	400			~100	
Lithium Cobalt Oxide	3.6	0.58	150-250	250-360	1800	2.8-5	5-10%	400–1200	2-6
Lithium-ion polymer	3.7	0.47-0.72	130-200	300	3000+	2.8-5.0	5%	500~1000	2-3

(Contd...)

(Contd...)

Type	Voltage[a]	Energy density[b]			Power[c]	E/$[e]	Disch.[f]	Cycles[g]	Life[h]
	(V)	(MJ/kg)	(Wh/kg)	(Wh/L)	(W/kg)	(Wh/$)	(%/month)	(#)	(years)
Lithium iron phosphate	3.25	0.32-0.4	80-120	170	1400	0.7-3.0		>10.000 90% DOD	>10
Lithium sulfur	2.0	0.94-1.44	400	350				~1400	
Lithium-ti-tanate	2.3	0.32	90		4000+	0.5-1.0		9000+	20+
Sodium-ion	1.7			30		3.3		5000+	Still testing
Thin film lithium	?		300	959	6000	?		40000	
Zinc bromide		0.27-0.31	75-85						
Vanadium redox	1.15-1.55	0.09-0.13	25-35				20%	14,000	10 (station-ary)
Sodium-sulfur		0.54	150						
Molten salt	2.58	0.25-1.04	70-290	160	150-220	4.54		3000+	<=20
Silver-oxide	1.86	0.47	130	240					
Quantum Battery (oxide semiconduc-tor)	1.5-3			500	8000(W/L)			100,000	

Notes :
- [a]Nominal cell voltage in V.
- [b]Energy density = energy/weight or energy/size, given in three different units
- [c]Specific power = power/weight in W/kg
- [e]Energy/consumer price in W·h/US$ (approximately)
- [f]Self-discharge rate in %/month
- [g]Cycle durability in number of cycles
- [h]Time durability in years
- [i]VRLA or recombinant includes gel batteries and absorbed glass mats
- [p]Pilot production
- [r]Depending upon charge rate

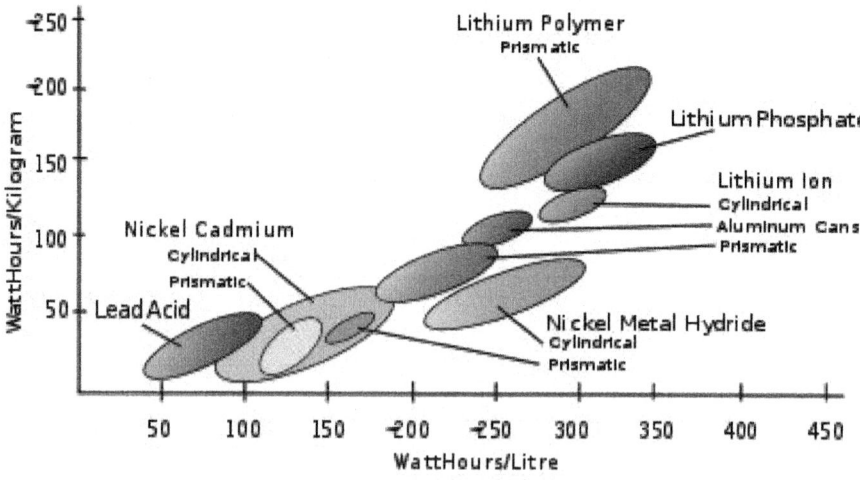

Fig. : Graph of mass and volume energy densities of several secondary cells.

Common Rechargeable Battery Types

Created by Waldemar Jungner of Sweden in 1899, it used nickel oxide hydroxide and metallic cadmium as electrodes. Cadmium is a toxic element, and was banned for most uses by the European Union in 2004. Nickel–cadmium batteries have been almost completely superseded by nickel–metal hydride (NiMH) batteries.

Nickel–metal Hydride Battery (NiMH)

First commercial types were available in 1989. These are now a common consumer and industrial type. The battery has a hydrogen-absorbing alloy for the negative electrode instead of cadmium.

Lithium-ion Battery

The technology behind the lithium-ion battery has not yet fully reached maturity. However, the batteries are the type of choice in many consumer electronics and have one of the best energy-to-mass ratios and a very slow loss of charge when not in use.

Lithium-ion Polymer Battery

These batteries are light in weight and can be made in any shape desired.

Less Common Types

Lithium Sulfur Battery

A new battery chemistry developed by Sion Power since 1994. Claims superior energy to weight than current lithium technologies on the market. Also lower material cost may help this product reach the mass market.

Thin Film Battery (TFB)

An emerging refinement of the lithium ion technology by Excellatron. The developers claim a very large increase in recharge cycles, around 40,000 cycles. Higher charge and discharge rates. At least $5C$ charge rate. Sustained $60C$ discharge, and $1000C$ peak discharge rate. And also a significant increase in specific energy, and energy density.

Also Infinite Power Solutions makes thin film batteries (TFB) for microelectronic applications, that are flexible, rechargeable, solid-state lithium batteries.

Smart Battery

A smart battery has the voltage monitoring circuit built inside.

Carbon Foam-based Lead Acid Battery

Firefly Energy has developed a carbon foam-based lead acid battery with a reported energy density of 30–40% more than their original 38 W·h/kg, with long life and very high power density.

Potassium-ion Battery

This type of rechargeable battery can deliver the best known cycleability, in order of a million cycles, due to the extraordinary electro-chemical stability of potassium insertion/extraction materials such as Prussian blue.

Sodium-ion Battery

This type is meant for stationary storage and competes with lead–acid batteries. It aims at a very low total cost ownership per kWh of storage. This is achieved by a long and stable lifetime. The number of cycles is above 5000 and the battery does not get damage by deep discharge. The energy density is rather low, somewhat lower than lead–acid.

Quantum Battery (Oxide Semiconductor)

A battery developed by the Japanese company MJC. It is a small, lightweight cell with a multi-layer film structure, high energy density and high power density. It is incombustible, has no electrolytes, and generates a low amount of heat during its charge cycle, making it a very safe and long-lasting battery. Its most groundbreaking feature is its ability to capture electrons physically rather than chemically.

Developments Since 2005

In 2007, Yi Cui and colleagues at Stanford University's Department of Materials Science and Engineering discovered that using silicon nanowires as the anode of a lithium-ion battery increases the volumetric charge density of the anode by up to a factor of 10, leading to the development of the nanowire battery.

Another development is the paper-thin flexible self-rechargeable battery combining a thin-film organic solar cell with an extremely thin and highly flexible lithium-polymer battery, which recharges itself when exposed to light.

Ceramatec, a research and development sub-company of CoorsTek, as of 2009 was testing a battery comprising a chunk of solid sodium metal mated to a sulfur compound by a paper-thin ceramic membrane which conducts ions back and forth to generate a current. The company claimed that it could fit about 40 kilowatt hours of energy into a package about the size of a refrigerator, and operate below 90°C; and that their battery would allow about 3,650 discharge/recharge cycles (or roughly 1 per day for one decade).

Researchers have developed a technique to microscopically view battery electrodes while they are bathed in wet electrolytes, which is very similar to actual conditions inside batteries. Although researchers have often used transmission electron microscopy to study wet environments, this is the first time is has been successfully applied to rechargeable battery research. This new research technique could lead to advances in rechargeable battery performance.

Alternatives

A rechargeable battery is only one of several types of rechargeable energy storage systems. Several alternatives to rechargeable batteries exist or are under development. For uses such as portable radios, rechargeable batteries may be replaced by clockwork mechanisms which are wound up by hand, driving dynamos, although this system may be used to charge a battery rather than to operate the radio directly. Flashlights may be driven by a dynamo directly. For transportation, uninterruptible power supply systems and laboratories, flywheel energy storage systems store energy in a spinning rotor for conversion to electric power when needed; such systems may be used to provide large pulses of power that would otherwise be objectionable on a common electrical grid.

Ultra-capacitors : Capacitors of extremely high value — are also used; an electric screwdriver which charges in 90 seconds and will drive about half as many screws as a device using a rechargeable battery was introduced in 2007, and similar flashlights have been produced. In keeping with the concept of ultra-capacitors, betavoltaic batteries may be utilized as a method of providing a trickle-charge to a secondary battery, greatly extending the life and energy capacity of the battery system being employed; this type of arrangement is often referred to as a "hybrid betavoltaic power source" by those in the industry.

Ultra-capacitors are being developed for transportation, using a large capacitor to store energy instead of the rechargeable battery banks used in hybrid vehicles. One drawback to capacitors compared with batteries is that the terminal voltage drops rapidly; a capacitor that has 25% of its initial energy left in it will have one-half of its initial voltage. By contrast, battery systems tend to have a terminal voltage that does not decline rapidly until nearly exhausted. The undesirable characteristic complicates the design of power electronics for use with ultra-capacitors. However, there are potential benefits in cycle efficiency, lifetime, and weight compared with rechargeable systems. China started using ultra-capacitors on two commercial bus routes in 2006; one of them is route 11 in Shanghai.

Flow batteries, used for specialized applications, are recharged by replacing the electrolyte liquid. A flow battery can be considered to be a type of rechargeable fuel cell.

BATTERY CHARGER

A **battery charger** or **recharger** is a device used to put energy into a secondary cell or rechargeable battery by forcing an electric current through it.

The charging protocol depends on the size and type of the battery being charged. Some battery types have high tolerance for over-charging and can be recharged by connection to a constant voltage source or a constant current source; simple chargers of this type require manual disconnection at the end of the charge cycle, or may have a timer to cut off charging current at a fixed time. Other battery types cannot withstand long high-rate over-charging; the charger may have temperature or voltage sensing circuits and a microprocessor controller to adjust the charging current, and cut off at the end of charge. A trickle charger provides a relatively small amount of current, only enough to counteract self-discharge of a battery that is idle for a long time. Slow battery chargers may take several hours to complete a charge; high-rate chargers may restore most capacity within minutes or less than an hour, but generally require monitoring of the battery to protect it from over-charge. Electric vehicles need high-rate chargers for public access; installation of such chargers and the distribution support for them is an issue in the proposed adoption of electric cars.

Charge Rate

Charge rate is often denoted as C or C-*rate* and signifies a charge or discharge rate equal to the capacity of a battery in one hour. For a 1.6Ah battery, $C = 1.6A$. A charge rate of $C/2 = 0.8A$ would need two hours, and a charge rate of $2C = 3.2A$ would need 30 minutes to fully charge the battery from an empty state, if supported by the battery. This also assumes that the battery is 100% efficient at absorbing the charge.

A battery charger may be specified in terms of the battery capacity or C rate; a charger rated $C/10$ would return the battery capacity in 10 hours, a charger rated at 4C would charge the battery in 15 minutes. Very rapid charging rates, 1 hour or less, generally require the charger to carefully monitor battery parameters such as terminal voltage and temperature to prevent over-charging and damage to the cells.

Types of Battery Chargers

Simple

A simple charger works by supplying a constant DC or pulsed DC power source to a battery being charged. The simple charger does not alter its output based on time or the charge on the battery. This simplicity means that a simple charger is inexpensive, but there is a tradeoff in quality. Typically, a simple charger takes longer to charge a battery to prevent severe over-charging. Even so, a battery left in a simple charger for too long will be weakened or destroyed due to over-charging. These chargers can supply either a constant voltage or a constant current to the battery.

Simple AC-powered battery chargers have much higher ripple current and ripple voltage than other kinds of battery supplies. When the ripple current is within the battery-manufacturer-recommended level, the ripple voltage will also be well within the recommended level. The maximum ripple current for a typical 12 V 100 Ah VRLA battery is 5 amps. As long as the ripple current is not excessive (more than 3 to 4 times the battery-manufacturer-recommended level), the expected life of a ripple-charged VRLA battery is within 3% of the life of a constant DC-charged battery.

Trickle

A trickle charger is typically a low-current (5–1,500 mA) battery charger. A trickle charger is generally used to charge small capacity batteries (2–30 Ah). These types of battery chargers are also used to maintain larger capacity batteries (> 30 Ah) that are typically found on cars, boats, RVs and other related vehicles. In larger applications, the current of the battery charger is sufficient only to provide a maintenance or trickle current (trickle is commonly the last charging stage of most battery chargers). Depending on the technology of the trickle charger, it can be left connected to the battery indefinitely. Some battery chargers that can be left connected to the battery without causing the battery damage are also referred to as smart or intelligent chargers.

Timer-based(HI)

The output of a timer charger is terminated after a pre-determined time. Timer chargers were the most common type for high-capacity Ni-Cd cells in the late 1990s for example (low-capacity consumer Ni-Cd cells were typically charged with a simple charger).

Often a timer charger and set of batteries could be bought as a bundle and the charger time was set to suit those batteries. If batteries of lower capacity were charged then they would be over-charged, and if batteries of higher capacity were charged they would be only partly charged. With the trend for battery technology to increase capacity year on year, an old timer charger would only partly charge the newer batteries.

Timer based chargers also had the drawback that charging batteries that were not fully discharged, even if those batteries were of the correct capacity for the particular timed charger, would result in over-charging.

Intelligent

A "smart charger" should not be confused with a "smart battery". A smart battery is generally defined as one containing some sort of electronic device or "chip" that can communicate with a smart charger about battery characteristics and condition. A smart battery generally requires a smart charger it can communicate with. A smart charger is defined as a charger that can respond to the condition of a battery, and modify its charging actions accordingly.

Some smart chargers are designed to charge :
• "Smart" batteries.
• "Dumb" batteries, which lack any internal electronic circuitry.

The term "smart battery charger" is thoroughly ambiguous, since it is not clear whether the adjective "smart" refers to the battery or only to the charger.

The output current of a smart charger depends upon the battery's state. An intelligent charger may monitor the battery's voltage, temperature or time under charge to determine the optimum charge current and to terminate charging.

For Ni-Cd and NiMH batteries, the voltage across the battery increases slowly during the charging process, until the battery is fully charged. After that, the voltage *decreases*, which indicates to an intelligent charger that the battery is fully charged. Such chargers are often labelled as a ΔV, "delta-V," or sometimes "delta peak", charger, indicating that they monitor the voltage change.

The problem is, the magnitude of "delta-V" can become very small or even non-existent if (very) high capacity rechargeable batteries are recharged. This can cause even an intelligent battery charger to not sense that the batteries are actually already fully charged, and continue charging. Over-charging of the batteries will result in some cases. However, many so called intelligent chargers employ a combination of cut off systems, which should prevent over-charging in the vast majority of cases.

A typical intelligent charger fast-charges a battery up to about 85% of its maximum capacity in less than an hour, then switches to trickle charging, which takes several hours to top off the battery to its full capacity.

Universal Battery Charger–analyzers

The most sophisticated types are used in critical applications *e.g.* : military or aviation batteries. These heavy-duty automatic "intelligent charging" systems can be programmed with complex charging cycles specified by the battery maker. The best are universal (*i.e.* : can charge all battery types), and include automatic capacity testing and analyzing functions too.

Fast

Fast chargers make use of control circuitry in the batteries being charged to rapidly charge the batteries without damaging the cells' elements. Most such chargers have a cooling fan to help keep the temperature of the cells under control. Most are also capable of acting as standard overnight chargers if used with standard NiMH cells that do not have the special control circuitry.

Pulse

Some chargers use *pulse technology* in which a series of voltage or current pulses is fed to the battery. The DC pulses have a strictly controlled rise time, pulse width, pulse repetition rate (frequency) and amplitude. This technology is said to work with any size, voltage, capacity or chemistry of batteries, including automotive and valve-regulated batteries. With pulse charging, high instantaneous voltages can be applied without overheating the battery. In a Lead–acid battery, this breaks down lead-sulfate crystals, thus greatly extending the battery service life.

Several kinds of pulse charging are patented. Others are open source hardware.

Some chargers use pulses to check the current battery state when the charger is first connected, then use constant current charging during fast charging, then use pulse charging as a kind of trickle charging to maintain the charge.

Some chargers use "negative pulse charging", also called "reflex charging" or "burp charging". Such chargers use both positive and brief negative current pulses. There is no significant evidence, however, that negative pulse charging is more effective than ordinary pulse charging.

Inductive

Inductive battery chargers use electromagnetic induction to charge batteries. A charging station sends electromagnetic energy through inductive coupling to an electrical device, which stores the energy in the batteries. This is achieved without the need for metal contacts between the charger and the battery. It is commonly used in electric toothbrushes and other devices used in bathrooms. Because there are no open electrical contacts, there is no risk of electrocution.

USB-based

Since the Universal Serial Bus specification provides for a five-volt power supply, it is possible to use a USB cable as a power source for recharging batteries. Products based on this approach include chargers for cellular phones, portable digital audio players, and tablet computers. They may be fully compliant USB peripheral devices adhering to USB power discipline, or uncontrolled in the manner of USB decorations.

Solar Chargers

Solar chargers convert light energy into DC current. They are generally portable, but can also be fixed mount. Fixed mount solar chargers are also known as solar panels. Solar panels are often connected to the electrical grid, whereas portable solar chargers are used off-the-grid (*i.e.* cars, boats, or RVs).

Although portable solar chargers obtain energy from the sun only, they still can (depending on the technology) be used in low light (*i.e.* cloudy) applications. Portable solar chargers are typically used for trickle charging, although some solar chargers (depending on the wattage), can completely recharge batteries. Other devices may exist, which combine this with other sources of energy for added recharging efficacy.

Motion-powered Charger

Several companies have begun making devices that charge batteries based on regular human motion. One example, made by Tremont Electric, consists of a magnet held between two springs that can charge a battery as the device is moved up and down, such as when walking. Such products have not yet achieved significant commercial success.

Applications

Since a battery charger is intended to be connected to a battery, it may not have voltage regulation or filtering of the DC voltage output. Battery chargers equipped with both voltage regulation and filtering may be identified as battery eliminators.

Mobile Phone Charger

Most mobile phone chargers are not really chargers, only power adapters that provide a power source for the charging circuitry which is almost always contained within the mobile phone. They are notoriously diverse, having a wide variety of DC connector-styles and voltages, most of which are not compatible with other manufacturers' phones or even different models of phones from a single manufacturer.

Users of publicly accessible charging kiosks must be able to cross-reference connectors with device brands/models and individual charge parameters and thus ensure delivery of the correct charge for their mobile device. A database-driven system is one solution, and is being incorporated into some designs of charging kiosks.

Mobile phones can usually accept a relatively wide range of voltages, as long as it is sufficiently above the phone battery's voltage. However, if the voltage is too high, it can damage the phone. Mostly, the voltage is 5 volts or slightly higher, but it can sometimes vary up to 12 volts when the power source is not loaded.

There are also human-powered chargers sold on the market, which typically consists of a dynamo powered by a hand crank and extension cords. A French startup offers a kind of dynamo charger inspired by the ratchet that can be used with only one hand. There are also solar chargers, including one that is a fully mobile personal charger and panel, which you can easily transport.

China, the European Commission and other countries are making a national standard on mobile phone chargers using the USB standard. In June 2009, 10 of the world's largest mobile phone manufacturers signed a Memorandum of Understanding to develop specifications for and support a microUSB-equipped common External Power Supply (EPS) for all data-enabled mobile phones sold in the EU. On October 22, 2009, the International Telecommunication Union announced a standard for a universal charger for mobile handsets (Micro-USB).

Battery Charger for Vehicles

There are two main types of charges for vehicles :
- To recharge a fuel vehicle's starter battery, where a modular charger is used.
- To recharge an electric vehicle (EV) battery pack.

Chargers rated only one or two amperes may be used to maintain charge on parked vehicle batteries or for small batteries on garden tractors or similar equipment. A motorist may keep a charger rated a few amperes to ten or fifteen amperes for maintenance of automobile batteries or to recharge a vehicle battery that has accidentally discharged. Service stations and commercial garages will have a large charger to fully charge a battery in an hour or two; often these chargers can briefly source the hundreds of amperes required to crank an internal combustion engine starter.

Stationary Battery Plants

Telecommunications, electric power, and computer uninterruptible power supply facilities may have very large standby battery banks (installed in battery rooms) to maintain critical loads for several hours during interruptions of primary grid power. Such chargers are permanently installed and equipped with temperature compensation, supervisory alarms for various system faults, and often redundant independent power supplies and redundant rectifier systems. Chargers for stationary battery plants may have adequate voltage regulation and filtration and sufficient current capacity to allow the battery to be disconnected for maintenance, while the charger supplies the DC system load. Capacity of the charger is specified to maintain the system load and recharge a completely discharged battery within, say, 8 hours or other interval.

Electric Vehicle Batteries

Electric vehicle battery chargers come in a variety of brands and characteristics. Zivan, Manzanita Micro, Elcon, Quick Charge, Rossco, Brusa, Delta-Q, Kelly, Lester and Soneil are the top 10 EV chargers in 2011 according to EVAlbum.com. These chargers vary from 1 KW to 7.5 KW maximum charge rate. Some use algorithm charge curves, others use constant voltage, constant current. Some are programmable by the end user through a CAN port, some have dials for maximum voltage and amperage, some are preset to specified battery pack voltage, amp-hour and chemistry. Prices range from $400 to $4500.

A 10 amp-hour battery could take 15 hours to reach a fully charged state from a fully discharged condition with a 1 amp charger as it would require roughly 1.5 times the battery's capacity.

Public EV charging stations provide 6 kW (host power of 208 to 240 VAC off a 40 amp circuit). 6 kW will recharge an EV roughly 6 times faster than 1 kW overnight charging.

Rapid charging results in even faster recharge times and is limited only by available AC power and the type of charging system.

Onboard EV chargers (change AC power to DC power to recharge the EV's pack) can be :

- *Isolated :* They make no physical connection between the A/C electrical mains and the batteries being charged. These typically employ some form of Inductive charging. Some isolated chargers may be used in parallel. This allows for an increased charge current and reduced charging times. The battery has a maximum current rating that cannot be exceeded.

- *Non-isolated :* The battery charger has a direct electrical connection to the A/C outlet's wiring. Non-isolated chargers cannot be used in parallel.

Power Factor Correction (PFC) chargers can more closely approach the maximum current the plug can deliver, shortening charging time.

Charge Stations

Non-contact Magnetic Charging

Researchers at the Korea Advanced Institute of Science and Technology (KAIST) have developed an electric transport system (called Online Electric Vehicle, OLEV) where the vehicles get their power needs from cables underneath the surface of the road *via* non-contact magnetic charging, (where a power source is placed underneath the road surface and power is wirelessly picked up on the vehicle itself.

Use in Experiments

A battery charger can work as a DC power adapter for experimentation. It may, however, require an external capacitor to be connected across its output terminals in order to "smooth" the voltage sufficiently, which may be thought of as a DC voltage plus a "ripple" voltage added to it. Note that there may be an internal resistance connected to limit the short circuit current, and the value of that internal resistance may have to be taken into consideration in experiments.

Prolonging Battery Life

What practices are best depend on the type of battery. NiCd cells need to be fully discharged occasionally, or else the battery loses capacity over time in a phenomenon known as "memory effect." Once a month (once every 30 charges) is sometimes recommended. This extends the life of the battery since memory effect is prevented while avoiding full charge cycles which are known to be hard on all types of dry-cell batteries, eventually resulting in a permanent decrease in battery capacity.

Most modern cell phones, laptops, and most electric vehicles use Lithium-ion batteries. These batteries last longest if the battery is frequently charged; fully discharging them will degrade their capacity relatively quickly. When storing however, lithium batteries degrade more while fully charged than if they are only 40% charged. Degradation also occurs faster at higher temperatures. Degradation in lithium-ion batteries is caused by an increased internal battery resistance due to cell oxidation. This decreases the efficiency of the battery, resulting in less net current available to be drawn from the battery. However, if Li-ION cells are discharged below a certain voltage a chemical reaction occurs that make them dangerous if recharged, which is why probably all such batteries in consumer goods now have an "electronic fuse" that permanently disables them if the voltage falls below a set level. The electronic fuse draws a small amount of current from the battery, which means that if a laptop battery is left for a long time without charging it, and with a very low initial state of charge, the battery may be permanently destroyed.

Motor vehicles, such as boats, RVs, ATVs, motorcycles, cars, trucks, and more use lead–acid batteries. These batteries employ a sulfuric acid electrolyte and can generally be charged and discharged without exhibiting memory effect, though sulfation (a chemical reaction in the battery which deposits a layer of sulfates on the lead) will occur over time. Typically sulfated batteries are simply replaced with new batteries, and the old ones recycled. Lead–acid batteries will experience substantially longer life when a maintenance charger is used to "float charge" the battery. This prevents the battery from ever being below 100% charge, preventing sulfate from forming. Proper temperature compensated float voltage should be used to achieve the best results.

CHARGE CYCLE

A **charge cycle** is the process of charging a rechargeable battery and discharging it as required into a load. The term is typically used to specify a battery's expected life, as the number of charge cycles affects life more than the mere passage of time. Discharging the battery fully before recharging may be called "deep discharge"; partially discharging then recharging may be called "shallow discharge".

In general, number of cycles for a rechargeable battery indicates how many times it can undergo the process of complete charging and discharging until failure or it starting to lose capacity.

Apple Inc. clarify that a charge cycle means using all the battery's capacity, but not necessarily by full charge and discharge; *e.g.*, using half the charge of a fully charged battery, charging it, and then using the same amount of charge again count as a single charge cycle.

SETTING BATTERY PERFORMANCE STANDARDS

A battery is a corrosive device that begins to fade the moment it comes off the assembly line. The stubborn behaviour of batteries has left many users in awkward situations. The British Army could have lost the Falklands War in 1982 on account of un-co-operative batteries. The officers assumed that a battery would always follow the rigid dictate of the military. Not so. When a key order was given to launch the British missiles, nothing happened. No missiles flew that day. Such battery-induced letdowns are common; some are simply a nuisance and others have serious consequences.

Even with the best of care, a battery only lives for a defined number of years. There is no distinct life span, and the health of a battery rests on its genetic makeup, environmental conditions and user patterns.

Lead acid reaches the end of life when the active material has been consumed on the positive grids; nickel-based batteries lose performance as a result of corrosion; and lithium-ion fades when the transfer of ions slows down for degenerative reasons. Only the super-capacitor achieves a virtually unlimited number of cycles, if this device can be called a battery, but it also has a defined life span.

Battery manufacturers are aware of performance loss over time, but there is a disconnect when educating buyers about the fading effect. Run-times are always estimated with a perfect battery delivering 100 per cent capacity, a condition that only applies when the battery is new.While a dropped phone call on a consumer product because of a weak battery may only inconvenience the cellular user, an unexpected power loss on a medical, military or emergency device can be more devastating.

Consumers have learned to take the advertised battery run-times in stride. The information means little and there is no enforcement. Perhaps no other specification is as loosely given as that of battery performance. The manufacturers know this and get away with minimal accountability. Very seldom does a user challenge the battery manufacturer for failing to deliver the specified battery performance, even when human lives are at stake. Less critical failures have been debated in court and punished in a harsh way.

The battery is an elusive scapegoat; it's as if it holds special immunity. Should the battery quit during a critical mission, then this is a situation that was beyond control and could not be prevented. It was an act of God and the fingers point in other directions to assign the blame. Even auditors of quality-control systems shy away from the battery and consider only the physical appearance; state-of-function appears less important to them.

HOW TO RATE BATTERY RUN-TIME

In the past, the battery industry got away with soft standards specifying battery run-times. Each manufacturer developed their own method, using the lightest load patterns possible to achieve good figures. This resulted in specifications that bore little resemblance to reality. Under pressure from consumer associations, manufacturers finally agreed to standardized testing procedures.

The Camera and Imaging Products Association (CIPA) succeeded in developing a standardized battery-life test for digital cameras. Under the test scheme, the camera takes a photo every 30 seconds, half with flash and the other without. The test zooms the lens in and out all the way before every shot and leaves the screen on. After every 10 shots, the camera is turned off for a while and the cycle is repeated. CIPA ratings replicate a realistic way a consumer would use a camera. Most new cameras adapt the CIPA protocol to rate the run-time.

The run-time on laptops is more complex to estimate than a digital camera as programs, type of activity, wireless features and screen brightness affect the load. To take these conditions into account, the computer industry developed a standard called Mobile Mark 2007. Not everyone agrees with this norm, and opponents say that the convention trims the applications down and ignores real-world habits. The setting of brightness is one example. The monitor is one of the most power-hungry components of a modern laptop and at full brightness the screen delivers 250 to 300 nits. Mobile Mark uses a setting that is less than half of this. Nor does Mobile Mark include Wi-Fi and Bluetooth; it leaves these peripherals up to the manufacturers to investigate. BAPCO (Business Applications Performance Corporation), the inventor of Mobile Mark 2007, is led by Intel and includes laptop and chip manufacturers, such as Advanced Micro Devices.

Cell phone manufacturers face similar challenges when estimating run-times. Standby and talk time are field-strength dependent and the closer you are to a repeater tower, the lower the transmit power will get and the longer the battery will last. CDMA (Code Division Multiple Access) takes slightly more power than GSM (Global System for Mobile Communications); however, the more critical power guzzlers are large colour displays, touch screens, video, web surfing, GPS, camera, voice dialing and Bluetooth. These peripherals drastically shorten the advertised run-time specifications if used frequently.

The insatiable appetite for information and entertainment on the go is devouring the excess energy enjoyed during the past 10 years when we used our cell phones for voice only. Although modern handsets draw considerably less power than older models and the battery capacity has doubled in 12 years, these improvements do not compensate for the modern peripherals, and a new energy crisis is in the making. Figure illustrates the lack of energy with analog cell phones during the 1990s, the sudden excess with the digital phones, and the looming energy shortage when making full use of modern features. These power needs are superimposed on a continuously improving battery.

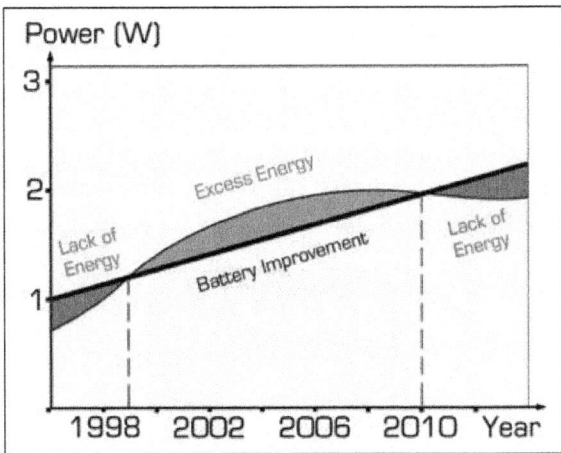

Fig. : Power needs of the past, present and future.

The capacity of Li-ion has doubled in 12 years and the circuits draw less power; however, these improvements do not compensate for the power demand of the new features, and a new energy crisis is in the making.
Courtesy of Cadex

Manufacturers of analog two-way radios test the run-time with a scheme called 5-5-90 and 10-10-80. The first number represents the transmit time at high current; the second denotes the receiving mode at a more moderate current; and the third refers to the long standby times between calls at low current. While 5-5-90 simulates the equivalent of a 5-second talk, 5-second receive and 90-seconds standby, the 10-10-80 schedule puts the intervals at a 10-second talk, 10-second receive and 80-second standby. The run-times of digital two-way radios are measured in a similar way, with the added complexity of tower distance and digital loading requirements that are reminiscent of a cellular phone.

HOW TO DEFINE BATTERY LIFE

Most new batteries go through a formatting process during which the capacity gradually increases and reaches optimal performance at 100–200 cycles. After this mid-life point, the capacity gradually begins decreasing and the depth of discharge, operating temperatures and charging method govern the speed of capacity loss. The deeper the batteries are discharged and the warmer the ambient temperature is, the shorter the service life. The effect of temperature on the battery can be compared with a jug of milk, which stays fresh longer when refrigerated.

Most portable batteries deliver between 300 and 500 full discharge/charge cycles. Fleet batteries in portable devices normally work well during the first year; however, the confidence in the portable equipment begins to fade after the second and third year, when some batteries begin to lose capacity. New packs are added and in time the battery fleet becomes a jumble of good and failing batteries. That's when the headaches begin. Unless date stamps or other quality controls are in place, the user has no way of knowing the history of the battery, much less the performance.

The green light on the charger does not reveal the performance of a battery. The charger simply fills the available space to store energy, and "ready" indicates that the battery is full. With age, the available space gradually decreases and the charge time becomes shorter. This can be compared to filling a jug with water. An empty jug takes longer because it can accept more water than one with rocks. Figure shows the "ready" light that often lies.

Fig. : The "ready" light lies.

The "ready" light on a charger only reveals that the battery is fully charged; there is no relationship to performance. A faded battery charges faster than a good one. Bad batteries gravitate to the top.
Courtesy of Cadex

Many battery users are unaware that weak batteries charge faster than good ones. Low performers gravitate to the top and become available by going to "ready" first. They form a disguised trap when unsuspecting users require a fully charged battery in a hurry. This plays havoc in emergency situations when freshly charged batteries are needed. The operators naturally grab batteries that show ready, presuming they carry the full capacity. Poor battery management is the common cause of system failure, especially during emergencies.

Failures are not foreign in our lives and to reduce breakdowns, regulatory authorities have introduced strict maintenance and calibration guidelines for important machinery and instruments. Although the battery can be an integral part of such equipment, it often escapes scrutiny. The battery as power source is seen as a black box, and for some inspectors correct size, weight and colour satisfies the requirements. For the users, however, state-of-function stands above regulatory discipline and arguments arise over what's more important, performance or satisfying a dogmatic mandate.

Ignoring the performance criteria of a battery nullifies the very reason why quality control is put in place. In defense of the quality auditor, batteries are difficult to check, and to this day there are only a few reliable devices that can check batteries with certainty.

HOW TO KNOW END-OF-BATTERY-LIFE

A critical concern among battery users is knowing *"readiness"* or how much energy a battery has at its disposal at any given moment. While installing a fuel gauge on a diesel engine is simple, estimating the energy reserve of a battery is more complex—we still struggle to read state-of-charge (SoC) with reasonable accuracy. Even if SoC were precise, this information alone has limited benefits without knowing the capacity, the storage capability of a battery. *Battery readiness,* or state-of-function (SoF), must also include internal resistance, or the "size of pipe" for energy delivery. Figure illustrates the bond between capacity and internal resistance on hand of a fluid-filled container that is being eroded as part of aging; the tap symbolizing the energy delivery.

Fig. : Relationship of CCA and capacity of a starter battery.

The liquid represents capacity, the leading health indicator; the tap symbolizes energy delivery or CCA. While the energy delivery remains strong, the capacity diminishes with age. Courtesy Cadex

Most batteries for critical missions feature a monitoring system, and stationary batteries were one of the first to receive supervision in the form of voltage check of individual cells. Some systems also include cell temperature and current measurement. Knowing the voltage drop of each cell at a given load provides cell resistance. Elevated resistance hints to cell failure caused by plate separation, corrosion and other malfunctions. Battery management systems (BMS) are also used in medical equipment, military devices, as well as the electric vehicle.

Although BMS serves an important role in supervising of batteries, such systems often falls short of expectations and here is why. The BMS device is matched to a new battery and does not adjust well to aging. As the battery gets older, the accuracy goes down and in extreme cases the data becomes meaningless. Most BMS also lack bandwidth in that they only reveal anomalies once the battery performance has dropped to 70 per cent. The all-important 70–100 per cent operating range is difficult to gauge and the BMS gives the battery a good bill-of-health. This prevents end-of-life prediction in that the operator must wait for the battery to show signs of wear before making a judgment. These shortcomings are not an oversight by the manufacturers, and engineers are trying to overcome them. The problem boils down to technology, or the lack thereof. Over-expectation is common and the user is stunned when stranded with a dead battery. Let's look how current systems work and examine new technologies.

The most simplistic method to determine end-of-battery-life is by applying a *date stamp* or observing *cycle count*. While this may work for military and medical instruments, such a routine is ill suited for commercial applications. A battery with less use has lower wear-and-tear than one in daily operation and to assure reliability of all batteries, the authorities may mandate that all batteries be replaced sooner. A system made to fit all sizes causes good batteries to be discarded too soon, leading to increased operational costs and environment concerns.

Laptops and other portable devices use coulomb counting for SoC readout. The theory goes back 250 years when Charles-Augustin de Coulomb first established the "Coulomb Rule." Coulomb counting works on the principle of measuring in-and out-flowing current of a battery. If, for example, a battery is charged for one hour at one ampere, the same energy should be available on discharge, but this is not the case. Internal losses and inaccuracies in capturing current flow add to an unwanted tracking error that must be corrected with periodic calibrations.

Calibration occurs naturally when running the equipment down. A full discharge sets the *discharge flag,* and the subsequent recharge establishes the *charge flag*. These two markers allow the calculation of state-of-charge by estimating the distance between the flags.

Coulomb counting should be self-calibrating, but in real life a battery does not always get a full discharge at a steady current. The discharge may be in form of a sharp pulse that is difficult to capture. The battery may then be partially recharged and be stored at high temperature, causing elevated self-discharge that cannot be tracked. To correct the tracking error, a "smart battery" in use should be calibrated once every three months or after 40 partial discharge cycles. This can be done by a deliberate discharge of the equipment or externally with a battery analyzer. Avoid too many intentional deep discharges as this stresses the battery.

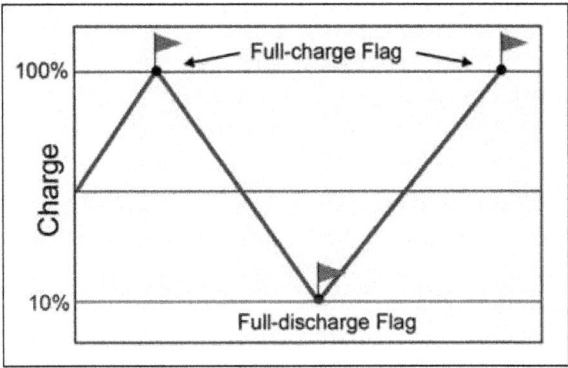

Fig. : Discharge and charge flags.

Calibration occurs by applying a full charge, discharge and charge. This can be done in the equipment or externally with a battery analyzer as part of battery maintenance.
Courtesy Cadex

Fifty years ago, the Volkswagen Beetle had few battery problems. The only battery management was ensuring that the battery was being charged while driving. Onboard electronics for safety, convenience, comfort and pleasure have added to the demands of the battery in modern cars. For the accessories to function reliably, the battery state-of-charge must be known at all times. This is especially critical with start-stop technologies, a future requirement in European cars to improve fuel economy.

When the engine of a start-stop vehicle turns off at a stoplight, the battery continues to draw 25–50 amperes to feed the lights, ventilators, windshield wipers and other accessories. The battery must have enough charge to crank the engine when the traffic light changes; cranking requires a brief 350A. To reduce engine loading during acceleration, the BMS delays charging for about 10 seconds.

Modern cars are equipped with a battery sensor that measures voltage, current and temperature. Packaged in a small housing and embedded into the positive battery clamp, the *electronic battery monitor* (EBM) provides a SoC accuracy of about +/–15 per cent on a new battery. As the battery ages, the EBM begins to drift and the accuracy drops to 20–30 per cent. This can result in a false warning message and some garage mechanics disconnect the EBM on an aging battery to stop annoyances. Disabling the control system responsible for the start-stop function immobilizes engine stop and reduces the legal clean air requirement of the vehicle.

Voltage, current and temperature readings are insufficient to assess battery SoF; the all-important capacity is missing. Until capacity can be measured with confidence on-board of a vehicle, the EBM will not offer reliable battery information. Capacity is the leading health indicator that in most cases determines the end-of-battery-life. Imagine measuring the liquid in a container that is continuously shrinking in size. State-of-charge alone has limited benefit if the storage has shrunk from 100 to 20 per cent and this change cannot be measured. Capacity fade may not affect engine cranking and the CCA can remain at a vigorous 70 per cent to the end of battery life. Because of reduced energy storage, a low capacity battery charges quickly and has normal vital signs, but failure is imminent. A bi-annual capacity check as part of service can identify low capacity batteries. Battery testers that read capacity are becoming available at garages.

A typical start-stop vehicle goes through about 2,000 micro cycles per year. Test data obtained from automakers and the Cadex laboratories indicate that the battery capacity drops to approximately 60 per cent in two years when in a start-stop configuration. The standard flooded lead acid is not robust enough for start-stop, and carmakers use a modified AGM (Absorbent Glass Mat) to attain longer life.

Automakers want to make sure that no driver gets stuck in traffic with a dead battery. To conserve energy when SoC is low, the BMS automatically turns unnecessary accessories off and the motor stays running at a stop-light. Even with this preventive measure, SoC can remain low when commuting in gridlock. Motor idling does not provide much charge and with essential accessories engaged, such as lights and windshield wipers, the net effect could be a small discharge.

Battery monitoring is also important in hybrid vehicles to optimize charge levels. The BMS prevents stressful over-charge above 80 per cent and avoids deep discharges below 30 per cent SoC. At low charge level, the internal combustion engine engages earlier and is left running for additional charge.

The driver of an electric vehicle (EV) expects similar accuracies on the energy reserve as is possible with a gasoline-powered car. Current technologies do not allow this and some EV drivers might get stuck with an empty battery when the fuel gauge still indicates reserve. Furthermore, the EV driver anticipates that a fully charged battery will travel the same distance, year after year. This is not possible and the range will decrease as the battery fades with age. Distances between charges will also be shorter than normal when driving in cold temperatures because of reduced battery performance.

Some lithium-ion batteries have a very flat discharge curve and the voltage method does not work well to provide SoC in the mid-range. An innovative new technology is being developed that measures battery SoC by magnetic susceptibility. *Quantum magnetism* (Q-Mag™) detects magnetic changes in the electrolyte and plates that correspond to state-of-charge. This provides accurate SoC detection in the critical 40-70 per cent mid-section. More impotently, Q-Mag™ allows measuring SoC while the battery is being charged and is under load.

The lithium iron phosphate battery in figure shows a clear decrease in *relative magnetic field units* while discharging and an increase while charging, which relates to SoC. We see no rubber band effect that is typical with the voltage method in which the weight of discharge lowers the terminal voltage and the charge lifts it up. Q-Mag™ also permits improved full-charge detection; however, the system only works with cells in plastic, foil or aluminum enclosures. Ferrous metals inhibit the magnetic field.

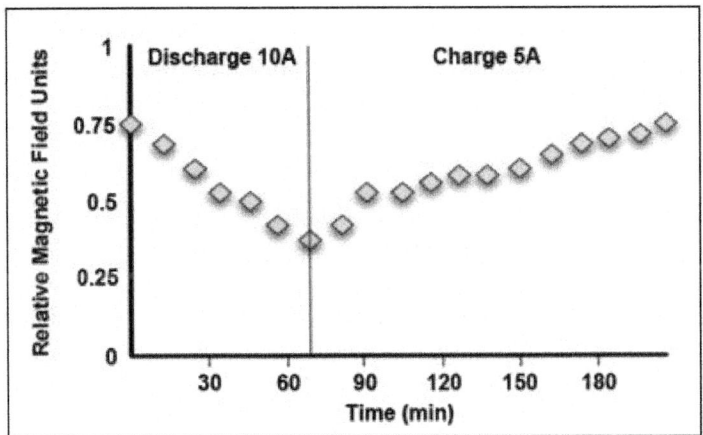

Fig.: Magnetic field measurements of a lithium iron phosphate during charge and discharge.

Relative magnetic field units provide accurate state-of-charge of lithium-and lead-based batteries.
Courtesy of Cadex (2011)

Q-Mag™ also works with lead acid. This opens the door to monitor starter batteries in vehicles. Figure illustrates the Q-Mag™ sensor installed in close proximity to the negative plate. Knowing the precise state-of-charge at any given moment optimizes charge methods and identifies battery deficiencies, including the end-of-battery-life with on-board capacity estimations.

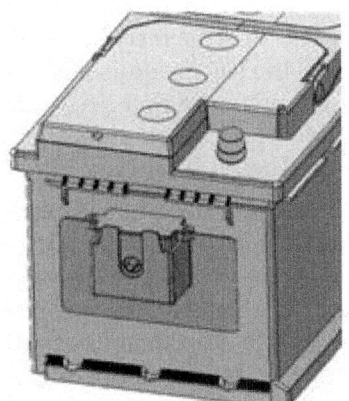

Fig. : Q-Mag™ sensor installed on the side of a starter battery.

The sensor measures the SoC of a battery by magnetic susceptibility. When discharging a lead acid battery, the negative plate changes from lead to lead sulfate. Lead sulfate has a different magnetic susceptibility than lead, which a magnetic sensor can measure.
Courtesy of Cadex (2009)

Q-Mag™ is also a candidate to monitor stationary batteries. The sensing mechanism does not need to touch the electrical poles for voltage measurements and this poses an advantage for high-voltage batteries. Furthermore, Q-Mag™ can assist EVs by providing SoF accuracies not possible with conventional BMS. Q-Mag™ may one day assist in the consumer market to test batteries by magnetism. It is conceivable that one day an iPhone or iPad can be placed on a test mat, similar to a charging mat, and read battery SoC and performance.

Summary

As a medical test at the doctor's office or a weather forecast broadcasted through the media requires multiple data to derive a result, so also can no single instrument fully assess a battery. Several methods are needed. Voltage as medium to estimate SoC works best if the battery type is known and the pack has rested for at least four hours. A charge or discharge falsifies the voltage and the battery needs several hours to neutralize. Temperature also alters the voltage; it is lower when hot and higher when cold. Coulomb counting offers better SoC accuracy but requires periodic calibration. The coulomb method is not immune to battery aging and the accuracy decreases with time. Measuring the internal resistance gives vital battery information but it presents only a snapshot and cannot predict end-of-battery-life. The resistance of most batteries stays low while the capacity fades as part of aging. Measuring battery state-of-charge with magnetic susceptibility is promising but the technology is only in research laboratories. In time, new technologies will evolve that promise clearer insight into the mystery of a battery. This moment cannot come quickly enough.

BATTERY FAILURE, REAL OR PERCEIVED

Battery manufactures use capacity to specify battery performance, and a new battery should have 100 per cent. This means that a 2Ah battery should deliver two amperes for one hour. If the battery quits after 30 minutes, then the capacity is only 50 per cent. Manufacturers use capacity to specify warranty obligations. Depending on chemistry and application, the warranty threshold is set between 70 and 80 per cent of the specified full capacity.

How does the user know when to claim warranty failure on a battery, or when to replace a pack that no longer performs as expected? Battery replacement has been an ongoing problem and the lack of easy-to-use testing procedures is in part to blame. On one hand, an aging battery may be kept too long until it begins affecting operation, while on the other hand perfectly good batteries are being replaced because of equipment problems or operator misapprehension. This commonly occurs with consumer products under warranty, especially cell phones. If the charge on a cell phone does not hold, the user naturally blames the battery when in many cases the fault lies in the device.

Cell phone manufacturers say that 90 per cent of batteries returned under warranty have no problem, and tests conducted in the Cadex laboratories confirm this finding. Many storefronts replace the batteries on the faintest complaint, and this frivolous battery return policy costs the manufacturers millions of dollars per year. Unrealistic expectations, perceived performance loss and lack of practical testing equipment contribute to this wasteful battery exchange behaviour.

Generous battery replacement policies are not limited to portable equipment alone : one German manufacturer of luxury cars points out that out of 400 starter batteries returned under warranty, 200 are working well and have no problem. Low charge and acid stratification are the most common causes of the apparent failure. This problem is more frequent with large luxury cars featuring power-hungry accessories than with the more basic models. A genuine factory defect is seldom the cause, and a leading European manufacturer of starter batteries says that factory defects cause less than seven per cent of the returned warranty batteries. Similar to the cell phone industry, the manufacturer of the starter battery must take responsibility for a problem that may be customer-induced.

Battery failure in Japan is the largest complaint among new owners. Motorists drive an average 13 km (8 miles) per day in congested cities. With the stop-and-go pattern, the battery has little chance to get fully charged and sulfation occurs. North America may be shielded from such battery problems in part because of the long-distance driving.

CAPACITY LOSS

The energy storage of a battery can be divided into three imaginary segments known as the available energy, theempty zone that can be refilled, and the unusable part, or rock content that has become inactive.

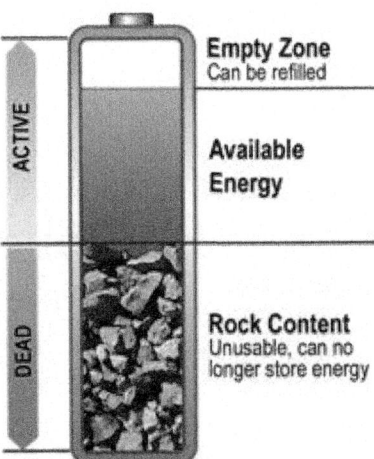

Fig. : Aging battery.

Batteries begin fading from the day they are manufactured. A new battery should deliver 100 per cent capacity; most packs in use operate at less.
Courtesy of Cadex

The manufacturer bases the run-time of a device on a battery that performs at 100 per cent; most packs in the field operate at less capacity. As time goes on, the performance declines further and the battery gets smaller in terms of energy storage. Most users are unaware of capacity fade and continue to use the battery. A pack should be replaced when the capacity drops to 80 per cent; however, the end-of-life threshold can vary according to application, user preference and company policy.

Besides age-related losses, sulfation and grid corrosion are the main killers of lead acid batteries. Sulfation is a thin layer that forms on the negative cell plate if the battery is allowed to dwell in a low state-of-charge. If sulfation is caught in time, an equalizing charge can reverse the condition. Read about Sulfation. Grid corrosion can be reduced with careful charging and optimization of the float charge. With nickel-based batteries, the so-called *rock content* is often the result of crystalline formation, also known as "memory," and a full discharge can sometimes restore the battery. The aging process of lithium-ion is cell oxidation, a process that occurs naturally as part of usage and aging and cannot be reversed.

RISING INTERNAL RESISTANCE

High battery capacity is of limited use if the pack cannot deliver the stored energy effectively. To supply power, the battery needs low internal resistance. Measured in milliohms (mΩ), resistance is the gatekeeper of the battery; the lower the resistance, the less restriction the pack encounters. This is especially important on heavy loads such as power tools and electric powertrains. High resistance causes the voltage to collapse on a load, triggering an early shutdown. Figure illustrates low and high resistance batteries in the form of free-flowing and restricted taps.

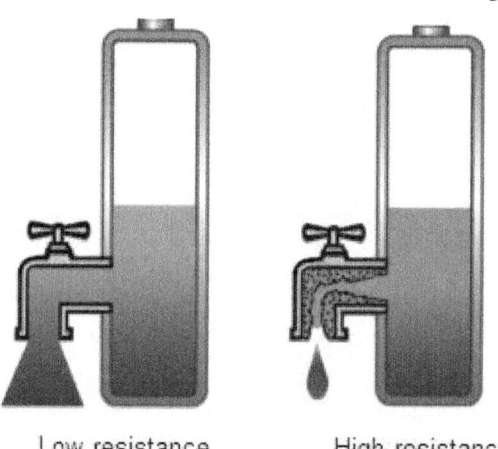

Low resistance   High resistance

Fig. : Effects of internal battery resistance.

A battery with low internal resistance delivers high current on demand. High resistance causes the battery voltage to collapse. The equipment cuts off, leaving energy behind. Courtesy of Cadex

Lead acid has a very low internal resistance, and the battery responds well to high current bursts that last for a few seconds. Due to inherent sluggishness, however, lead acid does not perform well on a sustained high current discharge and the battery needs a rest to recover. Sulfation and grid corrosion are the main contributor to the rise of the internal resistance. Temperature also affects the resistance; heat lowers it and cold raises it. Heating the battery will momentarily lower the internal resistance to provide extra run time.

Alkaline, carbon-zinc and other primary batteries have a relatively high internal resistance, and this limits its use to low-current applications such as flashlights, remote controls, portable entertainment devices and kitchen clocks. As these batteries discharge, the resistance increases further. This explains the relative short run-time when using alkaline cells in digital cameras.

ELEVATING SELF-DISCHARGE

All batteries are affected by self-discharge. Self-discharge is not a manufacturing defect but a battery characteristic, although poor manufacturing practices and improper handling can increase the problem. Figure illustrates self-discharge in the form of leaking fluid.

Fig. : Effects of high self-discharge.

Self-discharge increases with age, cycling and elevated temperature. Discard a battery if the self-discharge reaches 30 per cent in 24 hours.
Courtesy of Cadex

The amount of electrical self-discharge varies with battery type and chemistry. Primary cells such as lithium and alkaline retain the stored energy best and can be kept in storage for several years. Among rechargeable batteries, lead acid has the lowest self-discharge and loses only about five per cent per month. With age and usage, however, the flooded lead acid builds up sludge in the sediment trap, which causes a soft short when this semi-conductive substance reaches the plates.

Nickel-based rechargeable batteries leak the most and need recharging before use when placing them on the shelf for a few weeks. High-performance nickel-based batteries have a higher self-discharge than the standard versions. Furthermore, self-discharge increases with use and age, and the contributing factors are crystalline formation (memory), permitting the battery to "cook" in the charger or exposing it to repeated harsh deep discharge cycles.

Lithium-ion self-discharges about five per cent in the first 24 hours and then loses 1 to 2 per cent per month; the protection circuit adds another three per cent per month. A faulty separator can lead to a high self-discharge and if critical, the electrical current will generate enough heat that can in extreme cases lead to a thermal breakdown. Table shows the typical self-discharge of battery systems.

Table : Percentage of self-discharge in years and month. Primary batteries have considerably less self-discharge than secondary (rechargeable) batteries.

Battery System	Estimated Self-discharge
Primary lithium-metal	10% in 5 years
Alkaline	2-3% per year (7-10 years shelf life)
Lead-acid	5% per month
Nickel-based	10-15% in 24h, then 10-15% per month
Lithium-ion	5% in 24h, then 1-2% per month (plus 3% for safety circuit)

The energy loss is asymptotical, meaning that the self-discharge is highest right after charge and then tapers off. Nickel-based batteries lose 10 to 15 per cent of their capacity in the first 24 hours after charge, then 10 to 15 per cent per month. Figure shows the typical loss of a nickel-based battery while in storage.

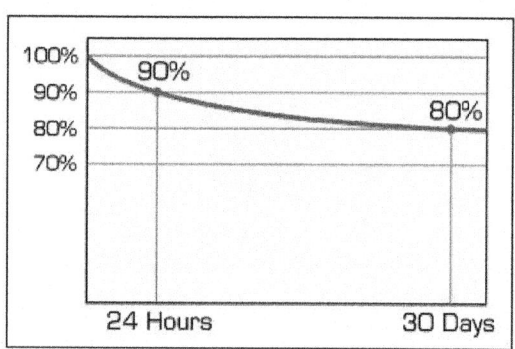

Fig. : Self-discharge as a function of time.

The discharge is highest right after charge and tapers off. The graph shows self-discharge of a nickel-based battery. Lead-and lithium-based systems have a lower self-discharge. Courtesy of Cadex

The self-discharge on all battery chemistries increases at higher temperature and the rate typically doubles with every 10°C (18°F). A noticeable energy loss occurs if a battery is left in a hot vehicle. High cycle count and aging also increase self-discharge. Nickel-metal-hydride is good for 300-400 cycles, whereas the standard nickel-cadmium lasts for over 1,000 cycles before elevated self-discharge starts interfering with performance. The self-discharge on an older nickel-based battery can get so high that the pack loses its energy through leakage rather than normal use.

Under normal circumstances the self-discharge of Li-ion is reasonably steady throughout its service life; however a full state-of-charge and elevated temperature increase the self-discharge. These very same factors also affect longevity, a phenomenon that applies to most batteries. Table shows the self-discharge per month of Li-ion at various temperatures and state-of-charge. The high self-discharge at full state-of-charge may come as a surprise to many. This explains in part the asymptotical self-discharge characteristic when removing a battery from the charger.

Table : Self-discharge per month of Li-ion at various temperatures and state-of-charge
Self-discharge increases with rising temperature and higher SoC.

Charge condition	0°C (32°F)	25°C (77°F)	60°C (140°F)
Full charge	6%	20%	35%
40–60% charge	2%	4%	15%

PREMATURE VOLTAGE CUT-OFF

Not all stored battery energy can or should be used on discharge, and some reserve is almost always left behind on purpose after the equipment cuts off. There are several reasons for this.

Most cell phones, laptops and other portable devices turn off when the lithium-ion battery reaches 3V/cell on discharge, and at this point the battery has about five per cent capacity left. Manufacturers choose this voltage threshold to enable some time before recharging. This grace period can be several months until self-discharge lowers the voltage to about 2.5V/cell, at which point the protection circuit opens. Most packs become unserviceable when this occurs.

A battery on a hybrid car is seldom fully charged or discharged; most operate between 20 to 80 per cent state-of-charge. This is the most effective working bandwidth of a battery; it also delivers the longest service life. A deep discharge causes undue stress, and the charge acceptance above 80 per cent diminishes. The emphasis on an electric powertrain and industrial applications is to maximize service life rather than optimize run-time, as it is the case with consumer products.

Power tools and medical devices with high current draw tend to push the battery voltage to an early cut-off. This is especially true if one of the cells has a high internal resistance, or if operating at cold temperature. These batteries may still have ample capacity left after the "cut-off;" discharging them with a battery analyzer at a moderate load will often give a residual capacity of 30 per cent. Figure illustrates the cut-off voltage graphically.

Alkaline batteries are not suitable for high load applications because of elevated internal resistance. Cold temperature and a partially depleted cell cause the internal resistance to rise further. This advances the cut-off and much of the energy is left behind.

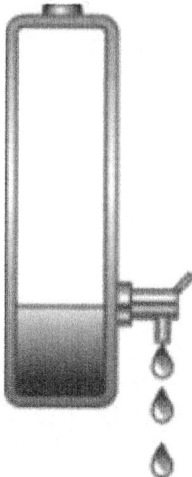

Fig. : Illustration of equipment with high cut-off voltage.

Portable devices do not utilize all available battery power and leave some energy behind.
Courtesy of Cadex

CAN BATTERIES BE RESTORED?

Battery users and entrepreneurs often ask, "Can batteries be restored?" The answer is, "It depends." Most battery failures are permanent and cannot be repaired, but there are exceptions. Sulfation on lead acid batteries can be removed if caught in time; crystalline formation, also known as "memory," on nickel-cadmium can be dissolved through deep-cycling. Read more about Memory : Myth or Fact?, and "sleeping" lithium-ion packs can be boosted if they have been over-discharged. Read more about Safety circuits for modern batteries.

Permanent battery defects include high internal resistance, elevated self-discharge, electrical short and capacity fade. Poorly designed chargers, exposure to excess heat, harsh charge and discharge cycles, and inappropriate storage contribute to early aging. Let's examine the cause of these non-correctable battery problems and explore what we can do to minimize them.

Low-capacity Cells

A manufacturer cannot predict the exact capacity when a battery comes off the production line, and this is especially true with lead acid batteries that involve manual assembly. Fully automated cell production in "clean rooms" also causes performance differences, and as part of quality control, each cell is measured and segregated into categories according to their inherent capacity levels. The high-capacity A-cells are reserved for special applications and sold at premium prices; the large mid-range B-group goes to commercial and industrial markets; and the low-grade C-cells may end up as consumer products in department stores. Cycling will not significantly improve the capacity of the low-end cell, and even though the cell may look good, the buyer must be aware of differences in capacity and quality, which often translate into life expectancy.

Cell Mismatch, Balancing

Matching of cells according to capacity is important, especially for industrial batteries. No perfect match is possible, and if slightly off, nickel-based cells adapt to each other after a few charge/discharge cycles similar to the players on a winning sports team. High-quality cells continue to perform longer than the lower-quality counterpart, and the cells degrade at a more even and controlled rate. Lower-grade cells, on the other hand, diverge more quickly with use and time, and failures due to cell mismatch are more widespread. Cell mismatch is a common cause of failure in industrial batteries. Manufacturers of professional power tools and medical equipment are careful in the choice of cells to attain good battery reliability and long life.

Let's look at what a weak cell does in a pack that is strung together with strong ones. The weak cell holds less capacity and is discharged more quickly than the strong brothers. Going empty first, the strong brothers overrun this feeble sibling and the resulting current on a continued discharge pushes the weak cell into reverse polarity. Nickel-cadmium can tolerate a reverse voltage of minus 0.2V and a reverse current of a few milliamps, but exceeding this level will cause a permanent electrical short. On charge, the weak cell reaches full charge first and it goes into heat-generating over-charge while the strong brothers still accept charge and stay cool. The low cell experiences a disadvantage on both charge and discharge. It continues to weaken until finally giving up the struggle.

The capacity tolerance between cells in an industrial battery should be +/-2.5 per cent. High-voltage packs designed for heavy loads and wide adverse temperature ranges should have lower tolerances. There is a strong correlation between cell balance and longevity.

Li-ion cells share similar deficiencies with nickel-based systems and need management. The mandatory protection circuit supervises the serially connected cells by clamping the voltage when exceeding 4.25 and 4.35V on charge, and disconnecting the pack from discharge when the weakest cell drops to between 2.50 and 2.80V/cell. This prevents the stronger cells from pushing the depleted cell into reverse polarization. The protection circuit acts like a guardian angel that shields the weaker siblings from being bullied by the stronger brothers. This may be help to explain why Li-ion packs for power tools last longer than nickel-based batteries, which normally do not have a protection circuit.

The capacity of quality Li-ion cells is consistent and the self-discharge is low. A problem arises when the cells exhibit a discrepancy in self-discharge. This can be attributed to lower-quality cells or high-temperature spots in a large automotive battery, which hastens aging. Balancing is required and there are two methods : *Passive balancing* bleeds the high-voltage cells; *active balancing* shuttles the extra-charge from higher-voltage cells to the lower-voltage cells without burning the energy. Active balancing is the preferred method on EVs.

With use and time all batteries become mismatched, and this also applies to lead acid. Shorted cells and those having high self-discharge are a common cause of cell imbalance and lead to subsequent failure. Manufacturers of golf cars, aerial work platforms, floor scrubbers and other battery-powered vehicles recommend an equalizing charge of 3–4 hours if the voltage difference between the cells is greater than +/–0.10V, or if the specific gravity varies more than 10 points (0.010 on the SG scale). An equalizing charge is a charge on top of a charge that brings all cells to full-charge saturation. This service must be administered with care because excessive charging can harm the battery. A difference in specific gravity of 40 points poses a performance problem and the cell is considered defective. A 40-point difference is one cell having an SG of 1.200 and another 1.240. A charge may temporarily cover the deficiency, but the flaw will resurface after a few hours of rest due to high self-discharge.

Shorted Cells

Manufacturers are at a loss to explain why some cells develop high electrical leakage or a short while still new. The culprit might be foreign particles that contaminate the cells during manufacture, or rough spots on the plates that damage the delicate separator. Clean rooms, improved quality control at the raw material level, and minimal human handling during the manufacturing process have reduced the "infant mortality rate."

Applying momentary high-current bursts to repair a shorted NiCd or NiMH cell has been tried but offers limited success. The short may temporarily evapourate but the damage in the separator remains. After service, the repaired cell may charge normally and reach correct voltages; however, high self-discharge will likely drain the battery and the short will return.

It is not advised to replace a shorted cell in an aging pack because of cell matching. The new cell will always be stronger than the others. Consider the biblical verses : "No one sews a patch of unshrunk cloth on an old garment. If he does, the new piece will pull away from the old, making the tear worse. And no one pours new wine into old wineskins. If he does, the wine will burst the skins, and both the wine and the wineskins will be ruined" (Mark 2 : 21, 22 NIV). Replacing faulty cells often leads to battery failures within six months. It's best not to disturb the cells. Instead, allow them to age naturally as an intact family.

Shorts or high leakage in a Li-ion cell are uncommon. If this occurs, the cell becomes unstable and a massive amount of power can dissipate, leading to a possible venting thermal breakdown. Such a leak can be compared to drilling a small pinhole into a high-pressure gas pipeline and holding a match to it. The resulting explosion could rupture the pipe. Similarly, the rushing current in the cell heats up the tiny malfunction, causes a major leak and releases all energy within seconds. (Read more about Safety circuits for modern batteries).

Cell disintegration caused by internal disturbances lies outside the safeguarding ability of the protection circuit. Most cell failures occur when the battery has been damaged by shock and vibration, has been over-charged or has been overheated. Li-ion cells for electric powertrains and demanding industrial applications use a heavy-duty separator to reduce the risk of an electrical short. These batteries are larger than consumer-type packs. Saying that Li-ion has twice the energy density of NiCd can be a misnomer; some long-lasting Li-ion cells have a specific energy as low as 60Wh/kg, the same as NiCd.

Caution : Applying a high current burst works best with nickel-based batteries. Do not use this method for lithium-ion cells.

Loss of Electrolyte

The loss of electrolyte in a flooded lead acid battery occurs through gassing, as hydrogen escapes during charging and discharging. Venting causes the electrolyte to become more concentrated and the balance must be restored by adding clean water. Do not add electrolyte, as this would upset the specific gravity and shorten battery life through excessive corrosion.

Permeation, or loss of electrolyte in sealed lead acid batteries, is a recurring problem that is often caused by over-charging. Careful adjustment of charging and float voltages, as well as operating at moderate temperatures, reduces this failure. Replenishing lost liquid in VRLA batteries by adding water has limited success. Although the lost capacity can often be regained with a catalyst, tampering with the cells turns the stack into a high-maintenance project that needs constant supervision.

Nickel-based batteries can lose electrolyte through venting due to excessive pressure during extreme charge or discharge. After repeated venting, the spring-loaded seal of the cells may not seal properly again, and the deposit of white powder around the seal opening is evidence of this. Losses of electrolyte may also occur as part of faulty manufacturing. Dry-up conditions result in a "soft" cell, a defect that cannot be corrected. On charge, the voltage of a "dry" cell goes high because the battery has no clamping action and does not draw current.

A properly designed and correctly charged lithium-ion cell should not generate gases, nor should it lose electrolyte through venting. In spite of what advocates have said, lithium-based cells can build up an internal pressure under certain conditions, and a bloated pouch cell is proof of this. Read more about The Pouch Cell. Some cells include an electrical switch that opens if the cell pressure reaches a critical level. Others feature a membrane that releases gases. Many of these safety features are one-way only, meaning that once activated, the cell becomes inoperable.

HOW TO RESTORE LEAD-ACID BATTERIES

A lead acid battery goes through three life phases, called *formatting, peak* and *decline*. In the formatting phase, try to imagine sponge-like lead plates that are being exposed to a liquid. Exercising the plates allows the absorption of more liquid, much like squeezing and releasing a sponge. This enables the electrolyte to better fill the usable areas, an exercise that increases the capacity.

Formatting is most important for deep-cycle batteries and requires 20 to 50 full cycles to reach peak capacity. Field usage achieves this. There is no need to apply added cycles for the sake of priming; however, manufacturers recommend to go easy on the battery until broken in. Starter batteries are less critical and do not need priming; the full cranking power is present right from the beginning, although the CCA reading will go up slightly with early use.

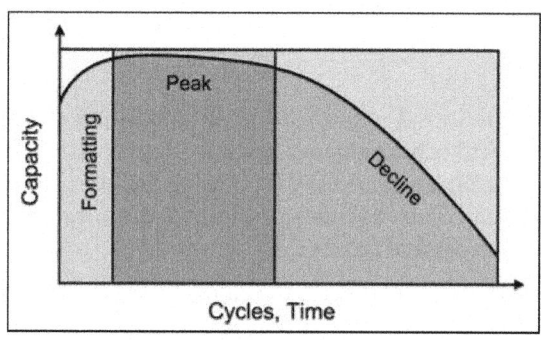

Fig. : Cycle life of a battery.

The three phases of a battery are formatting, peak and decline.
Courtesy of Cadex

A deep-cycle battery delivers 100–200 cycles before it starts the gradual decline. Replacement should occur when the capacity drops to 70 or 80 per cent. Some applications allow lower capacity thresholds but the time for retirement should not fall below 50 per cent because the aging occurs rapidly once the battery is past its prime. Apply a fully saturated charge of 14 to 16 hours. Operating at moderate temperatures assure the longest service times. If at all possible, avoid deep discharges; charge more often.

The primary reason for the relatively short cycle life of a lead acid battery is depletion of the active material. According to the *2010 BCI Failure Modes Study,** plate/grid-related breakdown has increased from 30 per cent five years ago to 39 per cent. The report does not give reasons for the increased wear-and-tear, other than to assume that higher demands of starter batteries in modern cars induce added stress.

While the depletion of the active material is well understood and can be calculated, a lead acid battery suffers from other infirmities long before plate-and grid-deterioration sound the death knell.

CORROSION, SHEDDING AND INTERNAL SHORT

Corrosion occurs primarily on the grid and is known as a softening and shedding of lead off the plates. This reaction cannot be avoided because the electrodes in a lead acid environment are always reactive. Lead shedding is a natural phenomenon that can only be slowed down and not eliminated. A battery that reaches the end of life through this failure mode has met or exceeded the anticipated life span. Limiting the depth of discharge, reducing the cycle count, operating at a moderate temperature and controlling over-charge are key in keeping corrosion in check. To reduce corrosion on long-life batteries, manufacturers keep the specific gravity at a moderate 1.200 when fully charged, compared to 1.265 and greater for high-performing lead acid batteries.

Applying prolonged over-charge is another contributor to grid corrosion. This is especially damaging to *sealed* lead acid systems. While the flooded lead acid has some resiliency to over-charge, sealed units must operate at a correct float charge. Chargers with variable float voltages adjust to the prevailing temperature to help to keep grid corrosion in check. Such chargers are in common use for stationary batteries.

To attain maximum surface area, the lead on a starter battery is applied in a sponge-like form. With time and use, chunks of lead fall off and reduce the performance. Figure illustrates the innards of a corroded lead acid battery.

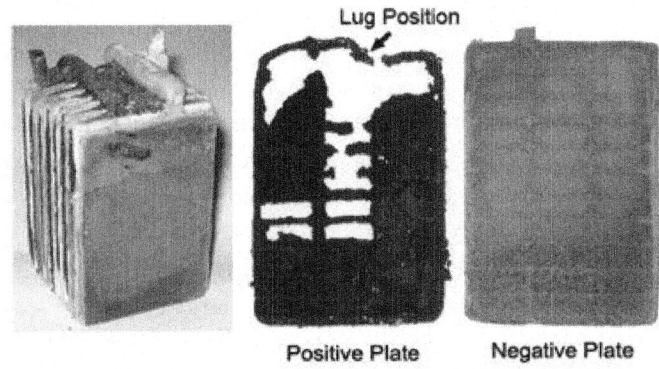

Lug Position

Positive Plate   Negative Plate

Fig. : Innards of a corroded lead acid battery.
Grid corrosion is unavoidable.

The terminals of a battery can also corrode, and this is often visible in the form of white powder. The phenomenon is a result of oxidation between two different metals connecting the poles. Terminal corrosion can eventually lead to an open electrical connection. Changing the connecting terminals to lead, the same material as the battery pole of a starter battery, will solve most corrosion problems.

Internal Short

The term "short" is commonly used to describe a general battery fault when no other definition is available. As the colloquial term "memory" was the cause for all battery ills in the NiCd days, so do battery users often judge non-functioning lead acid batteries simply as being "shorted."

The lead within a battery, especially in deep-cycle units, is mechanically active and when a battery discharges, the lead sulfate causes the plates to expand. This movement reverses during charge and the plates contract. The cells allow for some expansion but over time the growth of large sulfite crystals can result in a soft short that increases self-discharge. This mechanical action also causes *shedding* of the lead material. On a starter battery, the shedding is manageable because the lead plates are thin and the battery does not go through a deep discharge. On a deep-cycle battery, on the other hand, shedding is a concern.

As the battery sheds its lead to the bottom of the container, a conductive layer forms, and once the contaminated material fills the allotted space in the sediment trap, the now conductive liquid reaches the plates and creates a shorting effect. The term "short" is a misnomer and *elevated self-discharge* or *a soft short* would be a better term to describe the condition.

"Soft shorts" are difficult to detect because the battery appears normal immediately after a charge and everything seems to function as it should. In essence, the charge has wiped out all evidence of a soft short, except perhaps an elevated temperature on the battery housing. Once rested for 6–12 hours, the battery begins to show anomalies such as a lower open-circuit voltage and reduced specific gravity. The measured capacity will also be low because self-discharge has consumed some of the stored energy. According to the *2010 BCI Failure Modes Study*, shorted batteries accounted for 18 per cent of battery failures, a drop from 31 per cent five years earlier. Improved manufacturing methods may account for this reduction.

Another form of soft short is *mossing*. This occurs when the separators and plates are slightly misaligned as a result of poor manufacturing practices. This causes parts of the plates to become naked. The exposure promotes the formation of conductive crystal moss around the edges, which leads to elevated self-discharge.

Lead drop is another cause of short in which large chunks of lead break loose from the welded bars connecting the plates. Unlike a "soft" short that develops with wear-and-tear, a lead drop often occurs early in battery life. This causes a more serious short and is associated with a permanent voltage drop. The shorted cell may have little or no charge and the specific gravity of the electrolyte is close to 1.00. This is mostly a manufacturing defect and cannot be repaired.

The most radical and serious form of short is a mechanical failure in which the suspended plates become loose and touch each other. This results in a sudden high discharge current that can lead to excessive heat buildup and thermal runaway. Sloppy manufacturing as well as excessive shock and vibration are the most common contributors to this failure.

SULFATION AND HOW TO PREVENT IT

Sulfation occurs when a lead acid battery is deprived of a full charge. This is common with starter batteries in cars driven in the city with load-hungry accessories. A motor in idle or at low speed cannot charge the battery sufficiently.

Electric wheel chairs have a similar problem in that the users might not charge the battery long enough. An eight-hour charge during the night when the chair is free is not enough. Lead acid must periodically be charged 14–16 hours to attain full saturation. This may be the reason why wheel chair batteries last only two years, whereas golf car batteries deliver twice the service life. Longer leisure time allows golf car batteries to get the fully saturated charge.

Solar cells and wind turbines do not always provide sufficient charge, and lead acid banks succumb to sulfation. This happens in remote parts of the world where villagers draw generous amounts of electricity with insufficient renewable resources to charge the batteries. The result is a short battery life. Only a periodic fully saturated charge could solve the problem, but without an electrical grid at their disposal, this is almost impossible. An alternative is using lithium-ion, a battery that is forgiving to a partial charge, but this would cost about six-times as much as lead acid.

What is sulfation? During use, small sulfate crystals form, but these are normal and are not harmful. During prolonged charge deprivation, however, the amorphous lead sulfate converts to a stable crystalline that deposits on the negative plates. This leads to the development of large crystals, which reduce the battery's active material that is responsible for high capacity and low resistance. Sulfation also lowers charge acceptance. Sulfation charging will take longer because of elevated internal resistance.

There are two types of sulfation : *reversible* (or *soft sulfation*), and *permanent* (or *hard sulfation*). If a battery is serviced early, reversible sulfation can often be corrected by applying an over-charge to a fully charged battery in the form of a regulated current of about 200mA. The battery terminal voltage is allowed to rise to between 2.50 and 2.66V/cell (15 and 16V on a 12V mono block) for about 24 hours. Increasing the battery temperature to 50–60°C (122–140°F) further helps in dissolving the crystals. Permanent sulfation sets in when the battery has been in a low state-of-charge for weeks or months. At this stage, no form of restoration is possible.

There is a fine line between reversible and non-reversible sulfation, and most batteries have a little bit of both. Good results are achievable if the sulfation is only a few weeks old; restoration becomes more difficult the longer the battery is allowed to stay in a low SoC. A sulfated battery may improve marginally when applying a de-sulfation service. A subtle indication of whether a lead acid can be recovered is visible on the voltage discharge curve. If a fully charged battery retains a stable voltage profile on discharge, chances of reactivation are better than if the voltage drops rapidly with load.

Several companies offer anti-sulfation devices that apply pulses to the battery terminals to prevent and reverse sulfation. Such technologies tend to lower sulfation on a healthy battery but they cannot effectively reverse the condition once present. Manufacturers offering these devices take the "one size fits all" approach and the method is unscientific. A random service of pulsing or blindly applying an over-charge can harm the battery in promoting grid corrosion. Technologies are being developed that measure the level of sulfation and apply a calculated over-charge to dissolve the crystals. Chargers featuring this technique only apply de-sulfation if sulfation is present and only for the time needed.

WATER LOSS, ACID STRATIFICATION AND SURFACE CHARGE

During use, and especially on over-charge, the water in the electrolyte splits into hydrogen and oxygen. The battery begins to gas, which results in water loss. In flooded batteries, water can be added but in sealed batteries, water loss leads to an eventual dry-out and decline in capacity. Water loss from a sealed unit can eventually cause disintegration of the separator. The initial stages of dry-out can go undetected and a drop in capacity may not be immediately evident. Early detection of this failure is important. Read about Charging Lead Acid, under Watering.

On over-charge, a battery becomes a "water-splitting device" that turns water into oxygen and hydrogen. A parallel can be made with the *fuel cell*, but this device does the opposite; it turns oxygen and hydrogen back to electricity and produces water. Turning water into hydrogen needs energy and in a battery this is in the form of over-charge. Converting hydrogen and oxygen back to water regenerates energy. Read about the Fuel Cell.

Acid Stratification

The electrolyte of a stratified battery concentrates at the bottom, starving the upper half of the cell. Acid stratification occurs if the battery dwells at low charge (below 80 per cent), never receives a full charge and has shallow discharges. Driving a car for short distances with power-robbing accessories contributes to acid stratification because the alternator cannot always apply a saturated charge. Large luxury cars are especially prone to acid stratification. This is not a battery defect *per se* but the result of use. Figure illustrates a normal battery in which the acid is equally distributed from top to bottom.

Fig. : Normal battery.

The acid is equally distributed from the top to the bottom of the battery, providing good overall performance.
Courtesy of Cadex

Figure shows a stratified battery in which the acid concentration is light on top and heavy on the bottom. The light acid on top limits plate activation, promotes corrosion and reduces the performance, while the high acid concentration on the bottom makes the battery appear more charged than it is and artificially raises the open-circuit voltage. Because of unequal charge across the plates, CCA performance, or the ability to crank the engine, is also reduced.

Fig. : Stratified battery.

The acid concentration is light on top and heavy on the bottom. This raises the open circuit voltage and the battery appears fully charged. Excessive acid concentration induces sulfation on the lower half of the plates.
Courtesy of Cadex

Allowing the battery to rest for a few days, doing a shaking motion or tipping the battery on its side helps correct the problem. Applying an equalizing charge by raising the voltage of a 12-volt battery to 16 volts for one to two hours also helps by mixing the electrolyte through electrolysis. Avoid extending the topping charge beyond its recommended time.

Acid stratification cannot always be avoided. During cold winter months, starter batteries of most passenger cars dwell at a 75 per cent charge level. Knowing that motor idling and driving in gridlocked traffic does not sufficiently charge the battery, a charge with an external charger may be needed from time to time. If this is not practical, switch to an AGM battery. AGM does not suffer from acid stratification and is less subject to sulfation if undercharged than is the case with the flooded version. AGM is a little more expensive than the flooded starter battery but tends to last longer.

Surface Charge

Lead acid batteries are sluggish and cannot convert lead sulfate to lead and lead dioxide quickly enough during charge. As a result, most of the charge activities occur on the plate surfaces. This induces a higher state-of-charge on the outside than in the inner plate. A battery with surface charge has a slightly elevated voltage. To normalize the condition, switch on electrical loads to remove about one per cent of the battery's capacity, or allow the battery to rest for a few hours. Surface charge is not a battery defect but a reversible condition resulting from charging.

Simple Guidelines for Extending Battery Life

- Allow a fully saturated charge of 14–16 hours. Charge in a well-ventilated area.
- Always keep lead acid charged. Avoid storage below 2.10V/cell, or at a specific gravity level below 1.190.
- Avoid deep discharges. The deeper the discharge, the shorter the battery life will be. A brief charge on a 1 to 2 hour break during heavy use prolongs battery life.
- Never allow the electrolyte to drop below the tops of the plates. Exposed plates sulfate and become inactive. When low, add only enough water to cover the exposed plates before charging; fill to the correct level after charge.
- Never add acid. This would raise the specific gravity too high and cause excessive corrosion.
- Use distilled or ionized water. Tap water may be usable in some regions.
- When new, a deep-cycle battery may have a capacity of 70 per cent or less. Formatting as part of field use will gradually increase performance. Apply a gentle load for the first five cycles to allow a new battery to format.
- New batteries with low capacity many not perform as well as those that begin life with a high capacity. Low performers are known to have a short life. A capacity check as part of acceptance is advisable.

ADDITIVES TO BOOST FLOODED LEAD ACID

Adding chemicals to the electrolyte of flooded lead acid batteries can reduce the buildup of lead sulfate on the plates and improve the overall battery performance. This treatment has been in use since the 1950s (and perhaps longer) and provides a temporary performance boost for aging batteries. It's a stopgap measure because in most cases the plates have already been worn out through shedding. Chemical additives cannot replace the active material, nor can cracked plates, corroded connectors or damaged separators be restored with an outside remedy.

Extending the service life of an aging battery is a noble desire. The additives are cheap, readily available and worth the experiment of a handyman. Suitable additives are magnesium sulfate (Epsom salt), caustic soda and EDTA. (EDTA is a crystalline acid used in industry.) These salts may reduce the internal resistance of a sulfated battery to give it a few months of extra life. Using Epsom salt, follow these easy steps :

Heat up the water to about 66°C (150°F), mix 10 heaping tablespoons of Epsom salt into the water and stir until dissolved. The consistency of the brew should vary according to the extent of the sulfation. Avoid using too much salt because a heavy concentration will increase corrosion of the lead plates and internal connectors. Pour the warm solution into the battery.

Be careful not to overfill. Do not place un-dissolved Epsom salt directly into the battery because the substance does not dissolve well. In place of Epsom salt, try adding a pinch of caustic soda. Charge or equalize the battery after service. The results are not instantaneous and it may take a month for the treatment to work. The outcome is not guaranteed.

TRACKING BATTERY CAPACITY AND RESISTANCE AS PART OF AGING

Let's examine the aging mechanism of batteries in terms of fading capacity and increasing internal resistance. Figure shows a battery with high capacity and another that has aged. The capacity loss is illustrated with growing "rock content;" the rocks mark the unusable part of the battery. Figure looks at resistance and illustrates a good battery with a free-flowing tap and a high-resistance one with restricted flow.

New battery has high capacity   Aged battery has low capacity

Fig. : Battery capacity illustrated as liquid content. Both batteries are fully charged, but the "rock-content" limits the amount of energy being stored.

 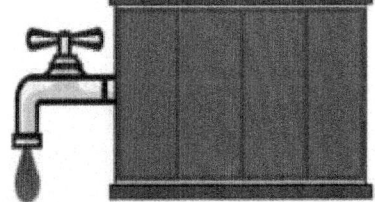

Battery with high CCA   Battery with low CCA

Fig. : Free-flowing and restricted taps representing CCA performance. The cranking current is about 300A. (A golf cart typically draws 56A.)

Automotive technicians are most familiar with CCA, but this reading reflects engine cranking only. Capacity, the energy storage component, remains mostly unknown. Figure illustrates the relationship between CCA and capacity on hand of a fluid-filled container. The liquid represents capacity and the taps symbolize CCA at different loading capabilities.

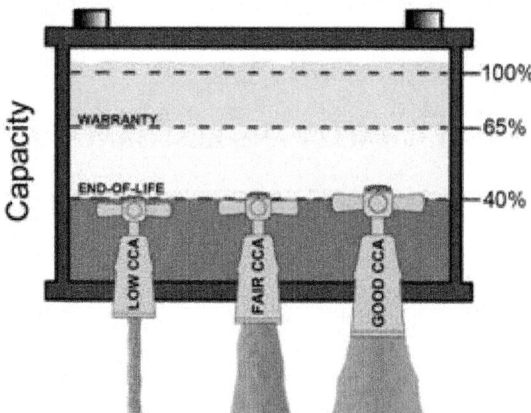

Fig. : Relationship of CCA and capacity of a starter battery.

Capacity represents energy content and CCA is power delivery. A battery with 40% capacity can still have a healthy crank but the low capacity indicates end-of-life.

Most rechargeable batteries maintain low internal resistance during the service life, and this reflects in a high CCA (cold cranking amps) on starter batteries. Capacity, on the other hand, begins to drop gradually as the battery ages. To study these changes, Cadex measured the capacity and CCA of 20 aging starter batteries. The results are laid out in figure, sorted according to capacity levels in percentage.

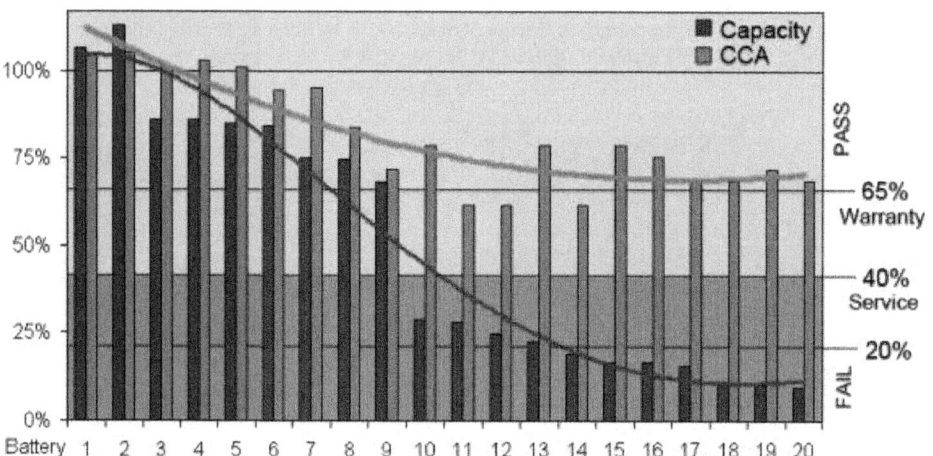

Fig. : Capacity and CCA readings of 20 aging batteries.

Batteries 1 to 9 have good CCA and high capacity; the CCA of batteries 10 to 20 remains reasonably strong but suffers from capacity loss. CCA tends to remain high while the capacity drops steadily as part of aging.

Test method : CCA was estimated with the Spectro CA-12 and the capacity was measured with an Agilent load bank by applying full discharges according to BCI standards.

Courtesy of Cadex

Batteries 1 to 9 perform well on capacity and CCA, but batteries 10 to 20 show notable capacity loss while the CCA remains strong. The motorist is unaware of the fading capacity until the car won't start one morning. This is especially critical during a cold spell, which further reduces the capacity.

Capacity is the leading health indicator of a battery, and car manufacturers often use 65 per cent as the pass/fail threshold for warranty replacement. This magic level forms a natural bend, a cliff between a high performing battery and one that is beginning to age. Service garages usually take 40 per cent as an end-of-life indication. Read more about How to Measure Capacity. Even though a starter battery with 40 per cent capacity may still crank well and have 6 to 12 months of service left before it will finally quit, the battery should be replaced. Thrifty drivers, (including this author) prefer to wait, but invariably get caught with a dead battery at the worst possible moment.

Evaluating the capacity of a starter battery gives the most accurate end-of-life prediction. Capacity sets the floor upon which CCA and other readings are compared. Without knowing capacity, other measurements mean little.

HOW HEAT AND LOADING AFFECT BATTERY LIFE

Heat is a killer of all batteries and high temperatures cannot always be avoided. This is the case with a battery inside a laptop, a starter battery under the hood of a car and stationary batteries in a tin shelter under the hot sun. As a guideline, each 8°C (15°F) rise in temperature cuts the life of a sealed lead acid battery in half. A VRLA battery for stationary applications that would last 10 years at 25°C (77°F) would only live for five years if operated at 33°C (92°F). The same battery would desist after 2½ years if kept at a constant desert temperature of 41°C (106°F). Once the battery is damaged by heat, the capacity cannot be restored. The life of a battery also depends on the activity and is shortened if the battery is stressed with frequent discharge.

According to the *2010 BCI Failure Mode Study*, starter batteries have become more heat-resistant over the past 10 years. In the 2000 study, a change of 7°C (12°F) affected battery life by roughly one year; in 2010 the heat tolerance has widened to 12°C (22°F). In 1962, a starter battery lasted 34 months, and in 2000 the life expectancy had increased to 41 months. In 2010, BCI reports an average age of 55 months of use. The cooler North attains 59 months and the warmer South 47 months.

Cranking the engine poses minimal stress on a starter battery. This changes in a start-stop function of a *micro hybrid*. The micro hybrid turns the IC engine off at a red traffic light and restarts it when the traffic flows. This results in about

2,000 micro cycles per year. Data obtained from car manufacturers show a capacity drop to about 60 per cent after two years of use in this configuration. To solve the problem, automakers are using specialty AGM and other variations that are more robust than the regular lead acid. Read more about Alternate Battery Systems. Figure shows the drop in capacity after 700 micro cycles. The simulated start-stop test was performed in Cadex laboratories. CCA remains high.

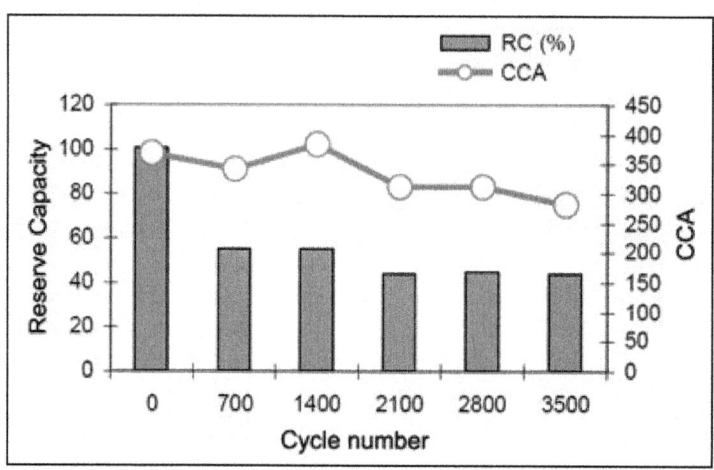

Fig. : Capacity drop of a flooded starter battery when micro cycling.

Start-stop function on a micro hybrid stresses the battery; the capacity drops to about 50 per cent after two years of use. AGM is more robust for this application.
Courtesy of Cadex, 2010.

Test method : The test battery was fully charged and then discharged to 70 per cent to resemble the SoC of a micro hybrid in real life. The battery was then discharged at 25A for 40 seconds to simulate engine off condition at stoplight with the headlight on, before cranking the engine at 400A and recharging. The CCA readings were taken with the Spectro CA-12.

The cell voltages on a battery string must be similar, and this is especially important for higher-voltage VRLA batteries. With time, individual cells fall out of line, and applying an equalizing charge every six months or so should theoretically bring the cells back to similar voltage levels. While equalizing will boost the needy cells, the healthy cell get stressed if the equalizing charge is applied carelessly. What makes this service so difficult is the inability to accurately measure the condition of each cell and provide the right dose of remedy. Gel and AGM batteries have lower over-charge acceptance than the flooded version and different equalizing conditions apply. Always refer to the manufacturer's specifications.

Water permeation, or loss of electrolyte, is a concern with sealed lead acid batteries, and over-charging contributes to this condition. While flooded systems accept water, a fill-up is not possible with VRLA. Adding water has been tried, but this does not offer a reliable fix. Experimenting with watering turns the VRLA into unreliable battery that needs high maintenance.

Flooded lead acid batteries are one of the most reliable systems. With good maintenance these batteries last up to 20 years. The disadvantages are the need for watering and providing good ventilation. When VRLA was introduced in the 1980s, manufacturers claimed similar life expectancy to flooded systems, and the telecom industry switched to these maintenance-free batteries. By mid 1990 it became apparent that the life for VRLA did not replicate that of a flooded type; the useful service life was limited to only 5–10 years. It was furthermore noticed that exposing the batteries to temperatures above 40°C (104°F) could cause a thermal runaway condition due to dry-out.

A new lead acid battery should have an open circuit voltage of 2.125V/cell. At this time, the battery is fully charged. During buyer acceptance, the lead acid may drop to between 2.120V and 2.125V/cell. Shipping, dealer storage and installation will decrease the voltage further but the battery should never go much below 2.10V/cell. This would cause sulfation. Battery type, applying a charge or discharge within 24 hours before taking a voltage measurement, as well as temperature will affect the voltage reading. A lower temperature lowers the OCV; warm ambient raises it.

HOW TO RESTORE NICKEL-BASED BATTERIES

During the nickel-cadmium years in the 1970s and 1980s, most battery ills were blamed on "memory." Memory is derived from "cyclic memory," meaning that a nickel-cadmium battery could remember how much energy was drawn on previous discharges and would not deliver more than was demanded before. On a discharge beyond regular duty, the voltage would abruptly drop as if to rebel against pending overtime. Improvements in battery technology have virtually eliminated the phenomenon of cycling memory.

Figure illustrates the stages of crystalline formation that occur on a nickel-cadmium cell if over-charged and not maintained with periodic deep discharges. The first enlargement shows the cadmium plate in a normal crystal structure; the middle image demonstrates full-blown crystalline formation; and the third reveals some form of restoration.

The modern nickel-cadmium battery is no longer affected by cyclic memory but suffers from *crystalline formation*. The active cadmium material is applied on the negative electrode plate, and with incorrect use a crystalline formation occurs that reduces the surface area of the active material. This lowers battery performance. In advanced stages, the sharp edges of the forming crystals can penetrate the separator, causing high self-discharge that can lead to an electrical short. The term "memory" on the modern NiCd refers to crystalline formation rather than the cycling memory of old.

When nickel-metal-hydride was introduced in the early 1990s, this chemistry was promoted as being memory-free but this claim is only partially true. NiMH is also subject to memory but to a lesser degree than NiCd. While NiMH has only the nickel plate to worry about, NiCd also includes the memory-prone cadmium negative electrode. This is a non-scientific explanation of why nickel-cadmium is more susceptible to memory than nickel-metal-hydride.

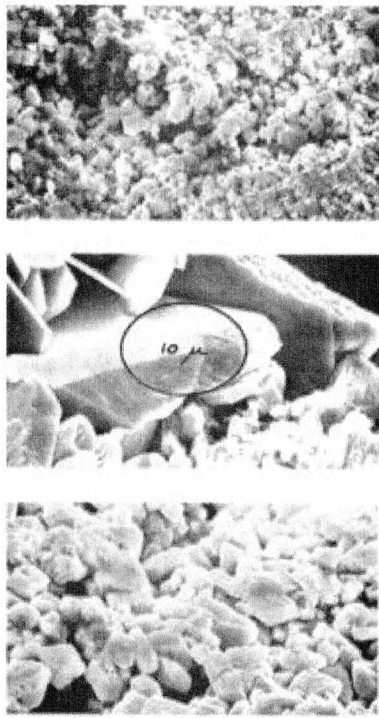

New nickel-cadmium cell. The anode (negative electrode) is in fresh condition. Hexagonal cadmium-hydroxide crystals are about 1 micron in cross-section, exposing large surface area to the electrolyte for maximum performance.

Cell with crystalline formation. Crystals have grown to 50 to 100 microns in cross-section, concealing large portions of the active material from the electrolyte. Jagged edges and sharp corners can pierce the separator, leading to increased self-discharge or electrical short.

Restored cell. After a pulsed charge, the crystals are reduced to 3–5 microns, an almost 100% restoration. Exercise or recondition is needed if the pulse charge alone is not effective.

Fig. : Crystalline formation on nickel-cadmium cell.

Crystalline formation occurs over a few months if battery is over-charged and not maintained with periodic deep discharges.

Courtesy of the US Army Electronics Command in Fort Monmouth, NJ.

Crystalline formation occurs if a nickel-based battery is left in the charger for days or repeatedly recharged without a periodic full discharge. Since most applications fall into this user pattern, NiCd requires a periodic discharge to one volt per cell to prolong service life. A discharge/charge cycle as part of maintenance, known as *exercise*, should be done every one to three months.Avoid over-exercising as this wears down the battery unnecessarily.

If regular exercise is omitted for six months and longer, the crystals ingrain themselves and a full restoration with a discharge to one volt per cell may no longer be sufficient. However, a restoration is often still possible by applying a secondary discharge called "recondition." Recondition is a slow discharge that drains the battery to a voltage cut-off point of about 0.4V/cell and lower. Tests done by the US Army indicate that a NiCd cell needs to be discharged to at least 0.6V to effectively break up the more resistant crystalline formations. During this corrective discharge, the current must be kept low to minimize cell reversal and, as discussed earlier, NiCd can tolerate a small amount of cell reversal. Figure illustrates the battery voltage during a discharge to 1V/cell, followed by the secondary discharge to 0.4V/cell.

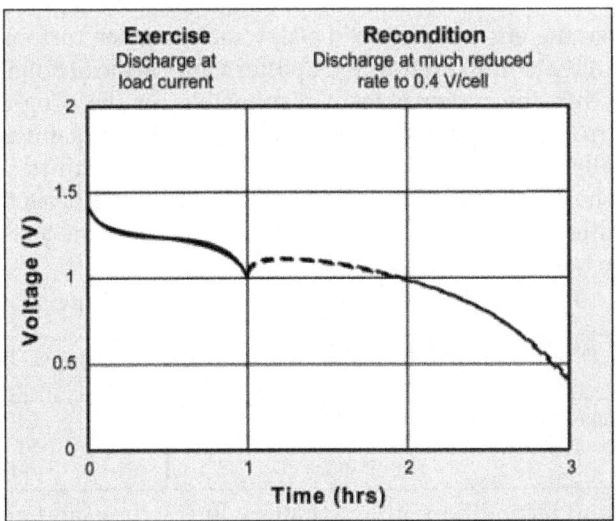

Fig. : Exercise and recondition features of a Cadex battery analyzer.

Recondition restores NiCd batteries with hard-to-remove memory. Recondition is a slow, deep dis-charge to 0.4V/cell.
Courtesy of Cadex.

Recondition is most effective with healthy batteries and the remedy is also known to improve new packs. Similar to a medical treatment, however, the service should only be applied when so needed because over-use will stress the battery. Automated battery analyzers (Cadex) only apply the recondition cycle if the user-set target capacity cannot be reached.

Recondition is only effective on working batteries. Best results in recovery are possible when applying a full discharge every 1–3 months. If exercise has been withheld for 6–12 months, the capacity may not recover fully, and if it does the pack might suffer from high self-discharge caused by a marred separator. Older batteries do not restore well and many get worse with recondition. When this happens, the battery is a ripe candidate for retirement.

Results of Battery Maintenance

After the Balkan War in the 1990s, the Dutch Army began servicing its arsenal of nickel-cadmium batteries that had been used for the two-way radios. The technicians in charge wanted to know the remaining capacity and how many batteries could be restored to full service using battery analyzers (Cadex). The army knew that allowing the batteries to sit in the chargers with only two to three hours of use per day during the war was not ideal, and the tests showed that the capacity on some packs had dropped to a low 30 per cent. With the recondition function, however, nine out of 10 batteries could be restored to 80 per cent and higher. The army uses 80 per cent as a threshold for usability. At time of service, the nickel-cadmium batteries were two to three years old.

To analyze the effectiveness of battery maintenance further, the US Navy carried out a study to find out how user pattern affects the life of nickel-cadmium batteries. For this, the research team responsible for the program established three battery groups. One group received charge only (no maintenance); another was periodically exercised (discharge to 1V/cell); and a third group received recondition. The 2,600 batteries studied were used for Motorola two-way radios deployed on three US aircraft carriers. Table summarizes the test results, including the cost factor.

Table : Replacement rates of nickel-cadmium batteries.

Maintenance method	Annual % of batteries requiring replacement	Annual battery cost (US$)
Charge-and-use only	45%	$40,500
Exercise	14%	$13,500
Recondition	5%	$4,500

Exercise and recondition prolong battery life by three-and nine-fold respectively.

GTE Government Systems, the organization that conducted the test, learned that with *charge-and-use* the annual percentage of battery failure was 45 per cent; with exercise the failure rate was reduced to 15 per cent; and with recondition only 5 per cent failed. The GTE report concludes that a battery analyzer featuring exercise and recondition costing US$2,500 would return the investment in less than one month on battery savings alone.

EFFECT OF ZAPPING NICKEL-CADMIUM

Remote control (RC) enthusiasts are experimenting with all imaginable methods to maximize battery performance. A race-car motor draws 30A for about four minutes delivered by a 7.2V battery. This amounts to over 200W of power, a large amount from a small battery. An experimental technique that seems to enhance power is zapping the cells with a very high pulse current. This is said to increase the cell voltage slightly and generate more power.

According to experts, zapping works best with NiCd cells. NiMH cells have been tried but do not produce consistent results. Companies specializing in zapping use a very high quality NiCd cell from Japan, and the sub-C is the most popular size for this application. The factory handpicks the cells and they come with a unique label in a fully discharged state. When measuring a totally empty cell (no charge), the open circuit voltage should read between 1.11 and 1.12V. If the voltage is lower than 1.06V, then the cell is considered suspect and zapping does not enhance the performance as well as with the others.

To zap a NiCd battery, a 47,000mF capacitor is charged to 90V, after which the raw power is discharged directly across a single NiCd cell of 1.2V. After the shock treatment, the cell is cycled and then zapped once more. Experts say that once a cell is treated and used in service, zapping will no longer improve performance, nor does it regenerate a weak cell.

The voltage increase on a successfully zapped battery is between 20 and 40mV when loaded with 30A. According to experts, the voltage gain is permanent, but there is a small drop in the gained voltage with usage and time.

There are no apparent side effects from zapping, however, the battery manufacturers remain silent about this treatment. No scientific explanations are available as to why zapping improves battery performance other than the belief that it lowers the internal resistance. There is little information available regarding the longevity of the cells after the treatment. Zapping only seems to work with high-quality standard NiCd cells, and in no way should this be used on lithium-based chemistries.

Another method to improve NiCd batteries is through a recondition program. Tests performed at the Cadex laboratories reveal a permanent capacity gain of about seven per cent when servicing new NiCd with recondition, a program that lowers the battery voltage to 0.4V/cell on a secondary discharge. This capacity gain is not fully understood, other than to assume that the battery improved through additional formatting. Another explanation is the presence of early memory. Since new batteries are stored with some charge, the self-discharge that occurs during storage may contribute to the buildup of crystalline formation, which recondition reverses. While NiCds once played a pivotal role for RC enthusiasts, the interest is shifting towards high-performance Li-ion.

HOW TO PROLONG LITHIUM-BASED BATTERIES

The lithium-ion battery works on ion movement between the positive and negative electrodes. In theory such a mechanism should work forever, but cycling, elevated temperature and aging decrease the performance over time. Since batteries are used in demanding environmental conditions, manufacturers take a conservative approach and specify the life of most Li-ion between 300 and 500 discharge/charge cycles.

Counting cycles is not conclusive because a discharge may vary in depth and there are no clearly defined standards of what constitutes a cycle. Read more about What Constitutes a Discharge Cycle?. In lieu of cycle count, some batteries in industrial instruments are date-stamped, but this method is not reliable either because it ignores environmental conditions. A battery may fail within the allotted time due to heavy use or unfavourable temperature conditions, but most quality packs will last considerably longer than what the stamp indicates.

The performance of a battery is measured in capacity, a leading health indicator. Internal resistance and self-discharge also play a role but with modern Li-ion these carry lower significance in predicting the end-of-battery-life. Figure illustrates the capacity drop of 11 Li-polymer batteries that have been cycled at a Cadex laboratory. The 1500mAh pouch cells for smartphones were first charged at a current of 1500mA (1C) to 4.20V/cell and allowed to saturate to 0.05C (75mA) as part of the full charge procedure. The batteries were then discharged at 1500mA to 3.0V/cell, and the cycle was repeated.

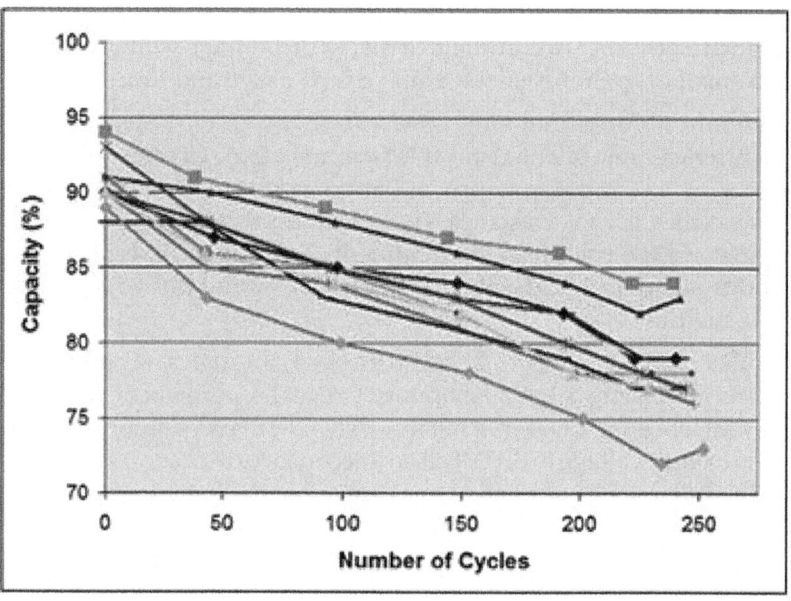

Fig. : Capacity drop as part of cycling.

A pool of new 1500mAh Li-ion-batteries for smartphones istested on a Cadex C7400 battery analyzer. All 11 pouch packs show a starting capacity of 88–94 per cent and decrease in capacity to 73–84 per cent after 250 full discharge cycles (2010).

Although a battery should deliver 100 per cent capacity during the first year of service, it is common to see lower than specified capacities, and shelf life may have contributed to this loss. In addition, manufacturers tend to overrate their batteries; knowing that very few customers would complain. In our test, the expected capacity loss of Li-ion batteries was uniform over the 250 cycles and the batteries performed as expected.

Similar to a mechanical device that wears out faster with heavy use, so also does the depth of discharge (DoD) determine the cycle count. The shorter the discharge (low DoD), the longer the battery will last. If at all possible, avoid full discharges and charge the battery more often between uses. Partial discharge on Li-ion is fine; there is no memory and the battery does not need periodic full discharge cycles to prolong life, other than to calibrate the fuel gauge on a smart battery once in a while. Read more about Battery Calibration.

Table compares the number of discharge/charge cycles Li-ion can deliver at various DoD levels before the battery capacity drops to 70 per cent. The number of discharge cycles depends on many conditions and includes charge voltage, temperature and load currents. Not all Li-ion systems behave the same.

Table : Cycle life as a function of depth of discharge.

A partial discharge reduces stress and prolongs battery life. Elevated temperature and high currents also affect cycle life.

Depth of discharge	Discharge cycles
100% DoD	300–500
50% DoD	1,200–1,500
25% DoD	2,000–2,500
10% DoD	3,750–4,700

Lithium-ion suffers from stress when exposed to heat, so does keeping a cell at a high charge voltage. A battery dwelling above 30°C (86°F) is considered *elevated temperature* and for most Li-ion, a voltage above 4.10V/cell is deemed as *high voltage*. Exposing the battery to high temperature and dwelling in a full state-of-charge for an extended time can be more stressful than cycling. Table demonstrates capacity loss as a function of temperature and SoC.

Table : Estimated recoverable capacity when storing Li-ion for one year at various temperatures.

Elevated temperature hastens capacity loss. The capacity cannot be restored. Not all Li-ion systems behave the same.

Temperature	40% charge	100% charge
0°C	98%	94%
25°C	96%	80%
40°C	85%	65%
60°C	75%	60%
		(after 3 months)

Most Li-ions are charged to 4.20V/cell and every reduction of 0.10V/cell is said to double cycle life. For example, a lithium-ion cell charged to 4.20V/cell typically delivers 300–500 cycles. If charged to only 4.10V/cell, the life can be prolonged to 600–1,000 cycles; 4.00V/cell should deliver 1,200–2,000 and 3.90V/cell 2,400–4,000 cycles. Table 4 summarizes these results. The values are estimate and depend on the type of li-ion-ion battery.

For safety reasons, lithium-ion cannot exceed 4.20V/cell. While a higher voltage would boost capacity, over-voltage shortens service life and compromises safety. Figure demonstrates cycle count as a function of charge voltage. At 4.35V, the cycle count is cut in half.

Table : Discharge cycles and capacity as a function of charge.

Every 0.10V drop below 4.20V/cell doubles the cycle; the retained capacity drops accordingly. Raising the voltage above 4.20V/cell stresses the battery and compromises safety.

Charge level (V/cell)	Discharge cycles	Capacity at full charge
[4.30]	[150–250]	~[110%]
4.20	300–500	100%
4.10	600–1,000	~90%
4.00	1,200–2,000	~80%
3.92	2,400–4,000	~75%

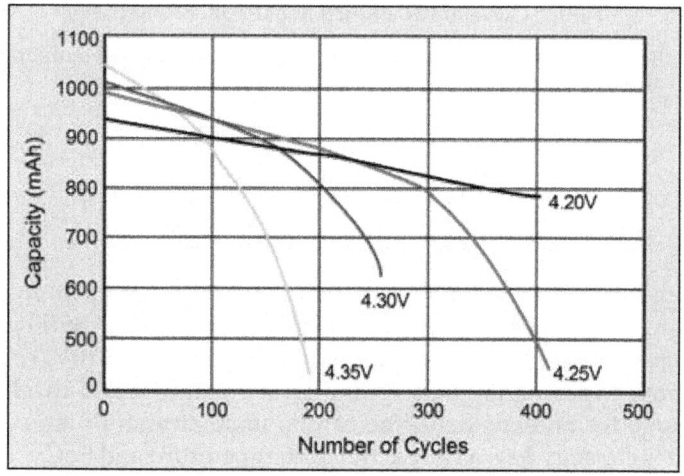

Fig. : Effects on cycle life at elevated charge voltages.

Higher charge voltages boost capacity but lowers cycle life and compromises safety.
Source : Choi *et. al.* (2002)

Chargers for cellular phones, laptops, tablets and digital cameras bring the Li-ion battery to 4.20V/cell. This allows maximum capacity, because the consumer wants nothing less than optimal run-time. Industry, on the other hand, is more concerned about longevity and may choose lower voltage thresholds. Satellites and electric vehicles are examples where longevity is more important than capacity.

Charging to 4.10V/cell the battery holds about 10 per cent less capacity than going all the way to 4.20V. In terms of optimal longevity, a voltage limit of 3.92V/cell works best but the capacity would only be about half compared to a 4.20V/cell charge (3.92V/cell is said to eliminate all voltage-related stresses).

Besides selecting the best-suited voltage thresholds for a given application, Li-ion should not remain at the high-voltage ceiling of 4.20V/cell for an extended time. When fully charged, remove the battery and allow to voltage to revert to a more natural level like relaxing after exercise. Although a properly functioning Li-ion charger will terminate charge when the battery is full, some chargers apply a topping charge if the battery terminal voltage drops to a given level.

What the User can Do

The author of this essay does not depend on the manufacturer's specifications alone but also listens to user comments. Battery University. comis an excellent sounding board to connect with the public and learn about reality. This approach might be unscientific, but it is genuine. When the critical mass speaks, the manufacturers listen. The voice of the multitude is in some ways stronger than laboratory tests performed in sheltered environments.

Tables look at cycle life as a function of discharge, temperature and charge level. A summary table should be added that also states the *Optimal Battery Energy Factor Over Life*. While this would help in selecting the optimal battery, battery makers are hesitant to release such a specification freely, and for good reason. A battery is in constant flux and capturing all of its data is exhaustive. A further criterion is price. Batteries can be built to perform better but this comes at a cost.

Let's look at real-life situations and examine what stresses lithium-ion batteries encounter. Most packs last three to five years. Environmental conditions, and not cycling alone, are a key ingredient to longevity, and the worst situation is keeping a fully charged battery at elevated temperatures. This is the case when running a laptop off the power grid. Under these conditions, a battery will typically last for about two years, whether cycled or not. The pack does not die suddenly but will give lower run-times with aging.

Even more stressful is leaving a battery in a hot car, especially if exposed to the sun. When not in use, store the battery in a cool place. For long-term storage, manufacturers recommend a 40 per cent charge. This allows for some self-discharge while still retaining sufficient charge to keep the protection circuit active. Finding the ideal state-of-charge is not easy; this would require a discharge with appropriate cut-off. Do not worry too much about the state-of-charge; a cool and dry place is more important than SoC.

Batteries are also exposed to elevated temperature when charging on wireless chargers. The energy transfer from a charging mat to a portable device is 70 to 80 per cent and the remaining 20 to 30 per cent is lost mostly in heat that is transferred to the battery through the mat. We keep in mind that the mat will cool down once the battery is fully charged.

Avoid charging a battery faster than 1C; a more moderate charge rate of 0.7C is preferred. Manufacturers of electric powertrains are concerned about super-fast charging of 20 minutes and less. Similarly, harsh discharges should be avoided as also this also adds to battery stress.

Commercial chargers do not allow changing the charge voltage limit. Adding this feature would have advantages, especially for laptops as a means to prolong battery life. When running on extended AC mode, the user could select the "long life" mode and the battery would charge to 4.00V/cell for a standby capacity of about 70 per cent. Before travelling, the user would apply the "full charge mode" to bring the charge to 100%. Some laptop manufacturers may offer this feature but often only computer geeks discover them.

Another way to extend battery life is to remove the pack from the laptop when running off the power grid. The *Consumer Product Safety Commission* advises to do this out of concern for overheating and causing a fire. Removing the battery has the disadvantage of losing unsaved work if a power failure occurs. Heat buildup is also a concern when operating a laptop in bed or on a pillow, as this may restrict airflow. Placing a ruler or other object under the laptop will improve air circulation and keep the device cooler.

"Should I disconnect my laptop from the power grid when not in use?" many ask. Under normal circumstances this should not be necessary because once the lithium-ion battery is full the charger discontinues charge and only engages when the battery voltage drops. Most users do not remove the AC power and I like to believe that this practice is safe.

Everyone wants to keep the battery as long as possible, but a battery must often operate in environments that are not conducive to optimal service life. Furthermore, the life of a battery may be cut short by an unexpected failure, and in this respect the battery shares human volatility.

To get a better understanding of what causes irreversible capacity loss in Li-ion batteries, several research laboratories are performing forensic tests. Scientists dissected failed batteries to find suspected problem areas on the electrodes. Examining an unrolled 1.5-meter-long strip (5 feet) of metal tape coated with oxide reveals that the finely structured nanomaterials have coarsened. Further studies revealed that the lithium ions responsible to shuttle electric charge between the electrodes had diminished on the cathode and had permanently settled on the anode. This results in the cathode having a lower lithium concentration than a new cell, a phenomenon that is irreversible. Knowing the reason for such capacity loss might enable battery manufacturers to prolong battery life in the future.

HOW TO AWAKEN SLEEPING LI-ION

Li-ion batteries contain a protection circuit that shields the battery against abuse. This important safeguard has the disadvantage of turning the battery off if over-discharged, and storing a discharged battery for any length of time can do this. The self-discharge during storage gradually lowers the voltage of a battery that is already discharged; the protection circuit will eventually cut off between 2.20 and 2.90V/cell.

Some battery chargers and analyzers, including those made by Cadex, feature a wake-up feature or "boost" to reactivate and charge batteries that have fallen asleep. Without this feature, a charger would render these batteries as unserviceable and the packs would be discarded. The boost feature applies a small charge current to first activate the protection circuit and then commence with a normal charge.

Do not boost lithium-based batteries back to life that have dwelled below 1.5V/cell for a week or longer. Copper shunts may have formed inside the cells that can lead to a partial or total electrical short. When recharging, such a cell might become unstable, causing excessive heat or showing other anomalies. The "boost" function by Cadex halts the charge if the voltage does not rise normally.

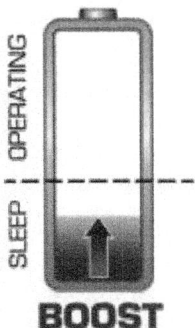

Fig. : Sleep mode of a lithium-ion battery.

Some over-discharged batteries can be "boosted" to life again. Discard pack if the voltage does not rise to a normal level within a minute while on boost.

A study done by Cadex to examine failed batteries reveals that three out of ten batteries are removed from service due to over-discharge. Furthermore, 90 per cent of returned batteries have no fault or can easily be serviced. Lack of test devices at the customer service level is in part to blame for the high exchange rate. Refurbishing batteries saves money and protects the environment.

HOW TO MAXIMIZE RUN-TIME

As the author of *Battery University. com*, I get many interesting enquiries from battery users. A man writes, "Hi, I am looking for an answer to a perplexing question. A co-worker and I have identical cell phones from the same provider. Moving into a new house, she complained of short battery run-time. I told her she was out of her mind, but then I noticed my battery behaving differently when I travel. Is there some mysterious force that's draining the battery?"

Yes, there is a hidden force that drains the battery but it's not mystical. When turned on, a cell phone is in constant communication with the tower, transmitting small bursts of power once every second or so to check for incoming calls. To save energy, the signal strength adjusts the transmission power to only what is needed. If the cell phone is close to a repeater tower, the energy required to communicate is very low. Move farther away or enter an area with high electrical noise, such as a shopping mall, hospital or factory, and the required energy increases. An analogy can be made to sitting in a restaurant. When the surroundings are quiet, the voices can be kept low, but as the crowd grows everyone needs to talk a bit louder.

Living in sight of a tower has advantages and your cellular battery will last longer between charges. Where you park your cell phone in the house also affects run-time. A manager of a large cellular provider in the UK said his son experienced shorter standby times after moving from the upstairs bedroom to the basement. If possible, leave your cell phone in an upstairs room facing a tower. When travelling by car place it near a window rather than on the floor but avoid direct exposure to the sun.

Similar rules apply to TETRA and P25 radio systems, cordless telephones, Wi-Fi and Bluetooth devices. A wireless headset that communicates with the cell phone from belt to ear provides longer run-times than when placing the handset on the dining-room table while cooking in the kitchen. Although the quality of communication stays the same, the Bluetooth headset needs to work harder when placed farther away from the user.

The energy savings only apply when the wireless device is in the "on" position. When "off," the load on the battery is very low and only provides power for housekeeping functions such as maintaining the clock and monitoring key commands. Housekeeping and self-discharge consume 5 to 10 per cent of the available battery energy per month.

During the last few years, standby and talk-times on cell phones have improved. Besides increases in the specific energy of lithium-ion, improvements in receiver and demodulator circuits have achieved notable energy savings. Figure illustrates the reduction of power consumption in these circuits since 2002. We must keep in mind that the savings apply mainly to standby and receiving circuits. Transmitting still requires about five times the power of the receiving and demodulation.

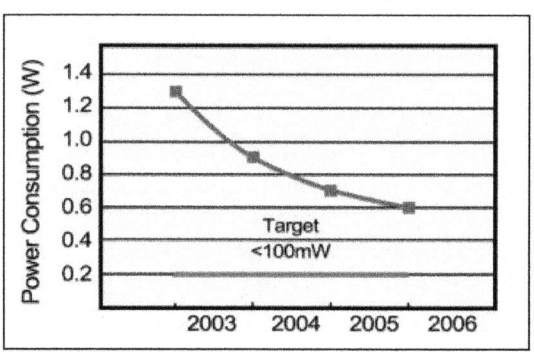

Fig. : Reduction in power consumption.

Cell phones have achieved notable power savings in the receiver and demodulator circuits. Transmitting needs the most power.
Souece : Sieber *et. al.* (2004).

Laptop batteries fare badly in terms of life span. Laptops are demanding bosses that request a steady stream of power under poor working conditions, toiling in an unbearable heat of 40–45°C (104–113°F). In addition, the battery is exposed to a high voltage by being kept at full charge. High heat and dwelling at full state-of-charge, not cycling, cause short battery life in laptops.

Laptop batteries have further demands — they must be small and lightweight. The laptop battery should be invisible to the user and deliver enough power to endure a five-hour flight. In reality, the battery runs for only about 90 minutes. Batteries are getting better; however, the request for higher performance counteracts the capacity gain, resulting in roughly the same run-time with more powerful features.

Although users want longer run-times, computer manufacturers are hesitant to add larger batteries because of increased size, weight and cost. A survey indicates that given the option of a larger size with added weight to gain longer run-times, most users would settle for what is offered today. For better or worse, we have learned to live with what we have.

Aftermarket Batteries

In the search for low-cost batteries, consumers may inadvertently purchase counterfeit batteries that are unsafe. The label appears *bona fide* and the buyer cannot distinguish between an original and a forged product. Cell phone manufacturers are concerned about these products flooding the market and advise customers to use approved brands; defiance could void the warranty. Manufacturers do not object to third-party suppliers as long as the aftermarket batteries are well built, safe and approved by a safety agency.

Caution also applies to purchasing counterfeit chargers. Some unsafe aftermarket chargers do not terminate the battery correctly and rely on the battery's internal protection circuit to cut off when full. The need for redundancy is important because a *bona fide* battery could have a malfunctioning protection circuit that was damaged by a static charge. If, for example, the maker of the counterfeit battery relies on the charger to terminate the charge, and the charger builder has full confidence that the battery will turn off when ready, the combination of these two products can have a lethal effect.

Some laptop manufacturers disallow aftermarket batteries by digitally locking the pack with a tamperproof security code. This is done in part for safety reasons, because the potential damage resulting from a faulty laptop battery is many times greater than that of a cell phone.

Simple Guidelines to Prolong Lithium-ion Batteries

- Do not discharge Li-ion too low; charge more often.
- A random or partial charge is fine. Li-ion does not need a full charge.
- Limit the time the battery resides at 4.20/cell (full charge), especially if warm.
- Moderate the charge current to between 0.5C and 0.8C for cobalt-based lithium-ion. Avoid ultra-fast charging and discharging.
- If the charger allows, lower the charge voltage limit to prolong battery life.
- Keep the battery cool. Move it away from heat-generating environments. Avoid hot cars and windowsills.
- High heat and full state-of-charge, not cycling, cause short battery life in laptops.
- Remove battery from laptop when used on the power grid.

AFTERMARKET BATTERIES AND WHAT EVERYONE SHOULD KNOW

In the search for low-cost batteries, consumers may inadvertently purchase counterfeit batteries that are unsafe. The label appears *bona fide* and the buyer cannot distinguish between an original and a forged product. Cell phone manufacturers are concerned about these products flooding the market and advise customers to use approved brands; defiance could void the warranty. Manufacturers do not object to third-party suppliers as long as the aftermarket batteries are well built, safe and approved by a safety agency.

• Caution also applies to purchasing counterfeit chargers. Some aftermarket chargers do not terminate the battery correctly and rely on the battery's internal protection circuit to cut off the charge when the battery is full. The need for redundancy in charging is important because a *bona fide* battery could have a malfunctioning protection circuit that was damaged by a static charge or other cause. If, for example, the maker of the counterfeit battery relies on the charger to terminate the charge, and the manufacturer of the charger depends fully on the battery's protection circuit, a combination exists that can have serious consequences.

• Some laptop manufacturers disallow aftermarket batteries by digitally locking the pack with a tamperproof security code. This is done in part for safety reasons; the potential damage resulting from a faulty laptop battery is many times greater than that of a cell phone.

HOW TO REPAIR A BATTERY PACK

Batteries for power tools and other industrial devices can often be repaired by replacing one or all cells. Finding a NiCd and NiMH cell is relatively easy; locating the correct Li-ion cell can be more difficult. Naked Li-ion cells are not readily available off the shelf and a reputable battery manufacturer may only sell to certified pack assemblers. Incorrect use or lack of an protection circuit could cause stress and disintegration of the replaced cell. When repairing a Li-ion pack make certain that each cell is properly connected to a protection circuit.

If a relatively new pack has only one defective cell, you may replace only the affected cell. On an aged battery, it's best to replace all cells. Adding a new cell with full capacity in between neighbouring cells that have faded would cause a cell mismatch. Matching the replacement cell with one of a lower rating may work but this fix is often of short duration. Always replace with the same chemistry cell.

A well-matched battery pack means that all cells have similar capacities. An anomaly can be drawn with a chain in which the weakest link determines the performance of the battery. Read more about Can Batteries Be Restored?

When replacing all cells, the rating is less important as long as the differences are not too large for the charger to handle. Cells with higher Ah will simply take a bit longer to charge. The state-of-charge of all cells being charged for the first time should have a similar charge level, and the open-circuit voltages should be within 10 per cent of each other.

Many visitors of Battery University. com ask if NiCd can be replaced with NiMH? Theoretically, this should be possible but charging may be an issue. NiMH uses a more defined charge algorithm than NiCd. A modern NiMH charger can charge both NiMH and NiCd; the old NiCd charger could over-charge NiMH by not properly detecting full charge state and applying a trickle charge that is too high.

Welding the cells is the only reliable way to get dependable connection. Limit the heat transfer to the cells during welding to prevent excess heat buildup.

Simple Guidelines When Repairing Battery Packs

- Only connect cells that are matched and have the identical state-of-charge. Do not connect cells of different chemistry, age or capacity.
- Never charge or discharge Li-ion batteries without a working protection circuit unattended. Each cell must be monitored individually.
- Include a temperature sensor that disrupts the current on high heat.
- Apply a slow charge only if the cells have different state-of-charge.
- Pay special attention when using an unknown brand of cells. Some may not contain a high level of intrinsic safety.
- Li-ion is sensitive to reverse polarization. Observe correct polarity.
- Do not charge a Li-ion battery that exhibits physical damage or has dwelled at a voltage of less than 1.5V/cell.
- When repairing Li-ion, assure that each cell is connected to a protection circuit.

HOW TO REPAIR A LAPTOP BATTERY

Most laptop batteries are smart. This means that the pack consists of two parts : the chemical cells and the digital circuit. If the cells are weak, cell replacement makes economic sense. While nickel-based cells are readily available, lithium-ion cells are commonly not sold over the counter, and most manufacturers only offer them to authorized pack assemblers. This precaution is understandable given liability issues. Read also about Safety Concerns and Protection Circuits. Always use the same chemistry; the mAh rating can vary if all cells are replaced.

A laptop battery may have only one weak cell, and the success rate of replacing the affected cell depends on the matching with the others. All cells in a pack must have a similar capacity because an imbalance shortens the life of the pack. Read more about Can Batteries Be Restored?. Furthermore, the state-of-charge of all cells being charged for the first time should have a similar charge level, and the open-circuit voltages should be within 10 per cent of each other. Welding the cells is the only reliable way to get dependable connection. Limit the heat transfer to the cells during welding to prevent excess heat buildup.

The typical SMBus battery has five or more battery connections consisting of positive and negative battery terminals, thermistor, clock and data. The connections are often unmarked; however, the positive and negative are commonly located at the outer edges of the connector and the inner contacts accommodate the clock and data. (The one-wire system combines clock and data.) For safety reasons, a separate thermistor wire is brought to the outside. Figure illustrates a battery with six connections.

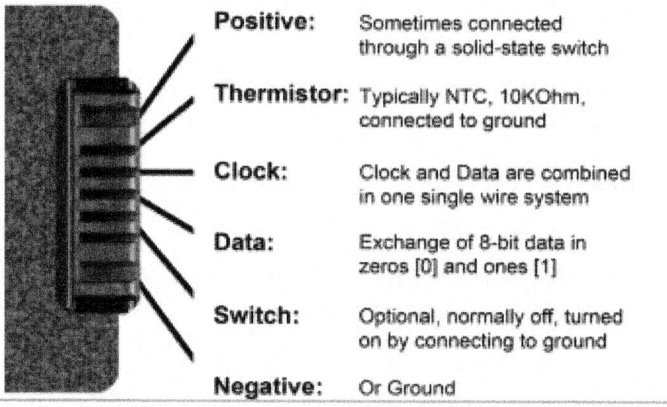

Positive: Sometimes connected through a solid-state switch

Thermistor: Typically NTC, 10KOhm, connected to ground

Clock: Clock and Data are combined in one single wire system

Data: Exchange of 8-bit data in zeros [0] and ones [1]

Switch: Optional, normally off, turned on by connecting to ground

Negative: Or Ground

Fig. : Terminal connection of a typical laptop battery.

The positive and negative terminals are usually placed on the outside; no norm exists on the arrangement of the other contacts.
Courtesy of Cadex

Some batteries are equipped with a solid-state switch that is normally in the "off" position and no voltage is present on the battery terminals; connecting the switch terminal to ground often turns the battery on. If this does not work, the pack may need a proprietary code for activation, and battery manufacturers keep these codes a well-guarded secret.

How can you find the correct terminals? Use a voltmeter to locate the positive and negative battery terminals and establish the polarity. If no voltage is available, a solid-state switch in the "off" position may need activating. Connecting the voltmeter to the outer terminals, take a 100-Ohm resistor (other values may also work), tie one end to ground, and with the other end touch each terminal while observing the voltmeter. If no voltage appears, the battery may be dead or the pack will require a security code. The 100-Ohm resistor is low enough to engage a digital circuit and high enough to protect the battery against a possible electrical short.

Establishing the connection to the battery terminals should now enable charging. If the charge current stops after 30 seconds, an activation code may be required, and this is often difficult if not impossible to obtain.

Some battery manufacturers add an end-of-battery-life switch that turns the battery off when reaching a certain age or cycle count. Manufacturers argue that customer satisfaction and safety can only be guaranteed by regularly replacing the battery. Such a policy tends to satisfy the manufacturer more than the user, and newer batteries do not include this feature.

If at all possible, connect the thermistor during charging and discharging to protect the battery against possible overheating. Use an ohmmeter to locate the internal thermistor. The most common thermistors are 10 Kilo Ohm NTC, which reads 10kΩ at 20°C (68°F). NTC stands for negative temperature coefficient, meaning that the resistance decreases with rising temperature. In comparison, a positive temperature coefficient (PTC) causes the resistance to increase. Warming the battery with your hand may be sufficient to detect a small change in resistor value when looking for the correct terminal on the battery.

In some cases the chemical battery can be restored, but the fuel gauge might not work, is inaccurate, or will provide wrong information. After re-packaging, the battery may need some sort of initialization/calibration process. Simply charging and discharging the pack to reset the flags might do the trick. A "flag" is a measuring point to mark and record an event.

The circuits of some smart batteries must be kept alive during cell replacement. Disconnecting the voltage for only a fraction of a second can erase vital data in the memory. The lost data could contain the resistor value of the digitized shunt that is responsible for the coulomb counter. Some integrated circuits (IC) responsible for fuel gauge function have wires going to each cell, and the sequence of assembly must to be done in the correct order.

To assure continued operation when changing the cells, supply a secondary voltage through a 100-Ohm resistor to the circuit before disconnection and remove the supply only after the circuit receives voltage again from the new cells. Cell replacement of a smart battery has a parallel with open-heart surgery, where doctors must keep all organs of the patient alive.

Anyone repairing an SMBus battery needs to be aware of compliance issues. Unlike other tightly regulated standards, the SMBus allows some variations, and this can cause problems when matching battery packs with existing chargers. The repaired SMBus battery should be checked for compatibility before use. More information on SMBus is available on.

Simple Guidelines When Repairing Battery Packs

- Only connect cells that are matched and have the identical state-of-charge. Do not connect cells of different chemistry, age or capacity.
- Never charge or discharge Li-ion batteries without a working protection circuit unattended. Each cell must be monitored individually.
- Include a temperature sensor that disrupts the current on high heat.
- Apply a slow charge only if the cells have different state-of-charge.
- Pay special attention when using an unknown brand of cells. Some may not contain a high level of intrinsic safety features.
- Li-ion is sensitive to reverse polarization. Observe correct polarity.
- Do not charge a Li-ion battery that exhibits physical damage or has dwelled at a voltage of less than 1.5V/cell.
- When repairing Li-ion, assure that each cell is connected to a protection circuit.

DIFFICULTIES WITH TESTING BATTERIES

A German manufacturer of luxury cars points out that one out of two starter batteries returned under warranty is working and has no problem. It is possible that battery testers used in service garages did not detect the batteries correctly before they were returned under warranty. ADAC reported in 2008 that 40 per cent of all roadside automotive failures are battery-related. In Japan, battery failure is the largest single complaint among new car owners. The average car is driven 13km (8 miles) per day and mostly in congested cities. The most common reason for battery failure is undercharge. Battery performance is important; problems during the warranty period tarnish customer satisfaction.

Battery malfunction during the warranty period is seldom a factory defect; driving habits are the culprits. A manufacturer of German-made starter batteries stated that factory defects account for only 5 to 7 per cent of warranty claims. The battery remains a weak link, and is evident when reviewing the *ADAC 2008 report* for the year 2007. The study examines the breakdowns of 1.95 million vehicles six years old or less, and Table provides the reasons.

Table : Most common car failures.
Batteries cause the most common failures requiring road assistance.

Percentage of Failure	Cause of Failure
52%	Battery
15%	Flat tire
8%	Engine
7%	Wheels
7%	Fuel injection
6%	Heating, cooling
5%	Fuel systems

The cellular phone industry experiences an even more astonishing battery return pattern. Nine out of 10 batteries returned under warranty have no problem or can easily be serviced. This is no fault of the manufacturers but they pay a price that is ultimately charged to the user.

Part of the problem lies in the difficulty of testing batteries at the consumer level, and this applies to storefronts and service garages alike. Battery rapid-test methods seem to dwell in medieval times, and this is especially evident when comparing advancements made on other fronts. We don't even have a reliable method to estimate state-of-charge — most of such measurements using voltage and coulomb counting are guesswork. Assessing capacity, the most reliable health indicator of a battery, dwells far behind.

The battery user may ask why the industry is lagging so far behind. The answer is simple : battery testing and monitoring is far more complex than outsiders perceive it. As there is no single diagnostic device that can assess the health of a person, so are there no instruments that can quickly check the state-of-health of a battery. Like the human body, batteries can have many hidden deficiencies that no single tester is able to identify with certainly. Yes, we can apply a discharge, but this takes the battery out of service and induces stress, especially on large systems. In some cases, even a discharge does not provide conclusive results either.

As doctors will examine a patient with different devices, so also does a battery need several approaches to find anomalies. A dead battery is easy to measure and all testers can do this. The challenge comes in evaluating a battery in the 80 to 100 per cent performance range. This chapter examines current and futuristic methods and how they stand up. One thing to remember is this : batteries cannot be measured; the appropriate instruments can only make predictions or estimations. This is synonymous with a doctor examining a patient, or the weatherman predicting the weather. All findings are estimations with various degrees of accuracies.

HOW TO MEASURE INTERNAL RESISTANCE

The resistance of a battery provides useful information about its performance and detects hidden trouble spots. High resistance values are often the triggering point to replace an aging battery, and determining resistance is especially useful in checking stationary batteries. However, resistance comparison alone is not effective, because the value between batches of lead acid batteries can vary by eight per cent. Because of this relatively wide tolerance, the resistance method only works effectively when comparing the values for a given battery from birth to retirement. Service crews are asked to take a snapshot of each cell at time of installation and then measure the subtle changes as the cells age. A 25 per cent increase in resistance over the original reading hints to an overall performance drop of 20 per cent.

Manufacturers of stationary batteries typically honour the warranty if the internal resistance increases by 50 per cent. Their preference is to get true capacity readings by applying a full discharge. It is their belief that only a discharge can provide reliable readings and they ask users to perform the service once a year. While this advice has merit, a full discharge requires a temporary disconnection of the battery from the system, and on a large battery such a test takes an entire day to complete. In the real world, very few battery installations receive this type of service and most measurements are based on battery resistance readings.

Measuring the internal resistance is done by reading the voltage drop on a load current or by AC impedance. The results are in ohmic values. There is a notion that internal resistance is related to capacity, and this is false. The resistance of many batteries stays flat through most of the service life. Figure shows the capacity fade and internal resistance of lithium-ion cells.

To estimate capacity and state-of-charge on the fly involves impedance trending by scanning a battery with frequencies ranging from less than one hertz to several thousand hertz.

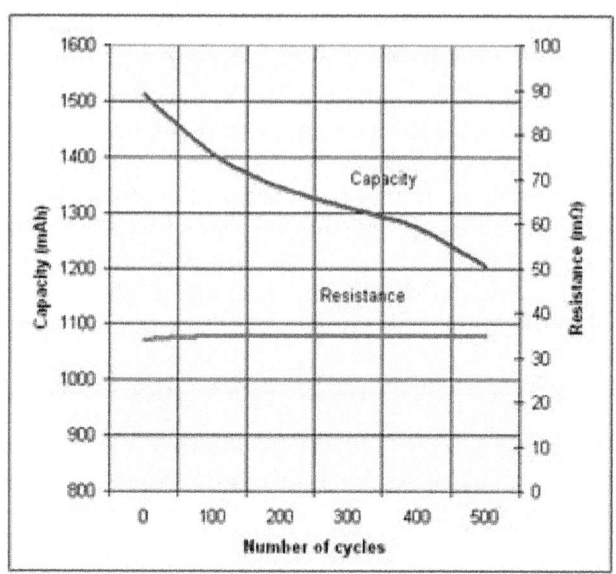

Fig. : Relationship between capacity and resistance as part of cycling.

Resistance does not reveal the state-of-health of a battery. The internal resistance often stays flat with use and aging.

Cycle test on Li-ion batteries at 1C : Charge : 1,500mA to 4.2V, 25°C Discharge : 1,500 to 2.75V, 25°C

Courtesy of Cadex

What is Impedance?

Before exploring the different methods of measuring the internal resistance of a battery, let's examine what electrical resistance means, and let's differentiate between a pure *resistance* (R) and *impedance* (Z) that includes reactive elements such as coils and capacitors. Both values are given in Ohms (W), a measure formulated by the German physicist Georg Simon Ohm, who lived from 1798 to 1854. (One Ohm produces a voltage drop of 1V with a current flow of 1A.) The difference between resistance and impedance lies in the *reactance*. Let me explain.

Most electrical loads, as well as batteries providing power, have internal impedance. Impedance consists of a capacitive reactance component (capacitor) and an inductive reactance component (coil). Capacitive reactance decreases with increasing frequency, while inductive reactance increases with increasing frequency. (To explain resistance change with frequency, we compare an oil damper that has a stiffer resistance when moved fast.).

A battery has resistive, capacitive and inductive resistance, and the term *impedance* includes all three in one. The inductive component only shows up at high frequencies when the inductive reactance of the conductors inside the battery becomes a factor in the overall impedance.

Impedance can best be illustrated with the Randles model. Figure illustrates the basic model of a lead acid battery, which reflects resistors and a capacitor (R1, R2 and C). The inductive reactance is commonly omitted because it plays a negligible role in a battery, especially at a low frequency.

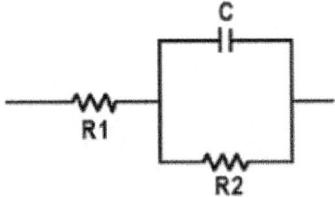

Fig. : Randles model of a lead acid battery.

The overall battery resistance consists of ohmic resistance, as well as inductive and capacitive reactance. The schematic and electrical values differ for every battery.

Now that we have learned the basics of internal battery resistance and how they can be applied to rapid-test batteries at different frequencies.

DC Load Method

Ohmic measurement is one of the oldest and most reliable test methods. The battery receives a brief discharge lasting a few seconds. A small pack gets an ampere or less and a starter battery is loaded with 50A and more. A voltmeter measures the voltage drop and Ohm's law calculates the resistance value (voltage divided by current equals resistance).

DC load measurements work well to check large stationary batteries, and the ohmic readings are very accurate and repeatable. Manufacturers of test instruments claim resistance readings in the 10 micro-ohm range. Many garages use the carbon pile to measure starter batteries, and with experience mechanics familiar with this loading device get a reasonably good assessment of the battery. The invasive test is in many ways more reliable than non-invasive methods.

The DC load method has a limitation in that it blends R1 and R2 of the Randles model into one combined resistorand ignores the capacitor. "C" is an important component of a battery that represents 1.5 farads per 100Ah capacity. In essence, the DC method sees the battery as a resistor and can only provide ohmic references.

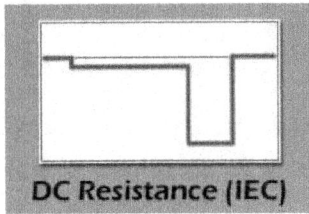

Fig. : DC load method.

The true integrity of the Randles model cannot be seen. R1 and R2 appear as one ohmic value. Courtesy of Cadex

The *two-tier DC load* method offers an alternative method by applying two sequential discharge loads of different currents and time durations. The battery first discharges at a low current for 10 seconds, followed by a higher current for three seconds, and Ohm's law calculates the resistance values. Evaluating the voltage signature under the two load conditions offers additional information about the battery, but the values are strictly resistive and do not reveal SoC and capacity estimations.

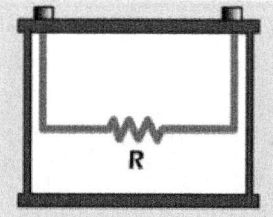

Fig. : Two-tier DC load.

The two-tier DC load follows the IEC 60285 and IEC 61436 standards and provides life-like test conditions for many battery applications. The load test is the preferred method for batteries powering DC loads.
Courtesy of Cadex

AC Conductance

The AC conductance method replaces the DC load and injects an alternating current into the battery. At a set frequency of between 80 and 90 hertz, the capacitive and inductive reactance converge, resulting in a negligible voltage lag that minimizes the reactance. Manufacturers of AC conductance equipment claim battery resistance readings in the 50 micro-ohm range, and these instruments are commonly used in North American car garages. The single-frequency technology as illustrated in figure sees the components of the Randles model as one complex impedance called the modulus of Z.

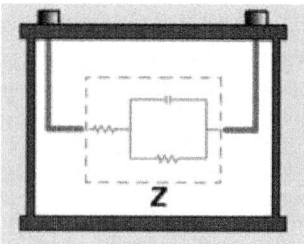

Fig. : AC conductance method.

The individual components of the Randles model are molten together and cannot be distinguished.

Smaller batteries often use the popular 1000-hertz (Hz) ohm test method. A 1000Hz signal excites the battery, and the Ohm's law calculates the resistance. It is important to note that the AC method shows different values to the DC load, and both are correct. For example, Li-ion in an 18650 cell produces about 36mOhm with a 1000Hz AC signal and roughly 110mOhm with a DC load. Since both readings are correct, and yet are so far apart, the user needs to consider the application. The pulse DC load method provides the best indication for a DC application such as driving a motor or powering a light, while the 1000Hz method better reflects the performance of a digital load, such as a cellular phone that relies to a large extent on the capacitor characteristics of a battery. Figure illustrates the 100Hz method.

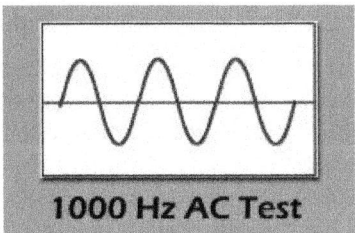

1000 Hz AC Test

Fig. : 1000-hertz method.

The IEC 1000-hertz is the preferred method to take impedance snapshots of batteries powering digital devices.

Electro-chemical Impedance Spectroscopy

Electro-chemical impedance spectroscopy (EIS) enables more than resistance readings; it can estimate state-of-charge and capacity. Research laboratories have been using EIS for many years to evaluate battery characteristics, but high equipment cost, slow test times and the need for trained professionals to decipher large volumes of data have limited this technology to laboratory environments. EIS is able to read each component of the Randles model individually; however, analyzing the value at different frequencies and correlating the data is an enormous task. Fuzzy logic and advanced digital signal processor (DSP) technology have simplified this task. Figure illustrates the battery component, which EIS technology is capable of reading.

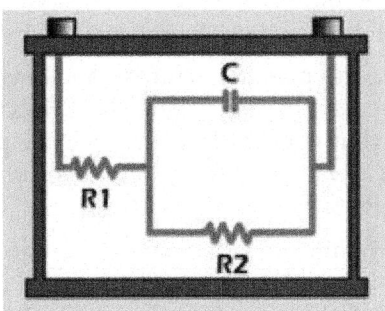

Fig. : Spectro™ method.

R1, R2 and C are measured separately, which enables state-of-charge and capacity measurements.

HOW TO MEASURE CCA (COLD CRANKING AMP)

Ever since Cadillac invented the starter motor in 1912, car mechanics explored ways to measure cold cranking amps. CCA assures that the battery has sufficient energy to crank the engine when cold. To do this without "freezing," testers look at internal resistance, the gatekeeper of a battery. A starter battery with low resistance assures reasonably good cranking, and a CCA reading of 400 to 500A is sufficient for most starter batteries. According to SAE J537, a CCA reading of 500A delivers 500A at -18°C (0°F) for 30 seconds without dropping below 7.2 volts.

Garages seldom do the full-fletched CCA test; this belongs to laboratories. Instead, device manufacturers offer alternatives and the *carbon pile* introduced in the 1980s is one of the oldest and most reliable methods. To do a pass/fail test, a fully charged starter battery is loaded with half the rated CCA for 15 seconds at a moderate temperature of 10°C (50°F) and higher. The battery will pass if the voltage stays above 9.6V. Colder temperatures cause the voltage to drop further. The DC load method has the advantage of detecting batteries with a partially shorted cell (low specific gravity) but the device cannot estimate battery capacity.

Mechanics prefer small sizes, and instead of applying the prolonged load that is typical of the carbon pile, device manufactures developed handheld testers that induce a high-current pulse. The Ohm's law calculates the internal resistance based on the load current and voltage drop. The test conditions and results of this device are similar to the carbon pile.

Meanwhile, non-invasive test methods emerged, meaning that the battery is no longer loaded for measurements. The *AC Conductance* method reads CCA by injecting a single frequency of 80–90 Hertz to the battery. The units are smaller than invasive devices and the battery does not need to be fully charged. AC Conductance meters cannot read capacity and a partially shorted cell may pass as good.

Critical progress has been made in *electro-chemical impedance spectroscopy* (EIS). Research centers have been using EIS for many years but high equipment cost, long test times and the need for trained professionals to decipher the data have kept this technology in laboratories. Fuzzy logic, advanced digital signal processors and a new algorithm to process the information have simplified this task.

Cadex took the EIS concept one step further and developed *multi-model electro-chemical impedance spectroscopy* or Spectroä for short. Spectro™ gives more accurate CCA estimation than what is possible with single-frequency AC Conductance, but the most important advantage is the ability to estimate capacity, the leading health indicator of a battery. Here is how it works :

A control signal ranging from 20 to 2000Hz is injected into the battery as if to capture the topography of a landscape. The scanned imprint is then compared against a *matrix* to derive at the reading. A matrix can be described as a multi-dimensional lookup table; and text recognition, fingerprint identification and visual imaging operate on a similar principle.

CCA works on a basic matrix that covers a broad range of starter batteries. Capacity, on the other hand, requires a complex model. To simplify testing, Cadex has developed a generic matrix that covers most starter batteries, flooded and AGM. The said generic matrix provides pass/fail information based on a capacity setting of 40 per cent, which serves as the end-of-life threshold. Battery-specific matrices can be made available that offer numeric capacity values in per cent. The test takes 15 seconds and works with a partially charged battery. Figure shows the Spectro CA-12 with Spectro™ technology.

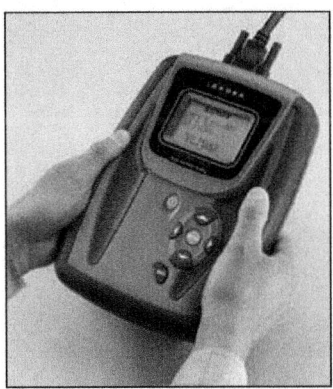

Fig. : Spectro CA-12 battery tester.

Multi-frequency concept Spectro™ concept displays capacity, CCA and state-of-charge; test time is 15 seconds.

"How accurate are the readings," car mechanics ask? This depends on the tester and method used. For example, the Spectro CA-12 with a generic matrix provides a CCA accuracy of 90 per cent; capacity is about 80 per cent. The single-frequecy AC Conductance, on the other hand, provides a CCA accuracy of roughly 70 per cent with no capacity readout. Such low accuracies may come as a surprise to many and service technicians ask for better than 90 per cent. This is impossible with lead acid batteries because of inherited inaccuracies. Capacity fluctuations of +/-12 per cent are common even with highly accurate discharge and charge equipment tested in a controlled laboratory environment.

State-of-charge (SoC) also affects accuracy. Figure compares CCA readings at different SoC, taken with the Spectro CA-12 and a device that is based on AC Conductance. While Spectro shows only a slight decrease with depleting charge, AC Conductance reflects a strong departure form the horizontal line; the readings are only similar at a 70 per cent SoC. Since most batteries hover at about 70 per cent when the car is brought in for service, the CCA readings of the two methods may appear similar.

Battery manufacturers are hesitant to endorse new test technologies. It is said that the first digital tester introduced in the early 1990s won approval by agreeing to give slightly higher CCA readings than what a lab test would provide. After all, who knows the true value ! Very few service garages would go through a full SAE

J537 verification that can take up to a week to complete for one battery. Showing a higher reading will indeed favour market acceptance, but this poses a problem when emerging technologies reveal correct readings that are at lower levels.

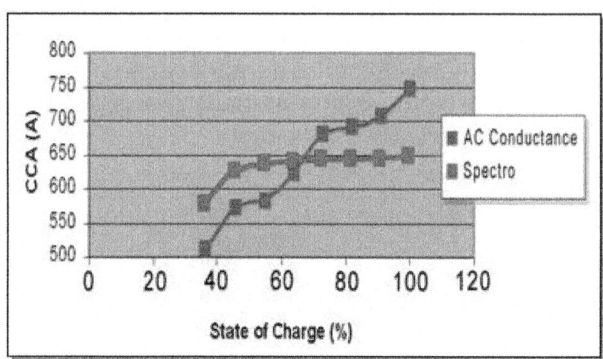

Fig. : CCA accuracy on state-of-charge.

The Spectro CA-12 provides stable CCA readings between a SoC of 100–40% (red); the values on AC Conductance drop rapidly with SoC (blue).

It so happened that the battery laboratory of a German luxury car manufacturer performed a comparison test as part of product qualification. The battery testers involved were the Spectro CA-12 and a device based on AC Conductance. With a dedicated matrix, the Spectro CA-12 achieved a CCA accuracy of 97%; capacity came in at 87 per cent. In comparison, the AC Conductance unit produced a correct CCA prediction of only 51 per cent with no capacity reference.

Fig. : Relationship of CCA and capacity of a starter battery.

Capacity represents energy and is shown as liquid. CCA relates to internal resistance and is responsible for energy delivery, best remembered as "pipe size." CCA tends to stay high while the capacity diminishes as part of aging.

One can clearly see that a CCA measurement at a low accuracy provides limited information regarding battery aging and end-of-life prediction. Furthermore, the driver can guess CCA on engine cranking. Capacity is the more reliable health indicator and there is some confusion in differentiating between the two. North America focuses on CCA, and RC (reserve capacity) is usually overlooked. Europe, on the other hand, is more in tune with capacity and their batteries are clearly marked with Ah. [Formula for RC to Ah conversion : RC divided by 2 plus 16]

Figure illustrates the bond between capacity and CCA on hand of a fluid-filled container. The liquid represents the capacity, and the tap symbolizes the energy delivery or CCA, best remembered as "pipe size." While CCA stays stable through most of the battery life, the capacity decreases steadily. The illustration represents the aging process with growing "rock content" that inhibits energy storage. The capacity gradually declines until there won't be enough "juice" one day to start the engine.

Conclusion

No single instrument can evaluate all battery anomalies and rapid testing only gives a rough estimation. There are battery defects that can only be revealed by applying a heavy load, and a micro short in a cell is such a case. A rapid-test might pass the battery as *good* even though the short has lowered the specific gravity to almost "empty" due to high self-discharge and the engine won't crank. A carbon pile or hydrometer is best able to find the anomaly but the test must be done after the battery has been removed from the charger for a few days. A charge will cover up the fault and everything will look normal.

There are no ideal battery test instruments; however, scientists predict that the battery industry is moving towards *electro-chemical impedance spectroscopy* to estimate battery performance. While advancements in battery rapid-testing are noteworthy, none is foolproof.

HOW TO MEASURE STATE-OF-CHARGE

Voltage Method

Measuring state-of-charge by voltage is the simplest method, but it can be inaccurate. Cell types have dissimilar chemical compositions that deliver varied voltage profiles. Temperature also plays a role. Higher temperature raises the open-circuit voltage, a lower temperature lowers it, and this phenomenon applies to all chemistries in varying degrees.

The most blatant error of voltage-based SoC occurs when disturbing the battery with a charge or discharge. This agitation distorts the voltage and no longer represents the true state-of-charge. To get accurate measurements, the battery needs to rest for at least four hours to attain equilibrium; battery manufacturers recommend 24 hours. Adding the element of time to neutralize voltage polarization does not sit well with batteries in active duty. One can see that this method is ill suited for fuel gauging.

Each battery chemistry delivers a unique discharge signature that requires a tailored model. While voltage-based SoC works reasonably well for a lead acid battery that has rested, the flat discharge curve of nickel-and lithium-based batteries renders the voltage method impracticable. And yet, voltage is commonly used on consumer products. A "rested" Li-cobalt of 3.80V/cell in open circuit indicates a SoC of roughly 50 per cent.

The discharge voltage curves of Li-manganese, Li-phosphate and NMC are very flat, and 80 per cent of the stored energy remains in this flat voltage profile. This characteristic assists applications requiring a steady voltage but presents a challenge in fuel gauging. The voltage method only indicates *full charge* and *low charge* and cannot estimate the large middle section accurately.

Lead acid has diverse plate compositions that must be considered when measuring SoC by voltage. Calcium, an additive that makes the battery mainte-nance-free, heat raises the voltage by 5–8 per cent. Temperature also affects the open-circuit voltage; heat raises it while cold causes it to decrease. Surface charge further fools SoC estimations by showing an elevated voltage immediately after charge; a brief discharge before measurement counteracts the error. Finally, AGM batteries produce a slightly higher voltage than the flooded equivalent.

When measuring SoC by open circuit voltage, the battery voltage must be truly "floating" with no load present. Installed in a car, the parasitic load present makes this a *closed circuit voltage (CCV)* condition that will falsify the readings. Adjustments must be made when measuring SoC in the CCV state by including the load current in the calculation. In spite of the notorious inaccuracies, most SoC measurements rely on the voltage method because it's simple. Voltage-based state-of-charge is popular for wheelchairs, scooters and golf cars.

Hydrometer

The hydrometer offers an alternative to measuring SoC, but this only applies to flooded lead acid and flooded nickel-cadmium. Here is how it works : As the bat-tery accepts charge, the sulfuric acid gets heavier, causing the specific gravity (SG) to increase. As the SoC decreases through discharge, the sulfuric acid removes itself from the electrolyte and binds to the plate, forming lead sulfate. The density of the electrolyte becomes lighter and more water-like, and the specific gravity gets lower. Table provides the BCI readings of starter batteries.

Table: BCI standard for SoC estimation of a maintenance-free starter battery with antimony. The readings are taken at room temperature of 26°C (78°F); the battery had rested for 24 hours after charge or discharge.

Approximate state-of-charge	Average specific gravity	Open circuit voltage			
		2V	6V	8V	12V
100%	1.265	2.10	6.32	8.43	12.65
75%	1.225	2.08	6.22	8. 30	12.45
50%	1.190	2.04	6.12	8.16	12.24
25%	1.155	2.01	6.03	8.04	12.06
0%	1.120	1.98	5.95	7.72	11.89

While BCI specifies the specific gravity of a fully charged starter battery at 1.265, battery manufacturers may go for 1.280 and higher. When increasing the specific gravity, the SoC readings on the look-up table will adjust upwards accordingly. Besides charge level and acid density, the SG can also vary due to low fluid levels, which raises the SG reading because of higher concentration. Alternatively, the battery can be over-filled, which lowers the number. When adding water, allow time for mixing before taking the SG measurement.

The specific gravity also varies according to battery type. Deep-cycle batteries use a dense electrolyte with an SG of up to 1.330 to get maximum run-time; aviation batteries have a SG of 1.285; traction batteries for forklifts are at 1.280; starter batteries come in at 1.265 and stationary batteries are at a low 1.225. Low specific gravity reduces corrosion. The resulting lower specific energy of stationary batteries is not as critical as longevity.

Nothing in the battery world is absolute. The specific gravity of fully charged deep-cycle batteries of the same model can range from 1.270 to 1.305; fully discharged, these batteries may vary between 1.097 and 1.201. Temperature is another variable that alters the specific gravity reading. The colder the temperature is, the higher (more dense) the SG value becomes. Table illustrates the SG gravity of a deep-cycle battery at various temperatures.

Table : Relation of specific gravity and temperature of deep-cycle battery.

Colder temperatures provide higher specific gravity readings.

Temperature of the Electrolyte		Gravity at full charge
40°C	104°F	1.266
30°C	86°F	1.273
20°C	68°F	1.280
10°C	50°F	1.287
0°C	32°F	1.294

Errors can also occur if the acid has stratified, meaning the concentration is light on top and heavy on the bottom. High acid concentration artificially raises the open circuit voltage, which can fool SoC estimations through false SG and voltage indication. The electrolyte needs to stabilize after charge and discharge before taking the SG reading.

Coulomb Counting

Laptops, medical equipment and other professional portable devices use coulomb counting as a SoC indication. This method works on the principle of measuring the current that flows in and out of the battery. If, for example, a battery was charged for one hour at one ampere, the same energy should be available on discharge. This is not the case. Inefficiencies in charge acceptance, especially towards the end of charge, as well as losses during discharge and storage reduce the total energy delivered and skew the readings. The available energy is always less than what had been fed to the battery, and compensation corrects the shortage.

Disregarding these irregularities, coulomb counting works reasonably well, especially for Li-ion. However, the one per cent accuracy some device manufacturers advertise is only possible in an ideal world and with a new battery. Independent tests show errors of up to 10 per cent when in typical use. Aging causes a gradual deviation from the working model on which the coulomb counter is based. The result is a laptop promising 30 minutes of remaining run-time and all of a sudden the screen goes dark. Periodic calibration by applying a full discharge and charge to reset the flags reduces the error.

There is a move towards electro-chemical impedance spectroscopy and even magnetism to measure state-of-charge. These new technologies get more accurate estimation than with voltage and can be used when the battery is under load. Furthermore, temperature, surface charge and acid stratification do not affect the readings noticeably.

BATTERY ENERGY—WHAT BATTERY PROVIDES MORE

Battery energy does not always remain quietly stored in the battery. Sometimes you must get that energy out in a hurry. Short bursts of power are required, for instance, when starting an automobile on a cold morning. Cold cranking requires high current from the battery. When you turn on the viewing screen of a digital camera, you also demand high current output from the camera's battery.

The Power Used Comes from Stored Energy

During every interval in which power is used, a quantity of energy is drained from the battery. That quantity of energy is equal to the amount of power, multiplied by the time the power flows. Energy has units of power and time, such as kilowatt-hours or watt-seconds. As the stored battery energy is used up, the available voltage and current drops lower and lower until finally the battery is exhausted. Then it is time to recharge or replace the battery. A good battery must supply two requirements. First, it must be able to meet the power demand by supplying adequate voltage and current when needed. Otherwise it is useless from the start. Secondly, there must be sufficient energy to last a long time, or else the battery must be economical, readily replaced, or easy to recharge.

Joules are Units of Energy or Work

The Joule is the International Standard unit of energy defined as one watt-second. One watt-second of *mechanical work* is the work done by a force of one Newton (or 0.2247 pound) pushing through a one-meter distance. 3600 Joules are contained in one watt-hour, since an hour contains 3600 seconds,. Batteries are often rated in milliampere-hours instead of watt-hours. This battery rating can be converted to energy if the average *voltage* of the battery during discharge is known. For instance, a 3.6-volt Lithium-ion battery rated at 850 mAh will maintain a voltage of 3.6 volts with little variation during discharge. Multiply the voltage of 3.6 volts times 850 mAh to yield 3060 mA-volt-hours, or 3060 milliwatt-hours. 3.06 watt-hours equal 11016 watt-seconds or Joules.

Joules may be converted to other familiar units using the numerical factors given below. Divide the number of Joules by 3.6 million to obtain kilowatt-hours. Divide the number of Joules by 1.356 to obtain the number of foot-pounds, a popular unit of work in the English system. Divide by 1055 to obtain the equivalent number of BTU (British Thermal Units). Divide by 4184 to obtain the number of food Calories ! Yes, food Calories are energy, of course. This comparison does not put batteries in a good light compared to peanut butter. Two tablespoons of smooth peanut butter contain 191 Calories, or almost 800,000 Joules! It takes a huge battery to contain this much energy.

What Battery Type Gives the Most Energy for the Price?

In the table below we present the cost per Watt-hour, Specific Energy, that is Watt-hours per kilogram, Joules per kg, and the Energy Density, Watt-hours/liter for various types of batteries. It is not surprising that the well-known Lead-acid storage batteries head the list. Fine for use in our cars, but a little inconvenient in a laptop. And why are Alkaline long-life and Carbon-zinc batteries in the list? Aren't they non-rechargeable? This was thought to be the case, previously, but now they can have their lifetimes extended by recharging. Ordinary alkaline cells may be recharged literally dozens of times using the new technology built into the *Battery Xtender.* Recharging alkaline, nickel-cadmium and nickel-metal hydride cells side-by-side in one automatic charger opens up new possibilities for battery selection economy.

Battery Type	Cost $ per Wh	Wh/kg	Joules/kg	Wh/liter
Lead-acid	$0.17	41	146,000	100
Alkaline long-life	$0.19	110	400,000	320
Carbon-zinc	$0.31	36	130,000	92
NiMH	$0.99	95	340,000	300
NiCad	$1.50	39	140,000	140
Lithium-ion	$0.47	128	460,000	230

Costs of lithium-ion batteries are falling rapidly in the race to develop new electric vehicles. The $0.47 price per watt-hour above is for the Nissan Leaf automobile, and they predict a target cost of $0.37 per watt-hour. Tesla Automobiles uses a smaller battery pack, and they are optimistic about reaching a price of $0.20 per watt-hour in the near future.

There is another type of battery that does not appear in the table above, since it is limited in the relative amount of current it can deliver. However, it has even higher energy storage per kilogram, and its temperature range is extreme, from-55 to + 150°C. That type is Lithium Thionyl Chloride. It is used in extremely hazardous or critical applications such as space flight and deep sea diving.

The specifications for Lithium Thionyl Chloride are $1.16 per watt-hour, 700 watts/kg, 2,000,000 Joules/kg, and 1100 watt-hours per liter. For more information of Lithium Thionyl Chloride please contact *Tadiran Batteries.*

Chapter 3

CORROSION AND STABILITY OF METALS

STABILITY CONSTANTS OF COMPLEXES

A **stability constant** (formation constant, binding constant) is an equilibrium constant for the formation of a **complex** in solution. It is a measure of the strength of the interaction between the reagents that come together to form the complex. There are two main kinds of complex : compounds formed by the interaction of a metal ion with a ligand and supramolecular complexes, such as host-guest complexes and complexes of anions. The stability constant(s) provide the information required to calculate the concentration(s) of the complex(es) in solution. There are many areas of application in chemistry, biology and medicine.

History

Jannik Bjerrum developed the first general method for the determination of stability constants of metal-ammine complexes in 1941. The reasons why this occurred at such a late date, nearly 50 years after Alfred Werner had proposed the correct structures for co-ordination complexes, have been summarised by Beck and Nagypál. The key to Bjerrum's method was the use of the then recently developed glass electrode and pH meter to determine the concentration of hydrogen ions in solution. Bjerrum recognised that the formation of a metal complex with a ligand was a kind of acid-base equilibrium : there is competition for the ligand, L, between the metal ion, M^{n+}, and the hydrogen ion, H^+. This means that there are two simultaneous equilibria that have to be considered. In what follows electrical charges are omitted for the sake of generality. The two equilibria are :

$$H + L \rightleftharpoons HL$$
$$M + L \rightleftharpoons ML$$

Hence by following the hydrogen ion concentration during a titration of a mixture of M and HL with base, and knowing the acid dissociation constant of HL, the stability constant for the formation of ML could be determined. Bjerrum went on to determine the stability constants for systems in which many complexes may be formed.

$$M + qL \rightleftharpoons ML_q$$

The following twenty years saw a veritable explosion in the number of stability constants that were determined. Relationships, such as the Irving-Williams series were discovered. The calculations were done by hand using the so-called graphical methods. The mathematics underlying the methods used in this period are summarised by Rossotti and Rossotti. The next key development was the use of a computer program, LETAGROP to do the calculations. This permitted the examination of systems too complicated to be evaluated by means of hand-calculations. Subsequently computer programs capable of handling complex equilibria in general, such as SCOGS and MINIQUAD were developed so that today the determination of stability constants has almost become a "routine" operation. Values of thousands of stability constants can be found in two commercial databases.

Theory

The formation of a complex between a metal ion, M, and a ligand, L, is in fact usually a substitution reaction. For example, in aqueous solutions, metal ions will be present as aqua-ions, so the reaction for the formation of the first complex could be written as :

$$[M(H_2O)_n] + L \rightleftharpoons [M(H_2O)_{n-1}L] + H_2O$$

The equilibrium constant for this reaction is given by

$$\beta' = \frac{[M(H_2O)_{n-1}L][H_2O]}{[M(H_2O)_n][L]}$$

[L] should be read as "the concentration of L" and likewise for the other terms in square brackets. The expression can be greatly simplified by removing those terms which are constant. The number of water molecules attached to each metal ion is constant. In dilute solutions the concentration of water is effectively constant. The expression becomes :

$$\beta = \frac{[ML]}{[M][L]}.$$

Following this simplification a general definition can be given, for the general equilibrium :

$$pM + qL \cdots \rightleftharpoons M_pL_q \cdots$$

$$\beta_{pq} \cdots = \frac{[M_pL_q \cdots]}{[M]p[L]q \cdots}$$

The definition can easily be extended to include any number of reagents. The reagents need not always be a metal and a ligand but can be any species which form a complex. Stability constants defined in this way, are *association* constants.

This can lead to some confusion as pK_a values are *dissociation* constants. In general purpose computer programs it is customary to define all constants as association constants. The relationship between the two types of constant is given in association and dissociation constants.

Step-wise and Cumulative Constants

A cumulative or overall constant, given the symbol β, is the constant for the formation of a complex from reagents. For example, the cumulative constant for the formation of ML_2 is given by :

$$M + 2L \rightleftharpoons ML_2; \beta_{12} = \frac{[ML_2]}{[M][L]^2}$$

The step-wise constants, K_1 and K_2 refer to the formation of the complexes one step at a time.

$$M + L \rightleftharpoons ML; K_1 = \frac{[ML]}{[M][L]}$$

$$ML + L \rightleftharpoons ML_2; K_2 = \frac{[ML_2]}{[ML][L]}$$

It follows that :

$$\beta_{12} = K_1 K_2$$

A cumulative constant can always be expressed as the product of step-wise constants. Conversely, any step-wise constant can be expressed as a quotient of two or more overall constants. There is no agreed notation for step-wise constants, though a symbol such as KL ML is sometimes found in the literature. It is best always to define each stability constant by reference to an equilibrium expression.

Hydrolysis Products

The formation of an hydroxo-complex is a typical example of an hydrolysis reaction. An hydrolysis reaction is one in which a substrate reacts with water, splitting a water molecule into hydroxide and hydrogen ions. In this case the hydroxide ion then forms a complex with the substrate.

$$M + OH \rightleftharpoons M(OH)$$

$$K = \frac{[M(OH)]}{[M][OH]}$$

In water the concentration of hydroxide is related to the concentration of hydrogen ions by the self-ionization constant, K_w.

$$K_w = [H^+][OH^-]; [OH^-] = K_w[H^+]^{-1}$$

The expression for hydroxide concentration is substituted into the formation constant expression :

$$K = \frac{[M(OH)]}{[M]K_W[H]^{-1}}$$

$$\beta^*_{1-1} = \frac{K}{K_W} = \frac{[M(OH)]}{[M][H]^{-1}}$$

The literature usually gives value of β^*.

Acid-base Complexes

A Lewis acid, A, and a Lewis base, B, can be considered to form a complex AB

$$A + B \rightleftharpoons AB : K = \frac{[AB]}{[A][B]}$$

There are three major theories relating to the strength of Lewis acids and bases and the interactions between them.

1. Hard and soft acid-base theory (HSAB). This is used mainly for qualitative purposes.

2. Drago and Wayland proposed a two-parameter equation which predicts the standard enthalpy of formation of a very large number of adducts quite accurately. $-\Delta H^\ominus (A - B) = E_A E_B + C_A C_B$. Values of the E and C parameters are available

3. Guttmann donor numbers : for bases the number is derived from the enthalpy of reaction of the base with antimony pentachloride in 1,2-Dichloroethane as solvent. For acids, an acceptor number is derived from the enthalpy of reaction of the acid with triphenylphosphine oxide.

THERMODYNAMICS

The thermodynamics of metal ion complex formation provides much significant information. In particular it is useful in distinguishing between enthalpic and entropic effects. Enthalpic effects depend on bond strengths and entropic effects have to do with changes in the order/disorder of the solution as a whole. The chelate effect, below, is best explained in terms of thermodynamics.

An equilibrium constant is related to the standard Gibbs free energy change for the reaction :

$\Delta G^\ominus = -2.303 \, RT \log_{10}\beta$.

R is the gas constant and T is the absolute temperature. At 25 °C, $\Delta G^\ominus = (-5.708$ kJ mol^{-1}) $\cdot$ log β. Free energy is made up of an enthalpy term and an entropy term.

$\Delta G^\ominus = \Delta H^\ominus - T\Delta S^\ominus$

The standard enthalpy change can be determined by calorimetry or by using the van 't Hoff equation, though the calorimetric method is preferable. When both the standard enthalpy change and stability constant have been determined, the standard entropy change is easily calculated from the equation above.

The fact that step-wise formation constants of complexes of the type ML_n decrease in magnitude as n increases may be partly explained in terms of the entropy factor. Take the case of the formation of octahedral complexes.

$$[M(H_2O)_m L_{n-1}] + L \rightleftharpoons [M(H_2O)_{m-1} L_n]$$

For the first step $m = 6$, $n = 1$ and the ligand can go into one of 6 sites. For the second step $m = 5$ and the second ligand can go into one of only 5 sites. This means that there is more randomness in the first step than the second one; $\Delta S^\ominus$ is more positive, so $\Delta G^\ominus$ is more negative and $\log K_1 > \log K_2$. The ratio of the step-wise stability constants can be calculated on this basis, but experimental ratios are not exactly the same because $\Delta H^\ominus$ is not necessarily the same for each step. The entropy factor is also important in the chelate effect, below.

Ionic Strength Dependence

The thermodynamic equilibrium constant, $K^\ominus$, for the equilibrium :

$$M + L \rightleftharpoons ML$$

can be defined as :

$$K^\ominus = \frac{\{ML\}}{\{M\}\{L\}}$$

where $\{ML\}$ is the activity of the chemical species ML etc. $K^\ominus$ is dimensionless since activity is dimensionless. Activities of the products are placed in the numerator, activities of the reactants are placed in the denominator.

Since activity is the product of concentration and activity coefficient (γ) the definition could also be written as :

$$K^\ominus = \frac{[ML]}{[M][L]} \times \frac{\gamma_{ML}}{\gamma_M \gamma_L} = \frac{[ML]}{[M][L]} \times \Gamma$$

where $[ML]$ represents the concentration of ML and Γ is a quotient of activity coefficients. This expression can be generalized as :

$$\beta^\ominus_{pq...} = \frac{[M_p L_q \cdots]}{[M]p[L]q \cdots} \times \Gamma$$

Fig. : Dependence of the stability constant for formation of [Cu(glycinate)]⁺ on ionic strength (NaClO₄).

To avoid the complications involved in using activities, stability constants are determined, where possible, in a medium consisting of a solution of a background electrolyte at high ionic strength, that is, under conditions in which Γ can be assumed to be always constant. For example, the medium might be a solution of 0.1 mol/dm^{-3} sodium nitrate or 3 mol/dm^{-3} potassium perchlorate. When Γ is constant it may be ignored and the general expression in theory, above, is obtained.

All published stability constant values refer to the specific ionic medium used in their determination and different values are obtained with different conditions, as illustrated for the complex CuL (L=glycinate). Furthermore, stability constant values depend on the specific electrolyte used as the value of Γ is different for different electrolytes, even at the same ionic strength. There does not need to be any chemical interaction between the species in equilibrium and the background electrolyte, but such interactions might occur in particular cases. For example, phosphates form weak complexes with alkali metals, so, when determining stability constants involving phosphates, such as ATP, the background electrolyte used will be, for example, a tetralkylammonium salt. Another example involves iron(III) which forms weak complexes with halide and other anions, but not with perchlorate ions.

When published constants refer to an ionic strength other than the one required for a particular application, they may be adjusted by means of specific ion theory (SIT) and other theories.

Temperature Dependence

All equilibrium constants vary with temperature according to the van 't Hoff equation :

$$\frac{d \ln K}{dT} \frac{\Delta H_m^{\ominus}}{RT^2}$$

R is the gas constant and T is the thermodynamic temperature. Thus, for exothermic reactions, (the standard enthalpy change, $\Delta H^{\ominus}$, is negative) K decreases with temperature, but for endothermic reactions ($\Delta H^{\ominus}$ is positive) K increases with temperature.

Factors Affecting the Stability Constants of Complexes

The Chelate Effect

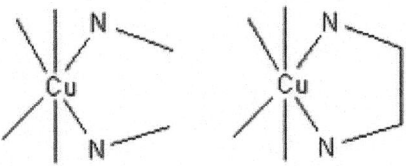

Fig. : Cu^{2+} complexes with methylamine (left) end ethylene diamine (right).

Consider the two equilibria, in aqueous solution, between the copper(II) ion, Cu^{2+} and ethylenediamine (en) on the one hand and methylamine, $MeNH_2$ on the other.

$$Cu^{2+} + en \rightleftharpoons [Cu(en)]^{2+} \tag{1}$$

$$Cu^{2+} + 2\ MeNH_2 \rightleftharpoons [Cu(MeNH_2)_2]^{2+} \tag{2}$$

In (1) the bidentate ligand ethylene diamine forms a chelate complex with the copper ion. Chelation results in the formation of a five–membered ring. In (2) the bidentate ligand is replaced by two monodentate methylamine ligands of approximately the same donor power, meaning that the enthalpy of formation of $Cu-N$ bonds is approximately the same in the two reactions. Under conditions of equal copper concentrations and when then concentration of methylamine is twice the concentration of ethylenediamine, the concentration of the complex (1) will be greater than the concentration of the complex (2). The effect increases with the number of chelate rings so the concentration of the EDTA complex, which has six chelate rings, is much higher than a corresponding complex with two mono-dentate nitrogen donor ligands and four monodentate carboxylate ligands. Thus, the phenomenon of the chelate effect is a firmly established empirical fact: under comparable conditions, the concentration of a chelate complex will be higher than the concentration of an analogous complex with monodentate ligands.

The thermodynamic approach to explaining the chelate effect considers the equilibrium constant for the reaction: the larger the equilibrium constant, the higher the concentration of the complex.

$$[Cu(en)] = \beta_{11}\ [Cu][en]$$

$$[Cu(MeNH_2)_2] = \beta_{12}[Cu][MeNH_2]^2$$

When the analytical concentration of methylamine is twice that of ethyl-enediamine and the concentration of copper is the same in both reactions, the concentration $[Cu(en)]^{2+}$ is much higher than the concentration $[Cu(MeNH_2)_2]^{2+}$ because $\beta_{11} \gg \beta_{12}$.

The difference between the two stability constants is mainly due to the difference in the standard entropy change, $\Delta S^{\ominus}$. In equation (1) there are two particles on the left and one on the right, whereas in equation (2) there are three particles on the left and one on the right. This means that less entropy of disorder is lost when the chelate complex is formed than when the complex with monodentate ligands is formed. This is one of the factors contributing to the entropy difference. Other factors include solvation changes and ring formation. Some experimental data to illustrate the effect are shown in the following table:

Equilibrium	$\log \beta$	$\Delta G^{\ominus}$	$\Delta H^{\ominus}$ /kJ mol^{-1}	$-T\Delta S^{\ominus}$ /kJ mol^{-1}
$Cd^{2+} + 4\ MeNH_2 \rightleftharpoons Cd(MeNH_2)_4^{2+}$	6.55	−37.4	−57.3	19.9
$Cd^{2+} + 2\ en \rightleftharpoons Cd(en)_2^{2+}$	10.62	−60.67	−56.48	−4.19

Fig. : An EDTA complex.

These data show that the standard enthalpy changes are indeed approximately equal for the two reactions and that the main reason why the chelate complex is so much more stable is that the standard entropy term is much less unfavourable, indeed, it is favourable in this instance. In general it is difficult to account precisely for thermodynamic values in terms of changes in solution at the molecular level, but it is clear that the chelate effect is predominantly an effect of entropy. Other explanations, Including that of Schwarzenbach, are discussed in Greenwood and Earnshaw.

The chelate effect increases as the number of chelate rings increases. For example the complex $[Ni(dien)_2)]^{2+}$ is more stable than the complex $[Ni(en)_3)]^{2+}$; both complexes are octahedral with six nitrogen atoms around the nickel ion, but dien (diethylenetriamine, 1,4,7-triazaheptane) is a tridentate ligand and en is bidentate. The number of chelate rings is one less than the number of donor atoms in the ligand. EDTA (ethylenediaminetetracetic acid) has six donor atoms so it forms very strong complexes with five chelate rings. Ligands such as DTPA, which have eight donor atoms are used to form complexes with large metal ions such as lanthanide or actinide ions which usually form 8-or 9-co-ordinate complexes.

5-membered and 6-membered chelate rings give the most stable complexes. 4-membered rings are subject to internal strain because of the small inter-bond angle is the ring. The chelate effect is also reduced with 7-and 8-membered rings, because the larger rings are less rigid, so less entropy is lost in forming them.

Ethylenediamine (en)

Diethylenetriamine (dien)

The Macrocyclic Effect

It was found that the stability of the complex of copper(II) with the macrocyclic ligand cyclam (1,4,8,11-tetraazacyclotetradecane) was much greater than expected in comparison to the stability of the complex with the corresponding open-chain amine. This phenomenon was named "the macrocyclic effect" and it was also interpreted as an entropy effect. However, later studies suggested that both enthalpy and entropy factors were involved.

An important difference between macrocyclic ligands and open-chain (che-lating) ligands is that they have selectivity for metal ions, based on the size of the cavity into which the metal ion is inserted when a complex is formed. For example, the crown ether 18-crown-6 forms much stronger complexes with the potassium ion, K^+ than with the smaller sodium ion, Na^+.

In hemoglobin an iron(II) ion is complexed by a macrocyclic porphyrin ring. The low-spin Fe^{2+} ion fits snugly into the cavity of the porhyrin ring, but high-spin iron(II) is significantly larger and the iron atom is forced out of the plane of the macrocyclic ligand. This effect contributes the ability of hemoglobin to bind oxygen reversibly under biological conditions. In Vitamin B12 a cobalt(II) ion is held in a corrin ring. Chlorophyll is a macrocyclic complex of magnesium(II).

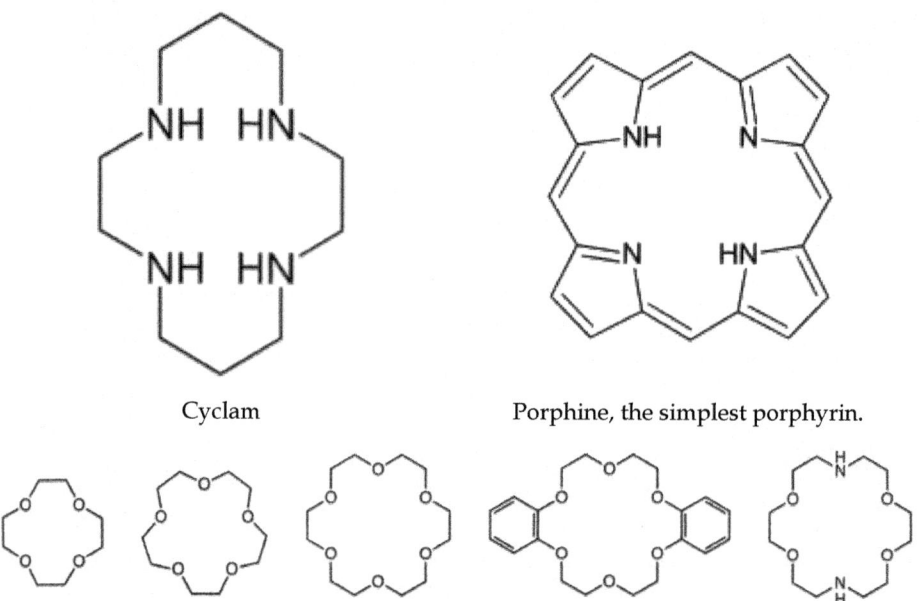

Cyclam   Porphine, the simplest porphyrin.

Fig. : Structures of common crown ethers : 12-crown-4, 15-crown-5, 18-crown-6, dibenzo-18-crown-6, and diaza-18-crown-6.

Geometrical Factors

Successive step-wise formation constants K_n in a series such as ML_n ($n = 1, 2,...$) usually decrease as n increases. Exceptions to this rule occur when the geometry of the ML_n complexes is not the same for all members of the series. The classic example is the formation of the diamminesilver(I) complex $[Ag(NH_3)_2]^+$ in aqueous solution.

$$Ag^+ + NH_3 \rightleftharpoons [Ag(NH_3)]^+ \quad K_1 = \frac{[[Ag(NH_3)]^+]}{[Ag^+][NH_3]}$$

$$Ag(NH_3)^+ + NH_3 \rightleftharpoons [Ag(NH_3)_2]^+ \quad K_2 = \frac{[[Ag(NH_3)_2]^+]}{[[Ag(NH_3)]^+[NH_3]}$$

In this case, $K_2 > K_1$. The reason for this is that, in aqueous solution, the ion written as Ag^+ actually exists as the four-co-ordinate tetrahedral aqua species $[Ag(OH_2)_4]^+$. The first step is then a substitution reaction involving the displacement of a bound water molecule by ammonia forming the tetrahedral complex $[Ag(NH_3)(OH_2)_3]^+$ (commonly abbreviated as $[Ag(NH_3)]^+$). In the second step, the aqua ligands are lost to form a linear, two-co-ordinate product $[H_3N-Ag-NH_3]^+$. Examination of the thermodynamic data shows that both enthalpy and entropy effects determine the result.

Equilibrium	$\Delta H^\ominus$ /kJ mol^{-1}	$\Delta S^\ominus$ /J K^{-1} mol^{-1}
$Ag^+ + NH_3 \rightleftharpoons [Ag(NH_3)]^+$	−21.4	8.66
$[Ag(NH_3)]^+ + NH_3 \rightleftharpoons [Ag(NH_3)_2]^+$	−35.2	−61.26

Other examples exist where the change is from octahedral to tetrahedral, as in the formation of $[CoCl_4]^{2-}$ from $[Co(H_2O)_6]^{2+}$.

Classification of Metal Ions

Ahrland, Chatt and Davies proposed that metal ions could be described as class A if they formed stronger complexes with ligands whose donor atoms are N, O or F than with ligands whose donor atoms are P, S or Cl and class B if the reverse is true. For example, Ni^{2+} forms stronger complexes with amines than with phosphines, but Pd^{2+} forms stronger complexes with phosphines than with amines. Later, Pearson proposed the theory of hard and soft acids and bases (HSAB theory). In this classification, class A metals are hard acids and class B metals are soft acids. Some ions, such as copper (i) are classed as borderline. Hard acids form stronger complexes with hard bases than with soft bases. In general terms hard-hard interactions are predominantly electrostatic in nature whereas soft-soft interactions are predominantly covalent in nature. The HSAB theory, though useful, is only semi-quantitative.

The hardness of a metal ion increases with oxidation state. An example of this effect is given by the fact that Fe^{2+} tends to form stronger complexes with N-donor ligands than with O-donor ligands, but the opposite is true for Fe^{3+}.

Effect of Ionic Radius

The Irving-Williams series refers to high-spin, octahedral, divalent metal ion of the first transition series. It places the stabilities of complexes in the order :

Mn < Fe < Co < Ni < Cu > Zn

This order was found to hold for a wide variety of ligands. There are three strands to the explanation of the series.

1. The ionic radius is expected to decrease regularly for Mn^{2+} to Zn^{2+}. This would be the normal periodic trend and would account for the general increase in stability.

2. The crystal field stabilisation energy (CFSE) increases from zero for manganese(II) to a maximum at nickel(II). This makes the complexes increasingly stable. CFSE returns to zero for zinc(II).

3. Although the CFSE for copper(II) is less than for nickel(II), octahedral copper(II) complexes are subject to the Jahn-Teller effect which results in a complex having extra-stability.

Another example of the effect of ionic radius the steady increase in stability of complexes with a given ligand along the series of trivalent lanthanide ions, an effect of the well-known lanthanide contraction.

Applications

Stability constant values are exploited in a wide variety of applications. Chelation therapy is used in the treatment of various metal-related illnesses, such as iron overload in β-thalassemia sufferers who have been given blood transfusions. The ideal ligand binds to the target metal ion and not to others, but this degree of selectivity is very hard to achieve. The synthetic drug *Deferiprone* achieves selectivity by having two oxygen donor atoms so that it binds to Fe^{3+} in preference to any of the other divalent ions that are present in the human body, such as Mg^{2+}, Ca^{2+} and Zn^{2+}. Treatment of poisoning by ions such as Pb^{2+} and Cd^{2+} is much more difficult since these are both divalent ions and selectivity is harder to accomplish. Excess copper in Wilson's disease can be removed by *penicillamine* or Triethylene tetramine (TETA). DTPA has been approved by the U.S. Food and Drug Administration for treatment of plutonium poisoning.

DTPA is also used as a complexing agent for gadolinium in MRI contrast enhancement. The requirement in this case is that the complex be very strong, as Gd^{3+} is very toxic. The large stability constant of the octadentate ligand ensures that the concentration of free Gd^{3+} is almost negligible, certainly well below toxicity threshold. In addition the ligand occupies only 8 of the 9 co-ordination sites on the gadolinium ion. The ninth site is occupied by a water molecule which exchanges rapidly with the fluid surrounding it and it is this mechanism that makes the paramagnetic complex into a contrast reagent.

EDTA forms such strong complexes with most divalent cations that it finds many uses. For example, it is often present in washing powder to act as a water softener by sequestering calcium and magnesium ions.

The selectivity of macrocyclic ligands can be used as a basis for the construction of an ion selective electrode. For example, potassium selective electrodes are available that make use of the naturally-occurring macrocyclic antibiotic valinomycin.

An ion-exchange resin such as chelex 100, which contains chelating ligands bound to a polymer, can be used in water softeners and in chromatographic separation techniques. In solvent extraction the formation of electrically-neutral complexes allows cations to be extracted into organic solvents. For example, in nuclear fuel reprocessinguranium(VI) and plutonium(VI) are extracted into kerosene as the complexes $[MO_2(TBP)_2(NO_3)_2]$ (TBP = tri-'n-*butyl phosphate*). In phase-transfer catalysis, a substance which is insoluble in an organic solvent can be made soluble by addition of a suitable ligand. For example, potassium permanganate oxidations can be achieved by adding a catalytic quantity of a crown ether and a small amount of organic solvent to the aqueous reaction mixture, so that the oxidation reaction occurs in the organic phase.

| Deferiprone | penicillamine | triethylenetet-ramine, TETA | Ethylenediamine tetracetic acid, EDTA |

| diethylenetri-aminepentacetic acid, DTPA | Valinomycin | tri-n-butylphos-phate |

In all these examples, the ligand is chosen on the basis of the stability constants of the complexes formed. For example, TBP is used in nuclear fuel reprocessing because (among other reasons) it forms a complex strong enough for solvent extraction to take place, but weak enough that the complex can be destroyed by nitric acid to recover the uranyl cation as nitrato complexes, such as $[UO_2(NO_3)_4]^{2-}$ back in the aqueous phase.

Supramolecular Complexes

Supramolecular complexes are held together by hydrogen bonding, hydrophobic forces, van der Waals forces, π-π interactions, and electrostatic effects, all of which can be described as noncovalent bonding. Applications include molecular recognition, host-guest chemistry and anion sensors.

A typical application in molecular recognition involved the determination of formation constants for complexes formed between a tripodal substituted urea molecule and various saccharides. The study was carried out using a non-aqueous solvent and NMR chemical shift measurements. The object was to examine the selectivity with respect to the saccharides.

An example of the use of supramolecular complexes in the development of chemosensors is provided by the use of transition-metal ensembles to sense for ATP.

Anion complexation can be achieved by encapsulating the anion in a suitable cage. Selectivity can be engineered by designing the shape of the cage. For example, dicarboxylate anions could be encapsulated in the ellipsoidal cavity in a large macrocyclic structure containing two metal ions.

DETERMINATION OF EQUILIBRIUM CONSTANTS

Equilibrium constants are determined in order to quantify chemical equilibria. When an equilibrium constant is expressed as a concentration quotient,

$$K = \frac{[S]^\sigma [T]^\tau \cdots}{[A]^\alpha [B]^\beta \cdots}$$

it is implied that the activity quotient is constant. In order for this assumption to be valid equilibrium constants should be determined in a medium of relatively high ionic strength. Where this is not possible, consideration should be given to possible activity variation.

The equilibrium expression above is a function of the concentrations [A], [B] etc. of the chemical species in equilibrium. The equilibrium constant value can be determined if any one of these concentrations can be measured. The general procedure is that the concentration in question is measured for a series of solutions with known analytical concentrations of the reactants. Typically, a titration is performed with one or more reactants in the titration vessel and one or more reactants in the burette. Knowing the analytical concentrations of reactants initially in the reaction vessel and in the burette, all analytical concentrations can be derived as a function of the volume (or mass) of titrant added.

The equilibrium constants may be derived by best-fitting of the experimental data with a chemical model of the equilibrium system.

Experimental Methods

There are four main experimental methods. For less commonly used methods.

Potentiometric Measurements

A free concentration [A] or activity {A} is measured by means of an ion selective electrode such as the glass electrode. If the electrode is calibrated using activity standards it is assumed that the Nernst equation applies in the form :

$$E = E^0 + \frac{RT}{nF} \ln A$$

where E^0 is the standard electrode potential. When buffer solutions of known pH are used for calibration the meter reading will be pH.

$$pH = \frac{nF}{RT}(E^0 - E)$$

At 298K, 1 pH unit is approximately equal to 59 mV.

When the electrode is calibrated with solutions of known concentration, by means of a strong acid/strong base titration, for example, a modified Nernst equation is assumed.

$$E = E^0 + s \log_{10} [A]$$

s an empirical slope factor. A solution of known hydrogen ion concentration may be prepared by standardization of a strong acid against borax. Constant-boilinghydrochloric acid may also be used as a primary standard for hydrogen ion concentration.

Spectrophotometric Measurements

Absorbance

It is assumed that the Beer–Lambert law applies.

$$A = \ell \sum \in c$$

where ℓ is the optical path length, ℓ is a molar absorbance at unit path length and c is a concentration. More than one of the species may contribute to the absorbance. In principle absorbance may be measured at one wavelength only, but in present-day practice it is common to record complete spectra.

Fluorescence (Luminescence) Intensity

It is assumed that the scattered light intensity is a linear function of species' concentrations.

$$I = \sum \phi c$$

where ℓ is a proportionality constant.

NMR Chemical Shift Measurements

Chemical exchange is assumed to be rapid on the NMR time-scale. An individual chemical shift δ is the mole-fraction-weighted average of the shifts δ of nuclei in contributing species.

$$\bar{\delta} = \frac{\sum c_i \delta_i}{\sum c_i}$$

Calorimetric Measurements

Simultaneous measurement of K and ΔH for 1 :1 adducts is routinely carried out using isothermal titration calorimetry. Extension to more complex systems is limited by the availability of suitable software.

Range and Limitations

1. *Potentiometry :* The most widely used electrode is the glass electrode which is selective for the hydrogen ion. This is suitable for all acid-base equilibria. $\log_{10}$

β values between about 2 and 11 can be measured directly by potentiometric titration using a glass electrode. This enormous range is possible because of the logarithmic response of the electrode. The limitations arise because the Nernst equation breaks down at very low or very high pH. The range can be extended by using the competition method. An example of the application of this method can be found in Palladium(II) cyanide

2. *Absorbance and Luminescence* : An upper limit on $\log_{10} \beta$ of 4 is usually quoted, corresponding to the precision of the measurements, but it also depends on how intense the effect is. Spectra of contributing species should be clearly distinct from each other

3. *NMR* : Limited precision of chemical shift measurements also puts an upper limit of about 4 on $\log_{10} \beta$. Limited to diamagnetic systems.

4. Calorimetry. Insufficient evidence is currently available.

Computational Methods

It is assumed that the experimental data which have been collected comprise a set of data points. At each i-th data point, the analytical concentrations of the reactants, $T_A(i)$, $T_B(i)$ etc. are known along with a measured quantity, y_i, that depends on one or more of these analytical concentrations. A general computational procedure has three main components.

1. Definion of a chemical model of the equilibria
2. Calculation of the concentrations of all the chemical species in each solution
3. Refinement of the equilibrium constants
4. Model selection.

The Chemical Model

The chemical model consists of a set of chemical species present in solution, both the **reactants** added to the reaction mixture and the **complex species** formed from them. Denoting the reactants by A, B..., each *complex species* is specified by the stoichiometric coefficients that relate the particular combination of *reactants* forming them.

$$pA + qB \cdots \rightleftharpoons A_p B_q \cdots : \beta_{pq} \cdots = \frac{[A_p B_q \cdots]}{[A]^p [B]^q \cdots}$$

When using general-purpose computer programs, it is usual to use cumulative, association constants, as shown above. Electrical charges are not shown in general expressions such as this and are often omitted from specific expressions, for simplicity of notation. In fact, electrical charges have no bearing on the equilibrium processes other that there being a requirement for overall electrical neutrality in all systems.

With aqueous solutions the concentrations of proton (hydronium ion) and hydroxide ion are constrained by the self-dissociation of water.

$$H_2O \rightleftharpoons H^+ + OH^- : K'_W = \frac{[H^+][OH^-]}{[H_2O]}$$

With dilute solutions the concentration of water can be assumed to be constant so the equilibrium expression is written in the familiar form of the ionic product of water.

$K_W = [H^+][OH^-]$

When both H^+ and OH^- must be considered as reactants, one of them is eliminated from the model by specifying that its concentration is to be derived from the concentration of the other. Usually the concentration of the hydroxide ion is given by :

$[OH^-] = K_W[H^+]^{-1}$

In this case the equilibrium constant for the formation of hydroxide has the stoichiometric coefficients -1 in regard to the proton and zero for the other reactants. This has important implications for all protonation equilibria in aqueous solution and for hydrolysis constants in particular.

It is quite usual to omit from the model those species whose concentrations are considered to be negligible. For example, it is usually assumed then there is no interaction between the reactants and/or complexes and the electrolyte used to maintain constant ionic strength or the buffer used to maintain constant pH. These assumptions may or may not be justified. Also, it is implicitly assumed that there are no other complex species present. When complexes are wrongly ignored a systematic error is introduced into the calculations.

Equilibrium constant values are usually estimated initially by reference to data sources.

Speciation Calculations

A speciation calculation is one in which the concentrations of all the species in an equilibrium system are calculated, knowing the analytical concentrations, T_A, T_B etc., of the reactants A, B etc. This means solving a set of non-linear equations of mass-balance :

$$T_A = [A] + \sum p\beta_{pq} \cdots [A]^p[B]^q \cdots$$

$$T_B = [B] + \sum q\beta_{pq} \cdots [A]^p[B]^q \cdots$$

$$\vdots$$

for the free concentrations [A], [B] etc. The concentrations of the complexes are derived from the free concentrations *via* the chemical model. Some authors include the free reactant terms in the sums by declaring *identity* (unit) β constants for which the stoichiometric coefficients are 1 for the reactant concerned and zero for all other reactants :

$$[A] = \beta_{10\ldots} [A], [B] = \beta_{01\ldots} [B] \cdots$$

$$\beta_{10\ldots} = \beta_{01\ldots} \cdots = 1$$

In this manner, all chemical species, *including the free reactants*, are treated in the same way, having been *formed* from the combination of reactants that is specified by the stoichiometric coefficients. The mass-balance equations assume the simpler form.

$$T_A = \sum p\beta_{pq\ldots} [A]^p [B]^q \cdots$$

$$T_B = \sum q\beta_{pq\ldots} [A]^p [B]^q \cdots$$

$$\vdots$$

In a titration system the analytical concentrations of the reactants at each titration point are obtained from the initial conditions, the burette concentrations and volumes. The analytical (total) concentration of a reactant R at the i'th titration point is given by :

$$T_R = \frac{R_0 + v_i [R]}{v_0 + v_i}$$

where R_0 is the initial **amount** of R in the titration vessel, v_0 is the initial volume, $[R]$ is the **concentration** of R in the burette and v_i is the volume added. The burette concentration of a reactant not present in the burette is taken to be zero.

In general, solving these non-linear equations presents a formidable challenge because of the huge range over which the free concentrations may vary. At the beginning, values for the free concentrations must be estimated. Then, these values are refined, usually by means of Newton–Raphson iterations. The logarithms of the free concentrations may be refined rather than the free concentrations themselves. Refinement of the logarithms of the free concentrations has the added advantage of automatically imposing a non-negativity constraint on the free concentrations. Once the free reactant concentrations have been calculated, the concentrations of the complexes are derived from them and the equilibrium constants.

Note that the free reactant concentrations can be regarded as implicit parameters in the equilibrium constant refinement process. In that context the values of the free concentrations are constrained by forcing the conditions of mass-balance to apply at all stages of the process.

Equilibrium Constant Refinement

The objective of the refinement process it to find equilibrium constant values that give the best fit to the experimental data. This is usually achieved by minimising an objective function, U, by the method of non-linear least-squares. First the residuals are defined as :

$$r_i = y_i^{obs} - y_i^{calc}$$

Then the most general objective function is given by :

$$U = \sum_i \sum_j r_i W_{ij} r_j$$

The matrix of weights, W, should be, ideally, the inverse of the variance-covariance matrix of the observations. It is rare for this to be known. However, when it is, the expectation value of U is one, which means that the data are fitted *within experimental error*. Most often only the diagonal elements are known, in which case the objective function simplifies to :

$$U = \sum_i W_{ii} r_i^2$$

with $W_{ij} = 0$ when $j \neq i$. Unit weights, $W_{ii} = 1$, are often used but, in that case, the expectation value of U is the root mean square of the experimental errors.

The minimization may be performed using the Gauss–Newton method. Firstly the objective function is linearised by approximating it as a first-order Taylor series expansion about an initial parameter set, p.

$$U = U^0 + \sum_i \frac{\partial U}{\partial p_i} \delta p_i$$

The increments δp_i are to be added to the corresponding initial parameters such that U is less than U^0. At the minimum the derivatives $\dfrac{\partial U}{\partial p_i}$, which are simply related to the elements of the Jacobian matrix, J

$$J_{jk} = \frac{\partial y_j^{calc}}{\partial p_k}$$

where p_k is the kth parameter of the refinement, are equal to zero. One or more equilibrium constants may be parameters of the refinement. However, the measured quantities represented by y are not expressed in terms of the equilibrium constants, but in terms of the species concentrations, which are implicit functions of these parameters. Therefore, the Jacobian elements must be obtained using implicit differentiation.

The parameter increments δp are calculated by solving the normal equations, derived from the conditions that $\dfrac{\partial U}{\partial p} = 0$ at the minimum.

$$(J^T W J)\, \delta p = J^T W r$$

The increments δp are added iteratively to the parameters :

$$p^{n+1} = p^n + \delta p$$

where n is an iteration number. The species concentrations and y^{calc} values are recalculated at every data point. The iterations are continued until no significant reduction in U is achieved, that is, until a convergence criterion is satisfied. If, however, the updated parameters do not result in a decrease of the objective function, that is, if divergence occurs, the increment calculation must be modified.

The simplest modification is to use a fraction, f, of calculated increment, so-called shift-cutting.

$$p^{n+1} = p^n + f\,\delta p$$

In this case, the direction of the shift vector, δp, is unchanged. With the more powerful Levenberg–Marquardt algorithm, on the other hand, the shift vector is rotated towards the direction of steepest descent, by modifying the normal equations,

$$(J^T W J + \lambda I)\,\delta p = J^T W r$$

where λ is the Marquardt parameter and I is an identity matrix. Other methods of handling divergence have been proposed.

A particular issue arises with NMR and *spectrophotometric* data. For the latter, the observed quantity is absorbance, A, and the Beer–Lambert law can be written as

$$A_i = \sum_i \epsilon_{pq\ldots}\, c_{pq\ldots},i$$

It can be seen that absorbance, A, is a linear function of the molar absorbptivities, ϵ, at the path length used. In matrix notation :

$$A = EC$$

There are two approaches to the calculation of the unknown molar absorptivities :

1. The ϵ values are considered to be parameters of the minimization and the Jacobian is constructed on that basis. However, the ϵ values themselves are calculated at each step of the refinement by linear least-squares :

    $$E = (C^T C)^{-1}\, C^T A$$

    using the refined values of the equilibrium constants to obtain the speciation. The matrix $(C^T C)^{-1}\, C^T$ is an example of a pseudo-inverse.

2. The Beer–Lambert law is written as :

    $$A = ((C^T C)^{-1}\, C^T A)C$$

    Golub and Pereyra showed how the pseudo-inverse can be differentiated so that parameter increments for both molar absorptivities and equilibrium constants can be calculated by solving the normal equations.

Parameter Errors and Correlation

In the region close to the minimum of the objective function, U, the system approximates to a linear least-squares system, for which :

$$p = (J^T W J)^{-1}\, J^T W y^{obs}$$

Therefore, the parameter values are (approximately) linear combinations of the observed data values and the errors on the parameters, p, can be obtained by error propagation from the observations, y^{obs}, using the linear formula. Let the variance-covariance matrix for the observations be denoted by Σ^y and that of the parameters by Σ^p. Then,

$$\Sigma^p = (J^TWJ)^{-1} \, J^TWSyWTJ(J^TWJ)^{-1}$$

When $W = (\Sigma^y)^{-1}$, this simplifies to

$$\Sigma^p = (J^TWJ)^{-1}$$

In most cases the errors on the observations are un-correlated, so that Σ^y is diagonal. If so, each weight should be the reciprocal of the variance of the corresponding observation. For example, in a potentiometric titration, the weight at a titration point, k, can be given by :

$$W_k = \frac{1}{\sigma_E^2 + \left(\dfrac{\partial E}{\partial v}\right)_k^2 \sigma_v^2}$$

where σ_E is the error in electrode potential or pH, $\left(\dfrac{\partial E}{\partial v}\right)_k$ is the slope of the titration curve and σ_v is the error on added volume.

When unit weights are used $(W = I, \ p = (J^TJ)^{-1} \, JTy)$ it is implied that the experimental errors are uncorrelated and all equal : $\Sigma^y = \sigma^2 I$, where σ^2 is known as the variance of an observation of unit weight, and I is an identity matrix. In this case σ^2 is approximated by $\dfrac{U}{n_d - n_p}$, where U is the minimum value of the objective function and n_d and n_p are the number of data and parameters, respectively.

$$\Sigma^p = \frac{U}{n_d - n_p}(J^T J)^{-1}$$

In all cases, the variance of the parameter p_i is given by Σ_{ii}^p and the covariance between parameters p_i and p_j is given by Σ_{ij}^p. Standard deviation is the square root of variance. These error estimates reflect only random errors in the measurements. The true uncertainty in the parameters is larger due to the presence of systematic errors which, by definition, cannot be quantified.

Note that even though the observations may be un-correlated, the parameters are always correlated.

Derived Constants

When cumulative constants have been refined it is often useful to derive stepwise constants from them. The general procedure is to write down the defining expressions for all the constants involved and then to equate concentrations. For example, suppose that one wishes to derive the pKa for removing one proton from a tribasic acid, LH_3, such as citric acid.

$$L^{3-} + 2H^+ \rightleftharpoons LH_2^- : [LH_2^-] = \beta_{12}[L^{3-}][H^+]^2$$

$L^{3-} + 3H^+ \rightleftharpoons LH_3 : [LH_3] = \beta_{13}[L^{3-}][H^+]^3$

The step-wise *association* constant for formation of LH_3 is given by :

$LH_2^- + H^+ \rightleftharpoons LH_3 : [LH_3] = K[LH_2^-][H^+]$

Substitute the expressions for the concentrations of LH_3 and LH_2^- into this equation :

$\beta_{13}[L^{3-}][H^+]^3 = K\beta_{12}[L^{3-}][H^+]^2[H^+]$

whence :

$$\beta_{13} = K\,\beta_{12} : K = \frac{\beta_{13}}{\beta_{12}}$$

and since $pKa = -\log(1/K)$ its value is given by :

$pKa_1 = \log\beta_{13} - \log\beta_{12}$

When calculating the error on the step-wise constant, the fact that the cumulative constants are correlated must be taken into account. By error propagation :

$$\sigma_K^2 = \sigma_{\beta 12}^2 + \sigma_{\beta 13}^2 - 2\sigma_{\beta 12}\sigma_{\beta 13}\rho_{12,13}$$

and

$$\sigma_{\log}K = \frac{\sigma_K}{K}$$

Model Selection

Once a refinement has been completed the results should be checked to verify that the chosen model is acceptable. generally speaking, a model is acceptable when the data are fitted within experimental error, but there is no single criterion with which to make the judgement. The following should be taken into consideration :

The Objective Function

When the weights have been correctly derived from estimates of experimental error, the expectation value of $\dfrac{U}{n_d - n_p}$ is 1. It is therefore very useful to estimate experimental errors and derive some reasonable weights from them as this is an absolute indicator of the goodness of fit.

When unit weights are used, it is implied that all observations have the same variance. $\dfrac{U}{n_d - n_p}$, is expected to be equal to that variance.

Parameter Errors

One would want the errors on the stability constants to be roughly commensurate with experimental error. For example, with pH titration data, if pH is measured to 2 decimal places, the errors of log β should not be much larger than 0.01. In exploratory work where the nature of the species present is not known in advance, several different chemical models may be tested and compared. There will be models where the uncertainties in the best estimate of an equilibrium constant may be somewhat or even significantly larger than σ_{pH}, especially with those constants governing the formation of comparatively minor species, but the decision as to how large is acceptable remains subjective. The decision process as to whether or not to include comparatively uncertain equilibria in a model, and for the comparison of competing models in general, can be made objective and has been outlined by Hamilton.

Distribution of Residuals

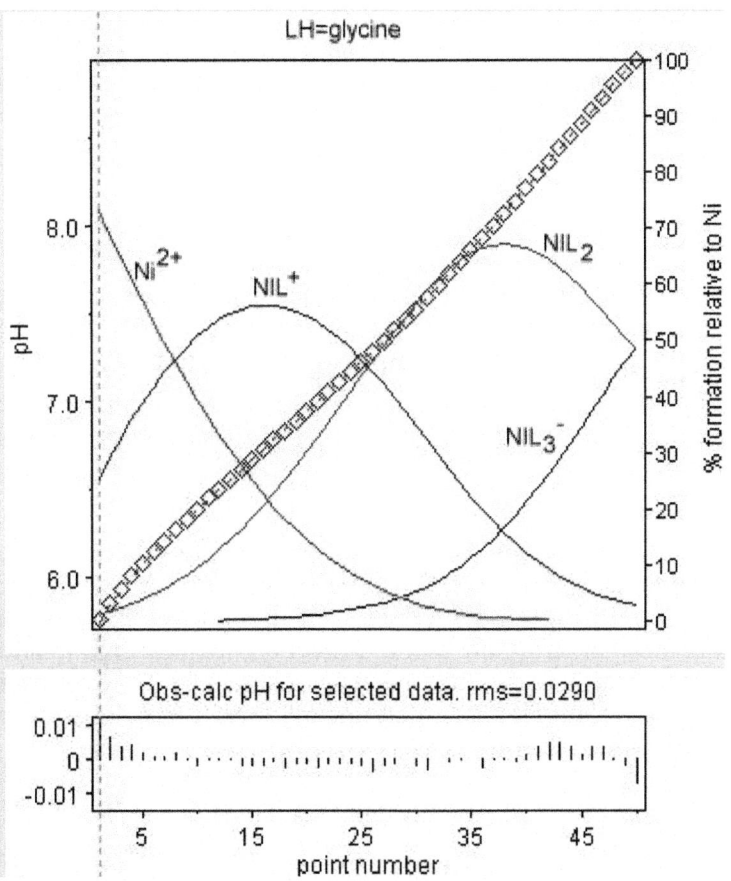

At the minimum in $\underline{U}$ the system can be approximated to a linear one, the residuals in the case of unit weights are related to the observations by :

$$r = y^{obs} - J \, (J^T T)^{-1} \, J^T y^{obs}$$

The symmetric, idempotent matrix $J(J^T T)^{-1} J$ is known in the statistics literature as the hat matrix, : H. Thus,

$$r = (I - H)y^{obs}$$

and

$$M^r = (I - H) M^y (I - H)$$

where I is an identity matrix and M^r and M^y are the variance-covariance matrices of the residuals and observations, respectively. This shows that even though the observations may be un-correlated, the residuals are always correlated.

The diagram at the right shows the result of a refinement of the stability constants of $Ni\ Gly^+$, $Ni(Gly)_2$ and $Ni(Gly)_3^-$ ($GlyH$ = glycine). The observed values are shown a blue diamonds and the species concentrations, as a percentage of the total nickel, are superimposed. The residuals are shown in the lower box. The presence of correlation is evident in the way sequences all have the same sign. Correlation notwithstanding, the magnitudes of the residuals show some randomness. Individual residuals are mostly commensurate with experimental error (about 0.002 in pH). This is about as good as it gets.

Physical Constraints

Some physical constraints are usually incorporated in the calculations. For example, all the concentrations of free reactants and species must have positive values and association constants must have positive values.

With spectrophotometric data the molar absorptivity (or emissivity) values should all be positive. Most computer programs do not impose this constraint on the calculations.

Other Models

If the model is not acceptable a variety of other models should be examined in order to find the model that best fits the experimental data, within experimental error. The main difficulty is with the so-called minor species. These are species whose concentration is so low that the effect on the measured quantity is at or below the level of error in the experimental measurement. The constant for a minor species may prove impossible to determine if there is no means to increase the concentration of the species.

POURBAIX DIAGRAM

In chemistry, a **Pourbaix diagram**, also known as a **potential/pH diagram**, $E_H\text{-}pH$ **diagram** or a pE/pH **diagram**, maps out possible stable (equilibrium) phases of an aqueous electro-chemical system. Predominant ion boundaries are represented by lines. As such a Pourbaix diagram can be read much like a standard phase diagram with a different set of axes. But like phase diagrams, they do not allow for reaction rate or kinetic effects.

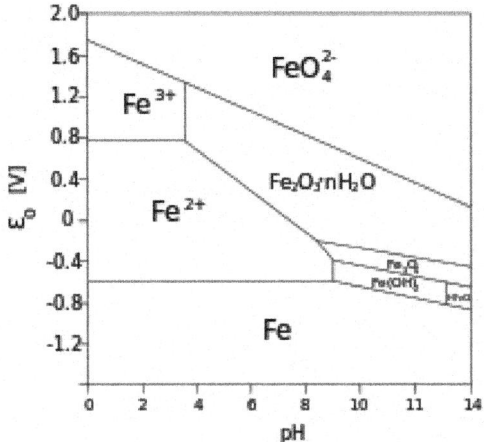

Fig. : Pourbaix diagram of iron.

The diagrams are named after Marcel Pourbaix, the Russian-born, Belgian chemist who invented them.

DIAGRAM

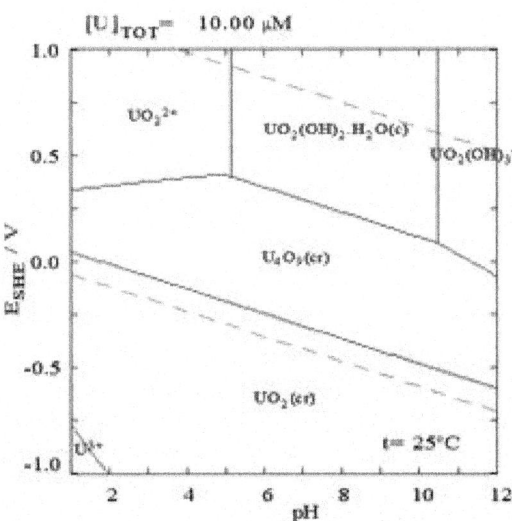

Fig. : The Pourbaix diagram for uranium in a non-complexing aqueous medium (*e.g.* perchloric acid/sodium hydroxide).

Pourbaix diagrams are also known as E_H-pH diagrams due to the labelling of the two axes. The vertical axis is labelled E_H for the voltage potential with respect to the standard hydrogen electrode (SHE) as calculated by the Nernst equation. The "H" stands for hydrogen, although other standards may be used, and they are for room temperature only.

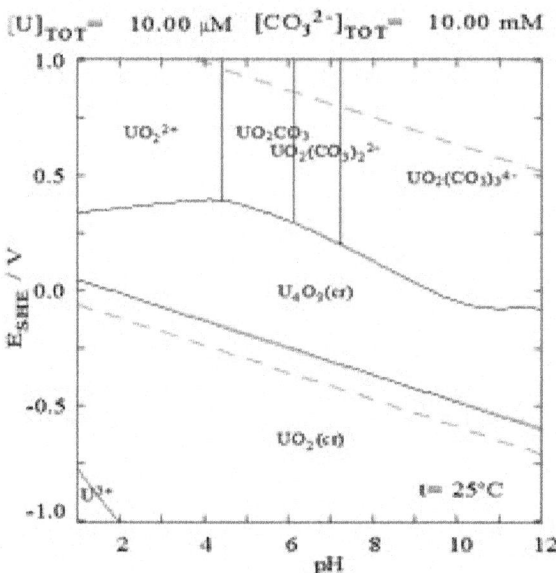

Fig. : The Pourbaix diagram for uranium in carbonate solution. The dashed green lines show the stability limits of water in the system.

$$EH = E^0 - \frac{0.0592}{n} \log \frac{[C]^c [D]^d}{[A]^a [B]^b} \{V\}$$

The horizontal axis is labelled pH for the-log function of the H^+ ion activity.

$$pH = - \log_{10}(a_{H+}) = \log_{10}\left(\frac{1}{a_{H+}}\right)$$

The lines in the Pourbaix diagram show the equilibrium conditions, that is, where the activities are equal, for the species on each side of that line. On either side of the line, one form of the species will instead be said to be predominant.

In order to draw the position of the lines with the Nernst equation, the activity of the chemical species at equilibrium must be defined. Usually, the activity of a species is approximated as equal to the concentration (for soluble species) or partial pressure (for gases). The same values should be used for all species present in the system.

For soluble species, the lines are often drawn for concentrations of $1\,M$ or 10^{-6} M. Sometimes additional lines are drawn for other concentrations.

If the diagram involves the equilibrium between a dissolved species and a gas, the pressure is usually set to $P^0 = 1$ atm $= 101{,}325$ Pa, the minimum pressure required for gas evolution from an aqueous solution at standard conditions.

While such diagrams can be drawn for any chemical system, it is important to note that the addition of a metal binding agent (ligand) will often modify the diagram. For instance, carbonate has a great effect upon the diagram for uranium.

The presence of trace amounts of certain species such as chloride ions can also greatly affect the stability of certain species by destroying passivating layers.

In addition, changes in temperature and concentration of solvated ions in solution will shift the equilibrium lines in accordance with the Nernst equation.

The diagrams also do not take kinetic effects into account, meaning that species shown as instable might not react to any significant degree in practice.

A simplified Pourbaix diagram indicates regions of "Immunity", "Corrosion" and "Passivity", instead of the stable species. They thus give a guide to the stability of a particular metal in a specific environment. Immunity means that the metal is not attacked, while corrosion shows that general attack will occur. Passivation occurs when the metal forms a stable coating of an oxide or other salt on its surface, the best example being the relative stability of aluminium because of the alumina layer formed on its surface when exposed to air.

The Stability Region of Water

In many cases, the possible conditions in a system are limited by the stability region of water. In the Pourbaix diagram for uranium, the limits of stability of water are marked by the two dashed green lines, and the stability region for water falls between these lines.

Under highly reducing conditions (low E_H/pE) water will be reduced to hydrogen according to :

$$2H_2O + 2e^- \rightarrow H_2(g) + 2OH^-$$

or

$$2H_3O^+ + 2e^- = H_2(g) + 2H_2O$$

Using the Nernst equation, setting $E^0 = -0.828\ V$ and the hydrogen gas fugacity (corresponding to activity) at 1, the equation for the lower stability line of water in the Pourbaix diagram will be :

$$E_H = -0.0591 * pH\ \{V\}$$

at standard temperature and pressure. Below this line, water will be reduced to hydrogen, and it will usually not be possible to pass beyond this line as long as there is still water present to be reduced.

Correspondingly, under highly oxidizing conditions (high E_H/pE) water will be oxidized to oxygen gas according to :

$$6H_2O \rightarrow 4H_3O^+ + O_2(g) + 4e^-$$

Using the Nernst equation as above, but with an E^0 of 1.229 V, gives an upper stability limit of water at :

$$E_H = 1.229V - 0.0591 * pH\ \{V\}$$

at standard temperature and pressure. Above this line, water will be oxidized to form oxygen gas, and it will usually not be possible to pass beyond this line as long as there is still water present to be oxidized.

Uses

Pourbaix diagrams have several uses, for example in corrosion studies, geology and in environmental studies.

In Environmental Chemistry

Pourbaix diagrams are widely used to describe the chemical behaviour of chemical species in the hydrosphere. In these cases, pE is used instead of E_H. pE is a dimensionless number and can easily be related to E_H by the equation :

$$pE = \frac{E_H}{0.0591V}$$

pE values in environmental chemistry ranges from -12 to 25, since at low or high potentials water will be reduced or oxidized, respectively. In environmental applications, the concentration of dissolved species is usually set to a value between 10^{-2} M and 10^{-5} M for the creation of the equilibrium lines.

BASICS OF ELECTRO-CHEMICAL CORROSION MEASUREMENTS

Introduction

Most metal corrosion occurs *via* electro-chemical reactions at the interface between the metal and an electrolyte solution. A thin film of moisture on a metal surface forms the electrolyte for atmospheric corrosion. Wet concrete is the electrolyte for reinforcing rod corrosion in bridges. Although most corrosion takes place in water, corrosion in non-aqueous systems is not unknown.

Corrosion normally occurs at a rate determined by an equilibrium between opposing electro-chemical reactions. The first is the anodic reaction, in which a metal is oxidized, releasing electrons into the metal. The other is the cathodic reaction, in which a solution species (often O_2 or H^+) is reduced, removing electrons from the metal. When these two reactions are in equilibrium, the flow of electrons from each reaction is balanced, and no net electron flow (electrical current) occurs. The two reactions can take place on one metal or on two dissimilar metals (or metal sites) that are electrically connected.

The vertical axis is potential and the horizontal axis is the logarithm of absolute current. The theoretical current for the anodic and cathodic reactions are shown as straight lines. The curved line is the total current—the sum of the anodic and cathodic currents. This is the current that you measure when you sweep the potential of the metal with your potentiostat. The sharp point in the curve is actually the point where the current changes signs as the reaction changes from anodic to cathodic, or *vice versa*. The sharp point is due to the use of a logarithmic axis. The use of a log axis is necessary because of the wide range of current values that must be displayed during a corrosion experiment. Because of the phenomenon of passivity, it is not uncommon for the current to change by six orders of magnitude during a corrosion experiment.

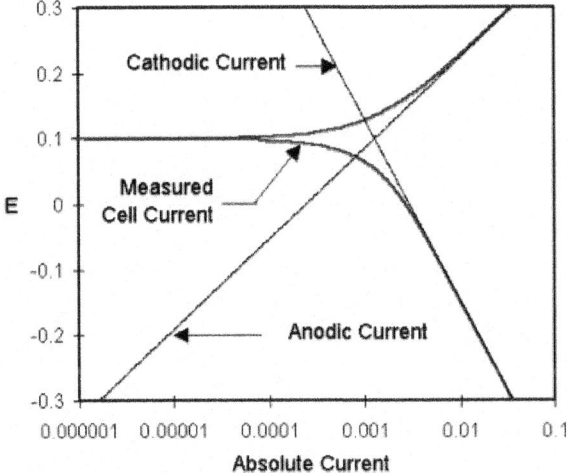

Fig. : Corrosion Process Showing Anodic and Cathodic Current Components.

The potential of the metal is the means by which the anodic and cathodic reactions are kept in balance. Notice that the current from each half reaction depends on the electro-chemical potential of the metal. Suppose the anodic reaction releases too many electrons into the metal. Excess electrons shift the potential of the metal more negative, which slows the anodic reaction and speeds up the cathodic reaction. This counteracts the initial perturbation of the system.

The equilibrium potential assumed by the metal in the absence of electrical connections to the metal is called the Open Circuit Potential, Eoc. In most electrochemical corrosion experiments, the first step is the measurement of Eoc. The terms Eoc (Open Circuit Potential) and Ecorr (Corrosion Potential) are usually interchangeable, but Eoc is preferred.

It is very important that the Corrosion Scientist measures the Eoc and allows sufficient time for the Eoc to stabilize before beginning the electro-chemical experiment. A stable Eoc is taken to indicate that the system being studied has reached "steady state", *i.e.*, the various corrosion reactions have assumed a constant rate. Some corrosion reactions reach steady state in a few minutes, while others may need several hours. Regardless of the time required, a computer-controlled system can monitor the Eoc and begin the experiment after it has stabilized.

The value of either the anodic or cathodic current at Eoc is called the Corrosion Current, Icorr. If we could measure *Icorr*, we could use it to calculate the corrosion rate of the metal. Unfortunately, *Icorr* cannot be measured directly. However, it can be estimated using electro-chemical techniques. In any real system, Icorr and Corrosion Rate are a function of many system variables including type of metal, solution composition, temperature, solution movement, metal history, and many others.

In practice, many metals form an oxide layer on their surface as they corrode. If the oxide layer inhibits further corrosion, the metal is said to passivate. In some cases, local areas of the passive film break down allowing significant metal corrosion to occur in a small area. This phenomena is called pitting corrosion or simply pitting.

Because corrosion occurs *via* electro-chemical reactions, electro-chemical techniques are ideal for the study of the corrosion processes. In electro-chemical studies, a metal sample with a surface area of a few square centimeters is used to model the metal in a corroding system. The metal sample is immersed in a solution typical of the metal's environment in the system being studied. Additional electrodes are immersed in the solution, and all the electrodes are connected to a device called a potentiostat. A potentiostat allows you to change the potential of the metal sample in a controlled manner and measure the current the flows as a function of potential.

Both controlled potential (potentiostatic) and controlled current (galvanostatic) polarization is useful. When the polarization is done potentiostatically, current is measured, and when it is done galvanostatically, potential is measured. This discussion will concentrate on controlled potential methods, which are much more common than galvanostatic methods. With the exception of Open Circuit Potential *vs.* Time, Electro-chemical Noise, Galvanic Corrosion, and a few others, potentiostatic mode is used to perturb the equilibrium corrosion process. When the potential of a metal sample in solution is forced away from Eoc, it is referred to as polarizing the sample. The response (current) of the metal sample is measured as it is polarized. The response is used to develop a model of the sample's corrosion behaviour.

Suppose we use the potentiostat to force the potential to an anodic region (towards positive potentials from Eoc). In figure, we are moving towards the top of the graph. This will increase the rate of the anodic reaction (corrosion) and decrease the rate of the cathodic reaction. Since the anodic and cathodic reactions are no longer balanced, a net current will flow from the electronic circuit into the metal sample. The sign of this current is positive by convention. For a discussion of electro-chemical sign conventions.

If we take the potential far enough from Eoc, the current from the cathodic reaction will be negligible, and the measured current will be a measure of the anodic reaction alone. In figure, notice that the curves for the cell current and the anodic current lie on top of each other at very positive potentials. Conversely, at strongly negative potentials, cathodic current dominates the cell current.

In certain cases as we vary the potential, we will first passivate the metal, then cause pitting corrosion to occur. With the astute use of a potentiostat, an experiment in which the current is measured *versus* potential or time may allow us to determine Icorr at Ecorr, the tendency for passivation to occur, or the potential range over which pitting will occur.

Because of the range of corrosion phenomena that can be studied with electrochemistry, the ability to measure very low corrosion rates, and the speed with which these measurements can be conducted, an electro-chemical corrosion measurement system has become a standard item in the modern corrosion laboratory.

Quantitative Corrosion Theory

In many cases, you can estimate it from current *versus* voltage data. You can measure a log current *versus* potential curve over a range of about one half volt. The voltage scan is centered on Eoc. You then fit the measured data to a theoretical model of the corrosion process.

The model we will use for the corrosion process assumes that the rates of both the anodic and cathodic processes are controlled by the kinetics of the electron transfer reaction at the metal surface. This is generally the case for corrosion reactions. An electro-chemical reaction under kinetic control obeys Equation 1-1, the Tafel Equation.

$$I = I_0 e^{(2.3\,(E-E^*)/\beta)} \quad \text{Equation 1-1}$$

where :

I = the current resulting from the reaction

I_0 = a reaction dependent constant called the Exchange Current

E = the electrode potential

E_o = the equilibrium potential (constant for a given reaction)

β = the reaction's Tafel Constant (constant for a given reaction). Beta has units of volts/decade.

The Tafel equation describes the behaviour of one isolated reaction. In a corrosion system, we have two opposing reactions–anodic and cathodic.

The Tafel equations for both the anodic and cathodic reactions in a corrosion system can be combined to generate the Butler-Volmer Equation (Equation 1-2).

$$I = Ia + I_c = I_{corr}(e^{(2.3(E-Eoc)/\beta a)} - e^{(-2.3(E-Eoc)/\beta c)}) \qquad \text{Equation 1-2}$$

where :

I = the measured cell current in amps

I_{corr} = the corrosion current in amps

E = the electrode potential

Eoc = the corrosion potential in volts

βa = the anodic Beta Tafel Constant in volts/decade

βc = the cathodic Beta Tafel Constant in volts/decade

What does Equation 1-2 predict about the current *versus* voltage curve? At Eoc, each exponential term equals one. The cell current is therefore zero, as you would expect. Near Eoc both exponential terms contribute to the overall current. Finally, as the potential is driven far from Eoc by the potentiostat, one exponential term predominates and the other term can be ignored. When this occurs, a plot of log current versus potential becomes a straight line.

A log I *versus* E plot is called a Tafel Plot.

In practice, many corrosion systems are kinetically controlled and thus obey Equation 1.2. A log current *versus* potential curve that is linear on both sides of E_{corr} is indicative of kinetic control for the system being studied. However, there can be complications, such as :

1.   Concentration polarization, where the rate of a reaction is controlled by the rate at which reactants arrive at the metal surface. Often cathodic reactions show concentration polarization at higher currents, when diffusion of oxygen or hydrogen ion is not fast enough to sustain the kinetically controlled rate. A more intuitive name for concentration polarization is "diffusion controlled".

2.   Oxide formation, which may or may not lead to passivation, can alter the surface of the sample being tested. The original surface and the altered surface may have different values for the constants in Equation 1–2.

3.   Other effects that alter the surface, such as preferential dissolution of one alloy component, can also cause problems.

4.   A mixed control process where more than one cathodic, or anodic, reaction occurs simultaneously may complicate the model. An example of mixed control is the simultaneous reduction of oxygen and hydrogen ion.

5.   Finally, potential drop as a result of cell current flowing through the resistance of your cell solution causes errors in the kinetic model. This last effect, if it is not too severe, may be correctable *via IR* compensation in the potentiostat.

In most cases, complications like those listed above will cause non-linearities in the Tafel Plot. The results derived from a Tafel Plot that does not have a well-defined linear region should be used with caution.

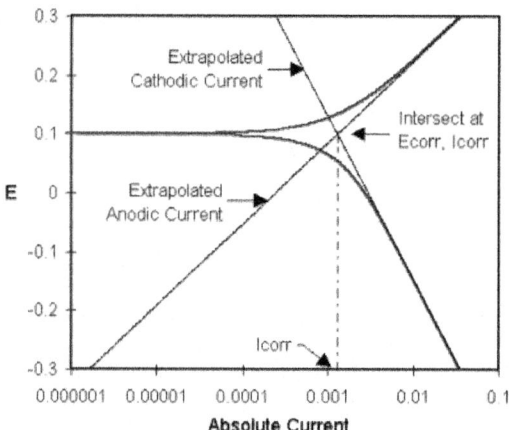

Fig. : Classic Tafel Analysis.

Classic Tafel analysis is performed by extrapolating the linear portions of a log current *versus* potential plot back to their intersection. The value of either the anodic or the cathodic current at the intersection is I_{corr}. Unfortunately, many real world corrosion systems do not provide a sufficient linear region to permit accurate extrapolation. Most modern corrosion test software, such as Gamry Instruments' DC105 DC Corrosion Techniques software, performs a more sophisticated numerical fit to the Butler-Volmer equation. The measured data is fit to Equation 1-2 by adjusting the values of E_{corr}, I_{corr}, βa, and βc. The curve fitting method has the advantage that it does not require a fully developed linear portion of the curve.

Polarization Resistance

Equation 1–2 can be further simplified by restricting the potential to be very close to E_{oc}. Near E_{oc}, the current versus voltage curve approximates a straight line. The slope of this line has the units of resistance (ohms). The slope is, therefore, called the Polarization Resistance, R_p. An R_p value can be combined with an estimate of the Beta coefficients to yield an estimate of the corrosion current.

If we approximate the exponential terms in Equation 1-2 with the first two terms of a power series expansion ($ex = 1 + x + x2/2...$) and simplify, we get one form of the Stern-Geary Equation :

$$I_{corr} = \frac{\beta_a \beta_c}{2.3 R_p (\beta_a + \beta_c)} \quad \text{Equation 1-3}$$

In a Polarization Resistance experiment, you record a current versus voltage curve as the cell voltage is swept over a small range of potential that is very near to E_{oc} (generally ± 10 mV). A numerical fit of the curve yields a value for the Polarization Resistance, R_p.

Polarization Resistance data does not provide any information about the values for the Beta coefficients. Therefore, to use Equation 1-3 you must provide Beta values. These can be obtained from a Tafel Plot or estimated from your experience with the system you are testing.

CALCULATION OF CORROSION RATE FROM CORROSION CURRENT

The numerical result obtained by fitting corrosion data to a model is generally the corrosion current. We are interested in corrosion rates in the more useful units of rate of penetration, such as millimeters per year. How is corrosion current used to generate a corrosion rate? Assume an electrolytic dissolution reaction involving a chemical species, S :

$S \rightarrow S^{n+} + ne^-$

You can relate current flow to mass *via* Faraday's Law.

$Q = nFM$ Equation 1-4

where :

Q = the charge in coulombs resulting from the reaction of species S

n = the number of electrons transferred per molecule or atom of S

F = Faraday's constant : 96,486.7 coulombs/mole

M = the number of moles of species S reacting.

A more useful form of Equation 1-4 requires the concept of equivalent weight. The equivalent weight (EW) is the mass of species S that will react with one Faraday of charge. For an atomic species, $EW = AW/n$ (where AW is the atomic weight of the species).

Recalling that $M = W/AW$ and substituting into Equation 1-4 we get :

$$W = \frac{EW \times Q}{F} \quad \text{Equation 1-5}$$

where W is the mass of species S that has reacted.

In cases where the corrosion occurs uniformly across a metal surface, the corrosion rate can be calculated in units of distance per year. Be careful — this calculation is only valid for uniform corrosion, it dramatically underestimates the problem when localized corrosion occurs !

For a complex alloy that undergoes uniform dissolution, the equivalent weight is a weighted average of the equivalent weights of the alloy components. Mole fraction, not mass fraction, is used as the weighting factor. If the dissolution is not uniform, you may have to measure the corrosion products to calculate EW.

Conversion from a weight loss to a corrosion rate (CR) is straightforward. We need to know the density, d, and the sample area, A. Charge is given by $Q = I\,T$, where T is the time in seconds and I is a current. We can substitute in the value of Faraday's constant. Modifying Equation 1-5 :

$$CR = \frac{I_{corr} \cdot K \cdot EW}{dA} \quad \text{Equation 1-6}$$

where :

CR = the corrosion rate, units are given by the choice of K

I_{corr} = the corrosion current in amps

K = a constant that defines the units for the corrosion rate

EW = the equivalent weight in grams/equivalent

d = density in grams/cm^3

A = sample area in cm^2

Table shows the value of K used in Equation 1-6 for corrosion rates in the units of your choice.

Table : Corrosion Rate Constants.

Corrosion Rate Units	K	Units
mm/year (mmpy)	3272	mm/(amp-cm-year)
milli-inches/year (mpy)	1.288 × 105	Milli-inches(amp-cm-year)

IR COMPENSATION

When you pass current between two electrodes in a conductive solution, there will always be regions of different potentials in the solution. Much of the overall change in potential occurs very close to the surface of the electrodes. Here the potential gradients are largely due to ionic concentration gradients set up near the metal surfaces. Also, there is always a potential difference (a potential drop) due to current flow through the resistance in the bulk of the solution.

In an electro-chemical experiment, the potential that you wish to control or measure is the potential of a metal specimen (called the Working Electrode) versus a Reference Electrode. You are normally not interested in the potential drops due to solution resistances.

The Gamry Instruments Series G or Reference 600 Potentiostat, like all modern electro-chemical instruments, is a three-electrode potentiostat. It measures and controls the potential difference between a non-current carrying Reference Electrode and one of the two current carrying electrodes (the Working Electrode). The potential drop near the other current carrying electrode (the Counter Electrode) does not matter when a three-electrode potentiostat is used.

Careful placement of the Reference Electrode can compensate for some of the IR drop resulting from the cell current, I, flowing through the solution resistance, R. You can think of the Reference Electrode as sampling the potential somewhere along the solution resistance. The closer it is to the Working Electrode, the closer you are to measuring a potential free from IR errors. However, complete IR compensation cannot be achieved in practice through placement of the reference electrode, because of the finite physical size of the electrode. The portion of the cell resistance that remains after placing the Reference Electrode is called the uncompensated resistance, Ru.

Gamry Potentiostats use current interrupt IR compensation to dynamically correct uncompensated resistance errors. In the current interrupt technique, the cell current is periodically turned off for a very short time. With no current flowing through the solution resistance, its IR drop disappears instantly. The potential drop at the electrode surface remains constant on a rapid time scale. The difference in potential with the current flowing and without is a measure of the uncompensated IR drop.

The potentiostat makes a current interrupt measurement immediately after each data point is acquired. It actually takes three potential readings : E_1 before the current is turned off, and E_2 and E_3 while it is off. Normally, the latter two are used to extrapolate the potential difference, delta E, back to the exact moment when the current was interrupted. The timing of the interrupt depends on the cell current. The interrupt time is 40 microseconds on the higher current ranges. On lower current ranges, the interrupt lasts longer.

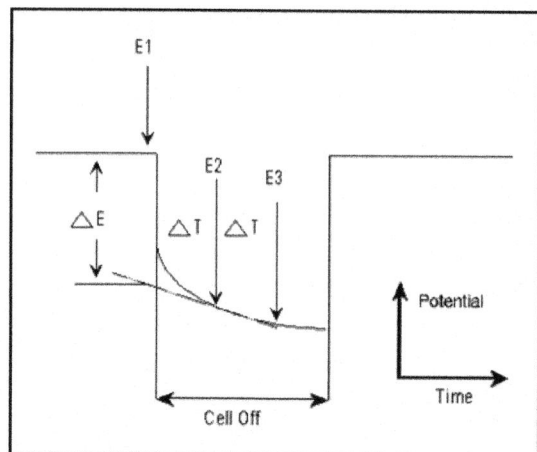

Fig. : Current Interrupt Potential *versus* Time.

In controlled potential modes, the applied potential can be dynamically corrected for the measured *IR* error in one of several ways. In the simplest of these, the *IR* error from the previous point is applied as a correction to the applied potential. For example, if an *IR* free potential of 1 volt is desired, and the measured *IR* error is 0.2 volts, the potentiostat will apply 1.2 volts. The correction is always one point behind, as the *IR* error from one point is applied to correct the applied potential for the next point. In addition to this normal mode, the Gamry PC4 offers more complex feedback modes in which the two points on the decay curve are averaged.

By default in the controlled potential modes, the potential error measured *via* current interrupt is used to correct the applied potential. In the controlled current modes, no correction is required. If *IR* compensation is selected, the measured *IR* error is subtracted from the measured potential. All reported potentials are therefore free from *IR* error.

Current and Voltage Conventions

Current polarities in electro-chemical measurements can be inconsistent. A current value of-1.2 mA can mean different things to workers in different areas of electro-chemistry or in different countries or even to different potentiostats. To an analytical electro-chemist it represents 1.2 mA of anodic current. To a corrosion scientist it represents 1.2 mA of cathodic current. A Gamry Potentiostat in default mode follows the corrosion convention for current in which positive currents are anodic and negative currents are cathodic. For the convenience of our users around the world, Gamry Potentiostats can provide the current polarity in your preferred polarity with a simple software command.

The polarity of the potential can also be a source of confusion. In electrochemical corrosion measurement, the equilibrium potential assumed by the metal in the absence of electrical connections to the metal is called the Open

Circuit Potential, Eoc. We have reserved the term Corrosion Potential, Ecorr, for the potential in an electro-chemical experiment at which no current flows, as determined by a numerical fit of current *versus* potential data. In an ideal case, the values for Eoc and Ecorr will be identical. One reason the two voltages may differ is that changes have occurred to the electrode surface during the scan.

With most modern potentiostats, all potentials are specified or reported as the potential of the working electrode with respect to either the reference electrode or the open circuit potential. The former is always labelled as "*vs.* Eref" and the later is labelled as "*vs.* Eoc". The equations used to convert from one form of potential to the other are :

E *vs.* Eoc = (E *vs.* Eref)-Eoc

E *vs.* Eref = (E *vs.* Eoc) + Eoc

Regardless off whether potentials are *versus* Eref or versus Eoc, one sign convention is used. The more positive a potential, the more anodic it is. More anodic potentials accelerate oxidation at the Working Electrode. Conversely, a negative potential accelerates reduction at the Working Electrode.

GALVANIC CORROSION

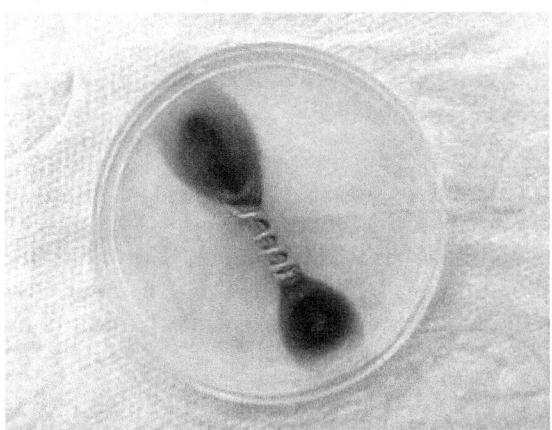

Fig. : Corrosion of an iron nail wrapped in bright copper wire, showing cathodic protection of copper, plus coloured chemical indicators of two types of ions diffusing through a moist agar medium.

Galvanic corrosion is an electro-chemical process in which one metalcorrodes preferentially to another when both metals are in electrical contact, in the presence of an electrolyte. This same galvanic reaction is exploited in primary batteries to generate an electrical voltage.

Overview

Dissimilar metals and alloys have different electrode potentials, and when two or more come into contact in an electrolyte, one metal acts as anode and the other as cathode. The electropotential difference between the dissimilar metals is the driving force for an accelerated attack on the anode member of the galvanic couple. The anode metal dissolves into the electrolyte, and deposit collects on the cathodic metal.

The electrolyte provides a means for ionmigration whereby metallic ions move from the anode to the cathode within the metal. This leads to the metal at the anode corroding more quickly than it otherwise would and corrosion at the cathode being inhibited. The presence of an electrolyte and an electrical conducting path between the metals is essential for galvanic corrosion to occur.

In some cases, this type of reaction is intentionally encouraged. For example, low-cost household batteries typically contain carbon-zinc cells. As part of a closed circuit (the electron pathway), the zinc within the cell will corrode preferentially (the ion pathway) as an essential part of the battery producing electricity. Another example is the cathodic protection of buried or submerged structures. In this case, sacrificial anodes work as part of a galvanic couple, promoting corrosion of the anode, while protecting the cathode metal.

In other cases, such as mixed metals in piping (for example, copper and cast iron), galvanic corrosion will contribute to accelerated corrosion of parts of the system. Corrosion inhibitors such as sodium nitrite or sodium molybdate can be injected into these systems to reduce the galvanic potential. However, the application of these corrosion inhibitors must be monitored closely. If the application of corrosion inhibitors increases the conductivity of the water within the system, the galvanic corrosion potential can be exponentially increased.

Acidity or alkalinity (pH) is also a major consideration with regard to closed loop bimetallic circulating systems. Should the pH and corrosion inhibition doses be incorrect, galvanic corrosion will be accelerated. In most HVAC systems, the use of sacrificial anodes and cathodes is not an option, as they would need to be applied within the plumbing of the system and, over time, would corrode and release particles that could cause potential mechanical damage to circulating pumps, heat exchangers, etc.

Examples of Corrosion

A common example of galvanic corrosion is the rusting of corrugated iron sheet, which becomes widespread when the protective zinc coating is broken and the underlying steel is attacked. The zinc is attacked preferentially because it is less noble, but once it has been consumed, rusting of the base metal can occur in earnest. By contrast, with a traditional tin can, the opposite of a protective effect occurs : because the tin is more noble than the underlying steel, when the tin coating is broken, the steel beneath is immediately attacked preferentially.

Statue of Liberty

A spectacular example of galvanic corrosion occurred in the Statue of Liberty when regular maintenance checks in the 1980s revealed that corrosion had taken place between the outer copper skin and the wrought iron support structure. Although the problem had been anticipated when the structure was built by Gustave Eiffel to Frédéric Bartholdi's design in the 1880s, the insulation layer of shellac between the two metals had failed over time and resulted in rusting of the iron supports. An extensive renovation requiring complete disassembly of the statue replaced the original insulation with PTFE. The structure was far from unsafe owing to the large number of unaffected connections, but it was regarded as a precautionary measure for what is considered a national symbol of the United States.

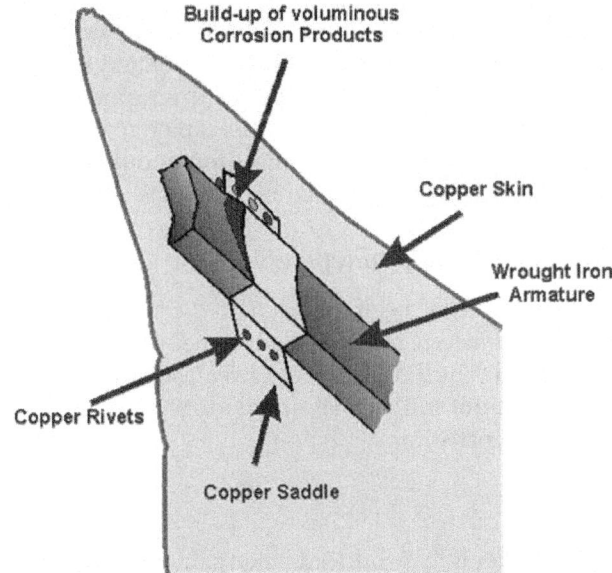

Fig. : Galvanic corrosion in the Statue of Liberty.

Fig. : Regular maintenance checks discovered that the Statue of Liberty suffered from galvanic corrosion.

Royal Navy and HMS Alarm

In 17th-century England, Samuel Pepys (then serving as Admiralty Secretary), agreed to the removal of lead sheathing from British Royal Navy vessels to prevent the mysterious disintegration of their rudder-irons and bolt-heads, though he confessed himself baffled as to the reason the lead caused the corrosion.

The problem recurred when vessels were sheathed in copper, to reduce marine weed accumulation and protect against shipworm. The protective copper on the Royal Navy frigate HMS *Alarm* detached from the wooden hull because iron nail fasteners had been "much rotted". Closer inspection revealed that water-resistant brown paper trapped under the nail head had protected some of the nails. The copper sheathing was delivered to the dockyard wrapped in the paper, which was not always removed before the sheets were nailed to the hull. The conclusion reported to the Admiralty in 1763, was that iron should not be allowed direct contact with copper in sea water to avoid corrosion.

US Navy Littoral Combat Ship Independence

Serious galvanic corrosion has been reported on the latest US Navy attack littoral combat vessel the USS *Independence* caused by steel water jet propulsion systems attached to an aluminium hull. Without electrical isolation between the steel and aluminium, the aluminium hull acts as an anode to the stainless steel, resulting in aggressive galvanic corrosion.

Lasagna Cell

A "lasagna cell" is accidentally produced when salty moist food such as lasagna is stored in a steel baking pan and is covered with aluminum foil. After a few hours the foil develops small holes where it touches the lasagna, and the food surface becomes covered with small spots composed of corroded aluminum.

In this example, the salty food (lasagna) is the electrolyte, the aluminum foil is the anode, and the steel pan is the cathode. If the aluminum foil only touches the electrolyte in small areas, the galvanic corrosion is concentrated, and corrosion can occur fairly rapidly.

Electrolytic Cleaning

The common technique of cleaning silverware by immersion of the silver and a piece of aluminum in an electrolytic bath (usually sodium bicarbonate) is an example of galvanic corrosion. (Care should be exercised because this will also strip silver oxide from the silverware, which may be there for decoration. Use on silverplate is inadvisable, as this may cause unwanted galvanic corrosion of the base metal.)

Fig. : Aluminum anodes mounted on a steel-jacketed structure.

PREVENTING GALVANIC CORROSION

There are several ways of reducing and preventing this form of corrosion.

- Electrically insulate the two metals from each other. If they are not in electrical contact, no galvanic couple will occur. This can be achieved by using non-conductive materials between metals of different electropotential. Piping can be isolated with a spool of pipe made of plastic materials, or made of metal material internally coated or lined. It is important that the spool be a sufficient length to be effective.

- Metal boats connected to a shore line electrical power feed will normally have to have the hull connected to earth for safety reasons. However the end of that earth connection is likely to be a copper rod buried within the marina, resulting in a steel-copper "battery" of about 0.5V. For such cases, the use of a galvanic isolator is essential, typically two semi-conductor diodes in series. This prevents any current flow while the applied voltage is **less** than 1.4V (*i.e.* 0.7V per diode), but allows a full flow of current in case of an electrical fault. There will still be a very minor leakage of current through the diodes, which may result in slightly faster corrosion than normal.

- Ensure there is no contact with an electrolyte. This can be done by using water-repellent compounds such as greases, or by coating the metals with an impermeable protective layer, such as a suitable paint, varnish, or plastic. If it is not possible to coat both, the coating should be applied to the more noble, the material with higher potential. This is advisable because if the coating is applied only on the more active material, in case of damage to the coating there will be a large cathode area and a very small anode area, and for the exposed anodic area the corrosion rate will be correspondingly high.

- Using antioxidant paste is beneficial for preventing corrosion between copper and aluminum electrical connections. The paste consists of a lower nobility metal than aluminum or copper.

- Choose metals that have similar electropotentials. The more closely matched the individual potentials, the lesser the potential difference and hence the lesser the galvanic current. Using the same metal for all construction is the easiest way of matching potentials.

• Electroplating or other plating can also help. This tends to use more noble metals that resist corrosion better. Chrome, nickel, silver and gold can all be used. Galvanizing with zinc protects the steel base metal by sacrificial anodic action.

• Cathodic protection uses one or more sacrificial anodes made of a metal which is more active than the protected metal. Alloys of metals commonly used for sacrificial anodes include zinc, magnesium, and aluminium. This is commonplace on many buried or immersed metallic structures.

• Cathodic Protection can also be applied by connecting a direct current (DC) electrical power supply to oppose the corrosive galvanic current.

GALVANIC SERIES

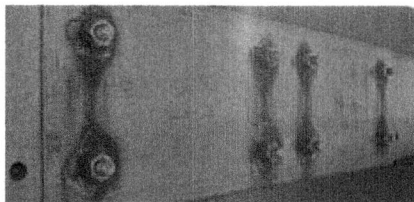

Fig. : Stainless steel cable ladder with mild steel bolts.

All metals can be classified into a galvanic series representing the electrical potential they develop in a given electrolyte against a standard reference electrode. The relative position of two metals on such a series gives a good indication of which metal is more likely to corrode more quickly. However, other factors such as water aeration and flow rate can influence the rate of the process markedly.

Anodic Index

Fig. : This new manifold for water meters has replaced the old one after 4 years of use, having been connected directly to a copper pipe in the building. With a PVC battery support, lifetime is unlimited.

Fig. : Zinc sacrificial anode (photo center) to protect a boat.

The compatibility of two different metals may be predicted by consideration of their anodic index. This parameter is a measure of the electro-chemical voltage that will be developed between the metal and gold. To find the relative voltage of a pair of metals it is only required to subtract their anodic indices.

For normal environments, such as storage in warehouses or non-temperature and humidity controlled environments, there should not be more than 0.25 V difference in the anodic index. For controlled environments, in which temperature and humidity are controlled, 0.50 V can be tolerated. For harsh environments, such as outdoors, high humidity, and salt environments, there should be not more than 0.15 V difference in the anodic index. For example; gold/silver would have a difference of 0.15V being acceptable

Often when design requires that dissimilar metals come in contact, the galvanic compatibility is managed by finishes and plating. The finishing and plating selected allows the dissimilar materials to be in contact, while protecting the base materials from corrosion. Note that it will always be the lower-down of the two metals which will ultimately suffer from corrosion when galvanic incompatibility is in play. This is why you should never place sterling silver and stainless steel tableware in a dishwasher at the same time, as the steel items will likely experience corrosion by the end of the cycle (soap and water having served as the chemical electrolyte, and heat having amplified the process).

Table : Anodic index.

Metal	Index (V)
Most Cathodic	
Gold, solid and plated, Gold-platinum alloy	–0.00
Rhodium plated on silver-plated copper	–0.05
Silver, solid or plated; monel metal. High nickel-copper alloys	–0.15
Nickel, solid or plated, titanium an s alloys, Monel	–0.30
Copper, solid or plated; low brasses or bronzes; silver solder; German silvery high copper-nickel alloys; nickel-chromium alloys	–0.35
Brass and bronzes	–0.40

(Contd...)

(*Contd...*)

Metal	Index (V)
High brasses and bronzes	–0.45
18% chromium type corrosion-resistant steels	–0.50
Chromium plated; tin plated; 12% chromium type corrosion-resistant steels	–0.60
Tin-plate; tin-lead solder	–0.65
Lead, solid or plated; high lead alloys	–0.70
2000 series wrought aluminum	–0.75
Iron, wrought, gray or malleable, plain carbon and low alloy steels	–0.85
Aluminum, wrought alloys other than 2000 series aluminum, cast alloys of the silicon type	–0.90
Aluminum, cast alloys other than silicon type, cadmium, plated and chromate	–0.95
Hot-dip-zinc plate; galvanized steel	–1.20
Zinc, wrought; zinc-base die-casting alloys; zinc plated	–1.25
Magnesium & magnesium-base alloys, cast or wrought	–1.75
Beryllium	–1.85
Most Anodic	

Mixed Potential Diagrams

Electrode kinetic data are typically presented in a graphical form called Evans diagrams, polarization diagrams or mixed potential diagrams. These diagrams are useful in describing and explaining many corrosion phenomena. According to the mixed-potential theory underlying these diagrams, any electro-chemical reaction can be algebraically divided into separate oxidation and reduction reactions with no net accumulation of electrical charge. In the absence of an externally applied potential, the oxidation of the metal and the reduction of some species in solution occur simultaneously at the metal/electrolyte interface. Under these circumstances the net measurable current is zero and the corroding metal is charge neutral, with no net accumulation of charge.

What did We Just Say?

The net measurable current is zero and the corroding metal is charge neutral, with no net accumulation of charge... **anodic current = cathodic current**

It is also important to realize that most textbooks present corrosion current data as *current densities*, since this normalized variable is more descriptive of the interfacial properties. However, one must use absolute current values to construct mixed potential diagrams, to properly balance the charges in question.

In order to construct mixed potential diagrams to model a corrosion situation, one must first gather the information concerning the (1) activation overpotential for each process that is potentially involved and (2) any additional information for processes that could be affected by concentration overpotential.

CORROSION

Corrosion is the gradual destruction of materials, (usually metals), by chemical reaction with its environment.

In the most common use of the word, this means electro-chemical oxidation of metals in reaction with an oxidant such as oxygen. Rusting, the formation of ironoxides, is a well-known example of electro-chemical corrosion. This type of damage typically produces oxide(s) or salt(s) of the original metal. Corrosion can also occur in materials other than metals, such as ceramics or polymers, although in this context, the term degradation is more common. Corrosion degrades the useful properties of materials and structures including strength, appearance and permeability to liquids and gases.

Many structural alloys corrode merely from exposure to moisture in air, but the process can be strongly affected by exposure to certain substances. Corrosion can be concentrated locally to form a pit or crack, or it can extend across a wide area more or less uniformly corroding the surface. Because corrosion is a diffusion-controlled process, it occurs on exposed surfaces. As a result, methods to reduce the activity of the exposed surface, such as passivation and chromate conversion, can increase a material's corrosion resistance. However, some corrosion mechanisms are less visible and less predictable.

Fig. : Galvanic corrosion of aluminium. A 5-mm-thick Al alloy plate is physically (and hence, electrically) connected to a 10-mm-thick mild steel structural support. Galvanic corrosion occurred on the Al plate along the joint with the steel. Perforation of Al plate occurred within 2 years.

Galvanic Corrosion

Galvanic corrosion occurs when two different metals have physical or electrical contact with each other and are immersed in a common electrolyte, or when the same metal is exposed to electrolyte with different concentrations. In a galvanic couple, the more active metal (the anode) corrodes at an accelerated rate and the more noble metal (the cathode) corrodes at a retarded rate. When immersed separately, each metal corrodes at its own rate. What type of metal(s) to use is readily determined by following the galvanic series. For example, zinc is often used as a sacrificial anode for steel structures. Galvanic corrosion is of major interest to the marine industry and also anywhere water (containing salts) contacts pipes or metal structures.

Factors such as relative size of anode, types of metal, and operating conditions (temperature, humidity, salinity, etc.) affect galvanic corrosion. The surface area ratio of the anode and cathode directly affects the corrosion rates of the materials. Galvanic corrosion is often prevented by the use of sacrificial anodes.

Galvanic Series

In a given environment (one standard medium is aerated, room-temperature seawater), one metal will be either more *noble* or more *active* than others, based on how strongly its ions are bound to the surface. Two metals in electrical contact share the same electrons, so that the "tug-of-war" at each surface is analogous to competition for free electrons between the two materials. Using the electrolyte as a host for the flow of ions in the same direction, the noble metal will take electrons from the active one. The resulting mass flow or electrical current can be measured to establish a hierarchy of materials in the medium of interest. This hierarchy is called a *galvanic series* and is useful in predicting and understanding corrosion. This method is expensive but offers maximum protection against corrosion.

Corrosion Removal

Often it is possible to chemically remove the products of corrosion. For example phosphoric acid in the form of naval jelly is often applied to ferrous tools or surfaces to remove rust. Corrosion removal should not be confused with electro-polishing, which removes some layers of the underlying metal to make a smooth surface. For example, phosphoric acid may also be used to electropolish copper but it does this by removing copper, not the products of copper corrosion.

Resistance to Corrosion

Some metals are more intrinsically resistant to corrosion than others. There are various ways of protecting metals from corrosion (oxidation) including painting, hot dip galvanizing, and combinations of these.

Intrinsic Chemistry

The materials most resistant to corrosion are those for which corrosion is thermodynamically unfavourable. Any corrosion products of gold or platinum tend to decompose spontaneously into pure metal, which is why these elements can be found in metallic form on Earth and have long been valued. More common "base" metals can only be protected by more temporary means.

Fig. : Gold nuggets do not naturally corrode, even on a geological time scale.

Some metals have naturally slow reaction kinetics, even though their corrosion is thermodynamically favourable. These include such metals as zinc, magnesium, and cadmium. While corrosion of these metals is continuous and ongoing, it happens at an acceptably slow rate. An extreme example is graphite, which releases large amounts of energy upon oxidation, but has such slow kinetics that it is effectively immune to electro-chemical corrosion under normal conditions.

Passivation

Passivation refers to the spontaneous formation of an ultrathin film of corrosion products known as passive film, on the metal's surface that act as a barrier to further oxidation. The chemical composition and micro-structure of a passive film are different from the underlying metal. Typical passive film thickness on aluminium, stainless steels and alloys is within 10 nanometers. The passive film is different from oxide layers that are formed upon heating and are in the micrometer thickness range–the passive film recovers if removed or damaged whereas the oxide layer does not. Passivation in natural environments such as air, water and soil at moderate pH is seen in such materials as aluminium, stainless steel, titanium, and silicon.

Passivation is primarily determined by metallurgical and environmental factors. The effect of pH is summarized using Pourbaix diagrams, but many other factors are influential. Some conditions that inhibit passivation include high pH for aluminium and zinc, low pH or the presence of chloride ions for stainless steel, high temperature for titanium (in which case the oxide dissolves into the metal, rather than the electrolyte) and fluoride ions for silicon. On the other hand, unusual conditions may result in passivation of materials that are normally unprotected, as the alkaline environment of concrete does for steelrebar. Exposure to a liquid metal such as mercury or hot solder can often circumvent passivation mechanisms. Passivation is primarily determined by metallurgical and environmental factors.

Corrosion in Passivated Materials

Passivation is extremely useful in mitigating corrosion damage, however even a high-quality alloy will corrode if its ability to form a passivating film is hindered. Proper selection of the right grade of material for the specific environment is important for the long-lasting performance of this group of materials. If breakdown occurs in the passive film due to chemical or mechanical factors, the resulting major modes of corrosion may include pitting corrosion, crevice corrosion and stress corrosion cracking.

Pitting Corrosion

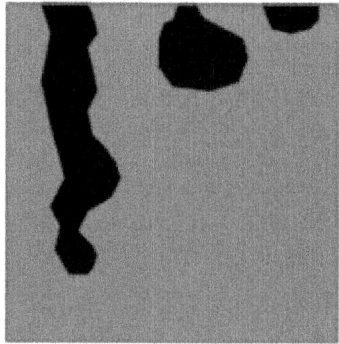

Fig. : The scheme of pitting corrosion.

Certain conditions, such as low concentrations of oxygen or high concentrations of species such as chloride which complete as anions, can interfere with a given alloy's ability to re-form a passivating film. In the worst case, almost all of the surface will remain protected, but tiny local fluctuations will degrade the oxide film in a few critical points. Corrosion at these points will be greatly amplified, and can cause *corrosion pits* of several types, depending upon conditions. While the corrosion pits only nucleate under fairly extreme circumstances, they can continue to grow even when conditions return to normal, since the interior of a pit is naturally deprived of oxygen and locally the pH decreases to very low values and the corrosion rate increases due to an autocatalytic process. In extreme cases, the sharp tips of extremely long and narrow corrosion pits can cause stress concentration to the point that otherwise tough alloys can shatter; a thin film pierced by an invisibly small hole can hide a thumb sized pit from view. These problems are especially dangerous because they are difficult to detect before a part or structure fails. Pitting remains among the most common and damaging forms of corrosion in passivated alloys, but it can be prevented by control of the alloy's environment.

Weld Decay and Knifeline Attack

Stainless steel can pose special corrosion challenges, since its passivating behaviour relies on the presence of a major alloying component (chromium, at least 11.5%).

Because of the elevated temperatures of welding and heat treatment, chromium carbides can form in the grain boundaries of stainless alloys. This chemical reaction robs the material of chromium in the zone near the grain boundary, making those areas much less resistant to corrosion. This creates a galvanic couple with the well-protected alloy nearby, which leads to *weld decay* (corrosion of the grain boundaries in the heat affected zones) in highly corrosive environments.

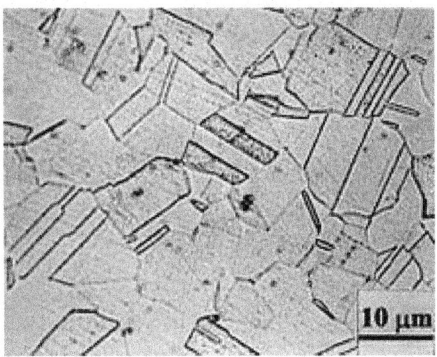

Fig. : Normal micro-structure.

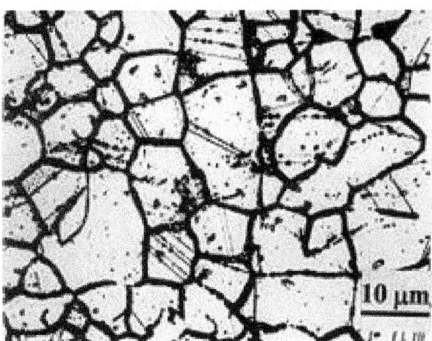

Fig. : Sensitized micro-structure.

A stainless steel is said to be sensitized if chromium carbides are formed in the micro-structure. A typical micro-structure of a normalized type-304 stainless steel shows no signs of sensitization while a heavily sensitized steel shows the presence of grain boundary precipitates. The dark lines in the sensitized micro-structure are networks of chromium carbides formed along the grain boundaries.

Special alloys, either with low carbon content or with added carbon "getters" such as titanium and niobium (in types 321 and 347, respectively), can prevent this effect, but the latter require special heat treatment after welding to prevent the similar phenomenon of *knifeline attack*. As its name implies, corrosion is limited to a very narrow zone adjacent to the weld, often only a few micrometers across, making it even less noticeable.

Crevice Corrosion

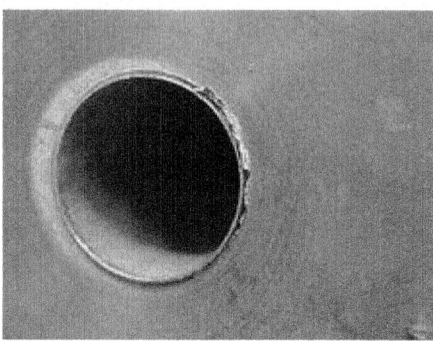

Fig. : Corrosion in the crevice between the tube and tube sheet (both made of type-316 stainless steel) of a heat exchanger in a seawater desalination plant.

Crevice corrosion is a localized form of corrosion occurring in confined spaces (crevices), to which the access of the working fluid from the environment is limited. Formation of a differential aeration cell leads to corrosion inside the crevices. Examples of crevices are gaps and contact areas between parts, under gaskets or seals, inside cracks and seams, spaces filled with deposits and under sludge piles.

Crevice corrosion is influenced by the crevice type (metal-metal, metal-non-metal), crevice geometry (size, surface finish), and metallurgical and environmental factors. The susceptibility to crevice corrosion can be evaluated with ASTM standard procedures. A critical crevice corrosion temperature is commonly used to rank a material's resistance to crevice corrosion.

Microbial Corrosion

Microbial corrosion, or commonly known as micro-biologically influenced corrosion (MIC), is a corrosion caused or promoted by micro-organisms, usually chemoautotrophs. It can apply to both metallic and non-metallic materials, in the presence or absence of oxygen. Sulfate-reducing bacteria are active in the absence of oxygen (anaerobic); they produce hydrogen sulfide, causing sulfide stress cracking. In the presence of oxygen (aerobic), some bacteria may directly oxidize iron to iron oxides and hydroxides, other bacteria oxidize sulfur and produce sulfuric acid causing biogenic sulfide corrosion. Concentration cells can form in the deposits of corrosion products, leading to localized corrosion.

Accelerated low-water corrosion (ALWC) is a particularly aggressive form of MIC that affects steel piles in seawater near the low water tide mark. It is characterized by an orange sludge, which smells of hydrogen sulfide when treated with acid. Corrosion rates can be very high and design corrosion allowances can soon be exceeded leading to premature failure of the steel pile. Piles that have been coating and have cathodic protection installed at the time of construction are not susceptible to ALWC. For unprotected piles, sacrificial anodes can be installed local to the affected areas to inhibit the corrosion or a complete retrofitted sacrificial anode system can be installed. Affected areas can also be treated electro-chemically by using an electrode to first produce chlorine to kill the bacteria, and then to produced a calcareous deposit, which will help shield the metal from further attack.

High-temperature Corrosion

High-temperature corrosion is chemical deterioration of a material (typically a metal) as a result of heating. This non-galvanic form of corrosion can occur when a metal is subjected to a hot atmosphere containing oxygen, sulfur or other compounds capable of oxidizing (or assisting the oxidation of) the material concerned. For example, materials used in aerospace, power generation and even in car engines have to resist sustained periods at high temperature in which they may be exposed to an atmosphere containing potentially highly corrosive products of combustion.

The products of high-temperature corrosion can potentially be turned to the advantage of the engineer. The formation of oxides on stainless steels, for example, can provide a protective layer preventing further atmospheric attack, allowing for a material to be used for sustained periods at both room and high temperatures in hostile conditions. Such high-temperature corrosion products, in the form of compacted oxide layer glazes, prevent or reduce wear during high-temperature sliding contact of metallic (or metallic and ceramic) surfaces.

Metal Dusting

Metal dusting is a catastrophic form of corrosion that occurs when susceptible materials are exposed to environments with high carbon activities, such as synthesis gas and other high-CO environments. The corrosion manifests itself as a break-up of bulk metal to metal powder. The suspected mechanism is firstly the deposition of a graphite layer on the surface of the metal, usually from carbon monoxide (CO) in the vapour phase. This graphite layer is then thought to form metastable M_3C species (where M is the metal), which migrate away from the metal surface. However, in some regimes no M_3C species is observed indicating a direct transfer of metal atoms into the graphite layer.

Protection from Corrosion

Fig. : US Army shrink wraps equipment such as helicopters to protect it from corrosion and thus save millions of dollars.

Surface Treatments

Applied Coatings

Plating, painting, and the application of enamel are the most common anti-corrosion treatments. They work by providing a barrier of corrosion-resistant material between the damaging environment and the structural material. Aside from cosmetic and manufacturing issues, there may be tradeoffs in mechanical flexibility *versus* resistance to abrasion and high temperature. Platings usually fail only in small sections, but if the plating is more noble than the substrate (for example, chromium on steel), a galvanic couple will cause any exposed area to corrode much more rapidly than an unplated surface would. For this reason, it is often wise to plate with active metal such as zinc or cadmium.

Painting either by roller or brush is more desirable for tight spaces; spray would be better for larger coating areas such as steel decks and waterfront applications. Flexible polyurethane coatings, like Durabak-M26 for example, can provide an anti-corrosive seal with a highly durable slip resistant membrane. Painted coatings are relatively easy to apply and have fast drying times although temperature and humidity may cause dry times to vary.

Reactive coatings

If the environment is controlled (especially in recirculating systems), corrosion inhibitors can often be added to it. These chemicals form an electrically insulating or chemically impermeable coating on exposed metal surfaces, to suppress electro-chemical reactions. Such methods make the system less sensitive to scratches or defects in the coating, since extra-inhibitors can be made available wherever metal becomes exposed. Chemicals that inhibit corrosion include some of the salts in hard water (Roman water systems are famous for their mineral deposits), chromates, phosphates, polyaniline, other conducting polymers and a wide range of specially-designed chemicals that resemble surfactants (*i.e.* long-chain organic molecules with ionic end groups).

Anodization

Aluminium alloys often undergo a surface treatment. Electro-chemical conditions in the bath are carefully adjusted so that uniform pores, several nanometers wide, appear in the metal's oxide film. These pores allow the oxide to grow much thicker than passivating conditions would allow. At the end of the treatment, the pores are allowed to seal, forming a harder-than-usual surface layer. If this coating is scratched, normal passivation processes take over to protect the damaged area.

Anodizing is very resilient to weathering and corrosion, so it is commonly used for building facades and other areas that the surface will come into regular contact with the elements. Whilst being resilient, it must be cleaned frequently. If left without cleaning, panel edge staining will naturally occur.

Bio-film Coatings

A new form of protection has been developed by applying certain species of bacterial films to the surface of metals in highly corrosive environments. This process increases the corrosion resistance substantially. Alternatively, anti-microbial-producing bio-films can be used to inhibit mild steel corrosion from sulfate-reducing bacteria.

Controlled Permeability Formwork

Controlled permeability formwork (CPF) is a method of preventing the corrosion of reinforcement by naturally enhancing the durability of the cover during concrete placement. CPF has been used in environments to combat the effects of carbonation, chlorides, frost and abrasion.

Cathodic Protection

Cathodic protection (CP) is a technique to control the corrosion of a metal surface by making that surface the cathode of an electro-chemical cell. Cathodic protection systems are most commonly used to protect steel, water, and fuel pipelines and tanks; steel pier piles, ships, and offshore oil platforms.

Sacrificial anode Protection

Fig. : Sacrificial anode on the hull of a ship.

For effective CP, the potential of the steel surface is polarized (pushed) more negative until the metal surface has a uniform potential. With a uniform potential, the driving force for the corrosion reaction is halted. For galvanic CP systems, the anode material corrodes under the influence of the steel, and eventually it must be replaced. The polarization is caused by the current flow from the anode to the cathode, driven by the difference in electro-chemical potential between the anode and the cathode.

Impressed Current Cathodic Protection

For larger structures, galvanic anodes cannot economically deliver enough current to provide complete protection. Impressed current cathodic protection (ICCP) systems use anodes connected to a DC power source (such as a cathodic protection rectifier). Anodes for ICCP systems are tubular and solid rod shapes of various specialized materials. These include high silicon cast iron, graphite, mixed metal oxide or platinum coated titanium or niobium coated rod and wires.

Anodic Protection

Anodic protection impresses anodic current on the structure to be protected (opposite to the cathodic protection). It is appropriate for metals that exhibit passivity (*e.g.*, stainless steel) and suitably small passive current over a wide range of potentials. It is used in aggressive environments, *e.g.*, solutions of sulfuric acid.

Rate of Corrosion

A simple test for measuring corrosion is the weight loss method. The method involves exposing a clean weighed piece of the metal or alloy to the corrosive environment for a specified time followed by cleaning to remove corrosion products and weighing the piece to determine the loss of weight. The rate of corrosion (R) is calculated as

$$R = KW/(\rho At)$$

where k is a constant, W is the weight loss of the metal in time t, A is the surface area of the metal exposed, and ρ is the density of the metal (in g/cm^3).

Economic Impact

Fig. : The collapsed Silver Bridge, as seen from the Ohio side.

In 2002, the US Federal Highway Administration released a study titled *Corrosion Costs and Preventive Strategies in the United States* on the direct costs associated with metallic corrosion in the U.S. industry. In 1998, the total annual direct cost of corrosion in the U.S. was ca. $276 billion (ca. 3.2% of the US gross domestic product).

Rust is one of the most common causes of bridge accidents. As rust has a much higher volume than the originating mass of iron, its build-up can also cause failure by forcing apart adjacent parts. It was the cause of the collapse of the Mianus river bridge in 1983, when the bearings rusted internally and pushed one corner of the road slab off its support. Three drivers on the roadway at the time died as the slab fell into the river below. The following NTSB investigation showed that a drain in the road had been blocked for road re-surfacing, and had not been unblocked; as a result, runoff water penetrated the support hangers. Rust was also an important factor in the Silver Bridge disaster of 1967 in West Virginia, when a steel suspension bridge collapsed within a minute, killing 46 drivers and passengers on the bridge at the time.

Similarly, corrosion of concrete-covered steel and iron can cause the concrete to spall, creating severe structural problems. It is one of the most common failure modes of reinforced concretebridges. Measuring instruments based on the half-cell potential can detect the potential corrosion spots before total failure of the concrete structure is reached.

Until 20–30 years ago; galvanized steel pipe was used extensively in the potable water systems for single and multi-family residents as well as commercial and public construction. Today, these systems have long ago consumed the protective zinc and are corroding internally resulting in poor water quality and pipe failures. The economic impact on home-owners, condo dwellers, and the public infrastructure is estimated at 22 billion dollars as insurance industry braces for a wave of claims due to pipe failures.

Corrosion in Nonmetals

Most ceramic materials are almost entirely immune to corrosion. The strong chemical bonds that hold them together leave very little free chemical energy in the structure; they can be thought of as already corroded. When corrosion does occur, it is almost always a simple dissolution of the material or chemical reaction, rather than an electro-chemical process. A common example of corrosion protection in ceramics is the lime added to soda-lime glass to reduce its solubility in water; though it is not nearly as soluble as pure sodium silicate, normal glass does form sub-microscopic flaws when exposed to moisture. Due to its brittleness, such flaws cause a dramatic reduction in the strength of a glass object during its first few hours at room temperature.

Corrosion of Polymers

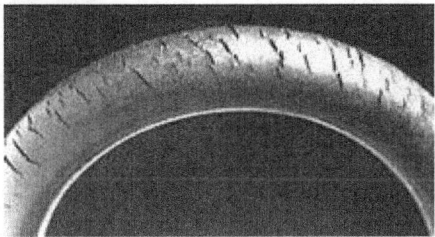

Fig. : Ozone cracking in natural rubber tubing.

Polymer degradation involves several complex and often poorly understood physio-chemical processes. These are strikingly different from the other processes discussed here, and so the term "corrosion" is only applied to them in a loose sense of the word. Because of their large molecular weight, very little entropy can be gained by mixing a given mass of polymer with another substance, making them generally quite difficult to dissolve. While dissolution is a problem in some polymer applications, it is relatively simple to design against. A more common and related problem is *swelling*, where small molecules infiltrate the structure, reducing strength and stiffness and causing a volume change. Conversely, many polymers (notably flexible vinyl) are intentionally swelled with plasticizers, which can be leached out of the structure, causing brittleness or other undesirable changes. The most common form of degradation, however, is a decrease in polymer chain length. Mechanisms which break polymer chains are familiar to biologists because of their effect on DNA : ionizing radiation (most commonly ultra-violet light), free radicals, and oxidizers such as oxygen, ozone, and chlorine. Ozone cracking is a well-known problem affecting natural rubber for example. Additives can slow these process very effectively, and can be as simple as a UV-absorbing pigment (*i.e.*, titanium dioxide or carbon black). Plastic shopping bags often do not include these additives so that they break down more easily as litter.

Corrosion of Glasses

Fig. : Glass corrosion.

Glass disease is the corrosion of silicate glasses in aqueous solutions. It is governed by two mechanisms : diffusion-controlled leaching (ion exchange) and hydrolytic dissolution of the glass network. Both mechanisms strongly depend on the pH of contacting solution : the rate of ion exchange decreases with pH as $10^{-0.5pH}$ whereas the rate of hydrolytic dissolution increases with pH as $10^{0.5pH}$.

Mathematically, corrosion rates of glasses are characterized by normalized corrosion rates of elements NR_i (g/cm² · d) which are determined as the ratio of total amount of released species into the water $M_i(g)$ to the water-contacting surface area $S(cm^2)$, time of contact t (days) and weight fraction content of the element in the glass f_i :

$$NR_i = \frac{M_i}{Sf_it}.$$

The overall corrosion rate is a sum of contributions from both mechanisms (leaching + dissolution) $NR_i = Nrx_i + NRh$. Diffusion-controlled leaching (ion exchange) is characteristic of the initial phase of corrosion and involves replacement of alkali ions in the glass by a hydronium (H_3O^+) ion from the solution. It causes an ion-selective depletion of near surface layers of glasses and gives an inverse square root dependence of corrosion rate with exposure time. The diffusion-controlled normalized leaching rate of cations from glasses (g/cm² · d) is given by :

$$NRx_i = 2\rho\sqrt{\frac{D_i}{\pi t}},$$

where t is time, D_i is the i-th cation effective diffusion coefficient (cm²/d), which depends on pH of contacting water as $D_i = D_{i0} \cdot 10^{-pH}$, and ρ is the density of the glass (g/cm³).

Glass network dissolution is characteristic of the later phases of corrosion and causes a congruent release of ions into the water solution at a time-independent rate in dilute solutions (g/cm² ·d) :

$$NRh = \rho r_h,$$

where r_h is the stationary hydrolysis (dissolution) rate of the glass (cm/d). In closed systems the consumption of protons from the aqueous phase increases the pH and causes a fast transition to hydrolysis. However, a further saturation of solution with silica impedes hydrolysis and causes the glass to return to an ion-exchange, *e.g.* diffusion-controlled regime of corrosion.

In typical natural conditions normalized corrosion rates of silicate glasses are very low and are of the order of 10^{-7}–10^{-5} g/(cm² ·d). The very high durability of silicate glasses in water makes them suitable for hazardous and nuclear waste immobilisation.

Glass Corrosion Tests

There exist numerous standardized procedures for measuring the corrosion (also called **chemical durability**) of glasses in neutral, basic, and acidic environments, under simulated environmental conditions, in simulated body fluid, at high temperature and pressure, and under other conditions.

The standard procedure ISO 719 describes a test of the extraction of water-soluble basic compounds under neutral conditions : 2 g of glass, particle size 300–500 μm, is kept for 60 min in 50 ml de-ionized water of grade 2 at 98°C; 25 ml of the obtained solution is titrated against 0.01 mol/l HCl solution.

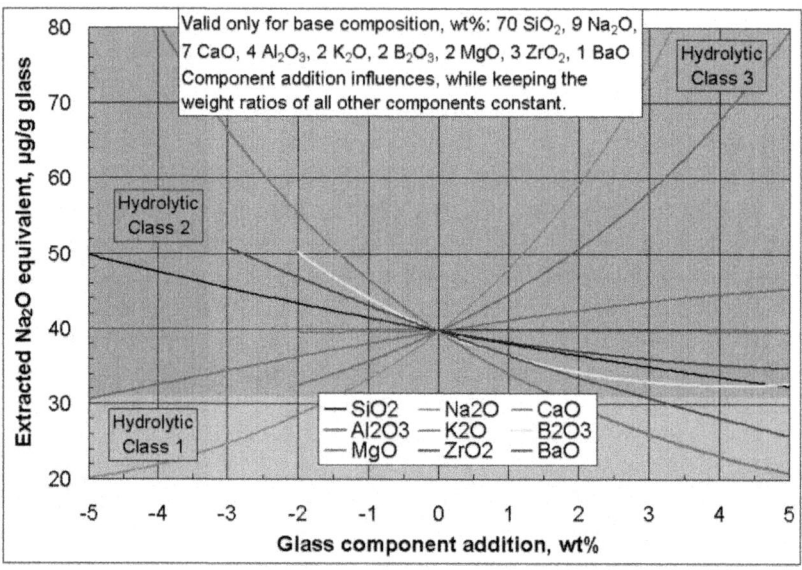

Fig. : Effect of addition of a certain glass component on the chemical durability against water corrosion of a specific base glass (corrosion test ISO 719).

Amount of 0.01M HCl needed to neutralize extracted basic oxides, ml	Extracted Na_2O equivalent, μg	Hydrolytic class
< 0.1	< 31	1
0.1-0.2	31-62	2
0.2-0.85	62-264	3
0.85-2.0	264-620	4
2.0-3.5	620-1085	5
> 3.5	> 1085	> 5

CATHODIC PROTECTION

Cathodic Protection (CP), also referred to as a sacrificial anode, is a technique used to control the corrosion of a metal surface by making it the cathode of an electro-chemical cell. A simple method of protection connects protected metal to a more easily corroded "sacrificial metal" to act as the anode. The sacrificial metal then corrodes instead of the protected metal. For structures such as long pipelines, where passive galvanic cathodic protection is not adequate, an external DC electrical power source is used to provide sufficient current.

Cathodic protection systems protect a wide range of metallic structures in various environments. Common applications are; steel water or fuel pipelines and storage tanks such as home water heaters, steel pier piles; ship and boat hulls; offshore oil platforms and onshore oil well casings and metal reinforcement bars in concrete buildings and structures. Another common application is in galvanized steel, in which a sacrificial coating of zinc on steel parts protects them from rust.

Cathodic protection can, in some cases, prevent stress corrosion cracking.

History

Cathodic protection was first described by Sir Humphry Davy in a series of papers presented to the Royal Society in London in 1824. The first application was to the HMS Samarang in 1824. Sacrificial anodes made from iron attached to the copper sheath of the hull below the waterline dramatically reduced the corrosion rate of the copper. However, a side effect of the cathodic protection was to increase marine growth. Copper, when corroding, releases copper ions which have an anti-fouling effect. Since excess marine growth affected the performance of the ship, the Royal Navy decided that it was better to allow the copper to corrode and have the benefit of reduced marine growth, so cathodic protection was not used further.

Davy was assisted in his experiments by his pupil Michael Faraday, who continued his research after Davy's death. In 1834, Faraday discovered the quantitative connection between corrosion weight loss and electric current and thus laid the foundation for the future application of cathodic protection.

Thomas Edison experimented with impressed current cathodic protection on ships in 1890, but was unsuccessful due to the lack of a suitable current source and anode materials. It would be 100 years after Davy's experiment before cathodic protection was used widely on oil pipelines in the United States — cathodic protection was applied to steel gas pipelines beginning in 1928 and more widely in the 1930s.

Types

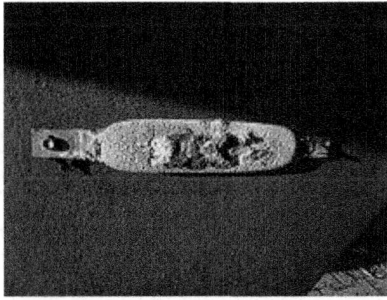

Fig. : Galvanic sacrificial anode attached to the hull of a ship, showing corrosion.

Galvanic

In the application of *passive* cathodic protection, a *galvanic anode*, a piece of a more electro-chemically "active" metal, is attached to the vulnerable metal surface where it is exposed to an electrolyte. Galvanic anodes are selected because they have a more "active" voltage (more negative electro-chemical potential) than the metal of the target structure (typically steel). For effective cathodic protection, the potential of the steel surface is polarized (pushed) more negative until the surface has a uniform potential. At that stage, the driving force for the corrosion reaction with the protected surface is removed. The galvanic anode continues to corrode, consuming the anode material until eventually it must be replaced. Polarization of the target structure is caused by the electron flow from the anode to the cathode, so the two metals must have a good electrically conductive contact. The driving force for the cathodic protection current is the difference in electro-chemical potential between the anode and the cathode.

Galvanic or sacrificial anodes are made in various shapes and sizes using alloys of zinc, magnesium and aluminium. ASTM International publishes standards on the composition and manufacturing of galvanic anodes.

In order for galvanic cathodic protection to work, the anode must possess a lower (that is, more negative) electro-chemical potential than that of the cathode (the target structure to be protected). The table below shows a simplified galvanic series which is used to select the anode metal. The anode must be chosen from a material that is lower on the list than the material to be protected.

Metal	Potential with respect to a $Cu:CuSO_4$ reference electrode in neutral pH environment (volts)
Carbon, Graphite, Coke	+0.3
Platinum	0 to –0.1
Mill scale on Steel	–0.2
High Silicon Cast Iron	–0.2
Copper, brass, bronze	–0.2
Mild steel in concrete	–0.2
Lead	–0.5
Cast iron (not graphitized)	–0.5
Mild steel (rusted)	–0.2 to –0.5
Mild steel (clean)	–0.5 to –0.8
Commercially pure aluminium	–0.8
Aluminium alloy (5% zinc)	–1.05
Zinc	–1.1
Magnesium Alloy (6% Al, 3% Zn, 0.15% Mn)	–1.6
Commercially Pure Magnesium	–1.75

Impressed Current Systems

For larger structures, or where electrolyte resistivity is high, galvanic anodes cannot economically deliver enough current to provide protection. In these cases, *impressed current cathodic protection* (ICCP) systems are used. These consist of anodes connected to a DC power source, often a transformer-rectifier connected to AC power. In the absence of an AC supply, alternative power sources may be used, such as solar panels, wind power or gas powered thermoelectric generators.

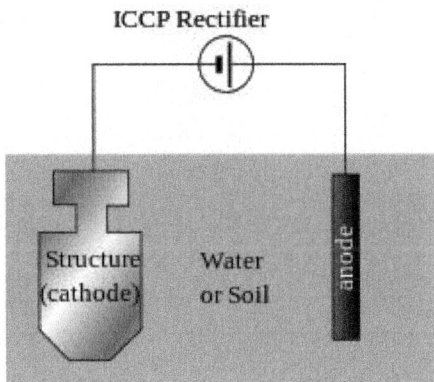

Fig. : Simple impressed current cathodic protection system. A source of DC electric current is used to help drive the protective electro-chemical reaction.

Anodes for ICCP systems are available in a variety of shapes and sizes. Common anodes are tubular and solid rod shapes or continuous ribbons of various materials. These include high silicon-cast iron, graphite, mixed metal oxide, platinum and niobium coated wire and other materials.

For pipelines, anodes are arranged in groundbeds either distributed or in a deep vertical holes depending on several design and field condition factors including current distribution requirements.

Cathodic protection transformer-rectifier units are often custom manufactured and equipped with a variety of features, including remote monitoring and control, integral current interrupters and various type of electrical enclosures. The output DC negative terminal is connected to the structure to be protected by the cathodic protection system. The rectifier output DC positive cable is connected to the anodes. The AC power cable is connected to the rectifier input terminals.

The output of the ICCP system should be optimised to provide enough current to provide protection to the target structure. Some cathodic protection transformer-rectifier units are designed with taps on the transformer windings and jumper terminals to select the voltage output of the ICCP system. Cathodic protection transformer-rectifier units for water tanks and used in other applications are made with solid state circuits to automatically adjust the operating voltage to maintain the optimum current output or structure-to-electrolyte potential. Analog or digital meters are often installed to show the operating voltage (DC and sometime AC) and current output. For shore structures and other large complex target structures, ICCP system are often designed with multiple independent zones of anodes with separate cathodic protection transformer-rectifier circuits.

Applications

Pipelines

Fig. : An air cooled cathodic protection rectifier connected to a pipeline.

Hazardous product pipelines are routinely protected by a coating supplemented with cathodic protection. An ICCP system for a pipeline consists of a DC power source, often an AC powered transformer rectifier and an anode, or array of anodes buried in the ground (the anode groundbed).

The DC power source would typically have a DC output of up to 50 amperes and 50 volts, but this depends on several factors, such as the size of the pipeline and coating quality. The positive DC output terminal would be connected *via* cables to the anode array, while another cable would connect the negative terminal of the rectifier to the pipeline, preferably through junction boxes to allow measurements to be taken.

Anodes can be installed in a groundbed consisting of a vertical hole backfilled with conductive coke (a material that improves the performance and life of the anodes) or laid in a prepared trench, surrounded by conductive coke and backfilled. The choice of groundbed type and size depends on the application, location and soil resistivity.

The output of the DC source is then adjusted to the optimum level after conducting various tests including measurements of electro-chemical potential.

It is sometimes more economically viable to protect a pipeline using galvanic anodes. This is often the case on smaller diameter pipelines of limited length.

Water pipelines of various pipe materials are also provided with cathodic protection where owners determine the cost is reasonable for the expected pipeline service life extension attributed to the application of cathodic protection.

Ships and Boats

Cathodic protection on ships is often implemented by galvanic anodes attached to the hull and ICCP for larger vessels. Since ships are regularly removed from the water for inspections and maintenance, it is a simple task to replace the galvanic anodes.

Fig. : The white patches visible on the ship's hull are zinc block sacrificial anodes.

Galvanic anodes are generally shaped to reduced drag in the water and fitted flush to the hull to also try to minimize drag.

Smaller vessels, with non-metallic hulls, such as yachts, are equipped with galvanic anodes to protect areas such as lower unit. As with all galvanic cathodic protection, this application relies on a solid electrical connection between the anode and the item to be protected.

For ICCP on ships, a DC power supply is provided within the ship and the anodes mounted on the outside of the hull. The anode cables are introduced into the ship *via* a compression seal fitting and routed to the DC power source. The negative cable from the power supply is simply attached to the hull to complete the circuit. Ship ICCP anodes are flush-mounted, minimizing the effects of drag on the ship, and located a minimum 5 ft below the light load line in an area to avoid mechanical damage. The current density required for protection is a function of velocity and considered when selecting the current capacity and location of anode placement on the hull.

Some ships may require specialist treatment, for example aluminium hulls with steel fixtures will create an electro-chemical cell where the aluminium hull can act as a galvanic anode and corrosion is enhanced. In cases like this, aluminium or zinc galvanic anodes can be used to offset the potential difference between the aluminium hull and the steel fixture. If the steel fixtures are large, several galvanic anodes may be required, or even a small ICCP system.

Marine

Marine cathodic protection covers many areas, jetties, harbors, offshore structures. The variety of different types of structure leads to a variety of systems to provide protection. Galvanic anodes are favored, but ICCP can also often be used. Because of the wide variety of structure geometry, composition, and architecture, specialized firms are often required to engineer structure-specific cathodic protection systems. Sometimes marine structures require retroactive modification to be effectively protected

Steel in Concrete

The application to concretereinforcement is slightly different in that the anodes and reference electrodes are usually embedded in the concrete at the time of construction when the concrete is being poured. The usual technique for concrete buildings, bridges and similar structures is to use ICCP, but there are systems available that use the principle of galvanic cathodic protection as well, although in the UK at least, the use of galvanic anodes for atmospherically exposed reinforced concrete structures is considered experimental.

For ICCP, the principle is the same as any other ICCP system. However, in a typical atmospherically exposed concrete structure such as a bridge, there will be many more anodes distributed through the structure as opposed to an array of anodes as used on a pipeline. This makes for a more complicated system and usually an automatically controlled DC power source is used, possibly with an option for remote monitoring and operation. For buried or submerged structures, the treatment is similar to that of any other buried or submerged structure.

Galvanic systems offer the advantage of being easier to retrofit and do not need any control systems as ICCP does.

For pipelines constructed from pre-stressed concrete cylinder pipe (PCCP), the techniques used for cathodic protection are generally as for steel pipelines except that the applied potential must be limited to prevent damage to the pre-stressing wire.

The steel wire in a PCCP pipeline is stressed to the point that any corrosion of the wire can result in failure. An additional problem is that any excessive hydrogen ions as a result of an excessively negative potential can cause hydrogen embrittlement of the wire, also resulting in failure. The failure of too many wires will result in catastrophic failure of the PCCP. To implement ICCP therefore requires very careful control to ensure satisfactory protection. A simpler option is to use galvanic anodes, which are self-limiting and need no control.

Internal Cathodic Protection

Vessels, pipelines and tanks which are used to store or transport liquids can also be protected from corrosion on their internal surfaces by the use of cathodic protection. ICCP and galvanic systems can be used. A common application of internal cathodic protection is water storage tanks.

Galvanized Steel

Galvanizing generally refers to hot-dip galvanizing which is a way of coating steel with a layer of metallic zinc. Galvanized coatings are quite durable in most environments because they combine the barrier properties of a coating with some of the benefits of cathodic protection. If the zinc coating is scratched or otherwise locally damaged and steel is exposed, the surrounding areas of zinc coating form a galvanic cell with the exposed steel and protect it from corrosion. This is a form of localized cathodic protection-the zinc acts as a sacrificial anode.

Galvanizing, while using the electro-chemical principle of cathodic protection, is not actually cathodic protection. Cathodic protection requires the anode to be separate from the metal surface to be protected, with an ionic connection through the electrolyte and an electron connection through a connecting cable, bolt or similar. This means that any area of the protected structure within the electrolyte can be protected, whereas in the case of galvanizing, only areas very close to the zinc are protected. Hence, a larger area of bare steel would only be protected around the edges.

Automobiles

Several companies market electronic corrosion control devices for automobiles and trucks. The systems are not effective and in 1996, the FTC in the USA fined David McCready and ordered him to pay $200,000 in consumer redress and stop marketing and selling his "Rust Evader" electronic corrosion control for cars. Systems marketed since that time are no more effective.

Testing

Electro-chemical potential is measured with reference electrodes. Copper-copper sulphate electrodes are used for structures in contact with soil or fresh water. Silver/sliver chloride/seawater electrodes or pure zinc electrodes are used for seawater applications. The methods are described in EN 13509 :2003 and NACE TM0497 along with the sources of error in the voltage that appears on the display of the meter. Interpretation of electro-chemical corrosion potential measurements to determine the potential at the interface between the anode of the corrosion cell and the electrolyte requires training and cannot be expected to match the accuracy of measurements done in laboratory work.

Problems

Production of Hydrogen Ions

A side effect of improperly applied cathodic protection is the production of atomic hydrogen, leading to its absorption in the protected metal and subsequent hydrogen embrittlement of welds and materials with high hardness. Under normal conditions, the atomic hydrogen will combine at the metal surface to create hydrogen gas, which cannot penetrate the metal. Hydrogen atoms, however, are small enough to pass through the crystalline steel structure, and lead in some cases to hydrogen embrittlement.

Cathodic Disbonding

This is a process of disbondment of protective coatings from the protected structure (cathode) due to the formation of hydrogen ions over the surface of the protected material (cathode). Disbonding can be exacerbated by an increase in alkali ions and an increase in cathodic polarization. The degree of disbonding is also reliant on the type of coating, with some coatings affected more than others. Cathodic protection systems should be operated so that the structure does not become excessively polarized, since this also promotes disbonding due to excessively negative potentials. Cathodic disbonding occurs rapidly in pipelines that contain hot fluids because the process is accelerated by heat flow.

Cathodic Shielding

Effectiveness of cathodic protection (CP) systems on steel pipelines can be impaired by the use of solid film backed dielectric coatings such as polyethylene tapes, shrinkable pipeline sleeves, and factory applied single or multiple solid film coatings. This phenomenon occurs because of the high electrical resistivity of these film backings. Protective electric current from the cathodic protection system is blocked or shielded from reaching the underlying metal by the highly resistive film backing. Cathodic shielding was first defined in the 1980s as being a problem, and technical papers on the subject have been regularly published since then.

A 1999 report concerning a 20,600 bbl (3,280 m^3) spill from a Saskatchewan-crude oil line contains an excellent definition of the cathodic shielding problem :

"The triple situation of disbondment of the (corrosion) coating, the dielectric nature of the coating and the unique electro-chemical environment established under the exterior coating, which acts as a shield to the electrical CP current, is referred to as CP shielding. The combination of tenting and disbondment permits a corrosive environment around the outside of the pipe to enter into the void between the exterior coating and the pipe surface. With the development of this CP shielding phenomenon, impressed current from the CP system cannot access exposed metal under the exterior coating to protect the pipe surface from the consequences of an aggressive corrosive environment. The CP shielding phenomenon induces changes in the potential gradient of the CP system across the exterior coating, which are further pronounced in areas of insufficient or sub-standard CP current emanating from the pipeline's CP system. This produces an area on the pipeline of insufficient CP defense against metal loss aggravated by an exterior corrosive environment."

Standards

- 49 CFR 192.112 — Requirements for Corrosion Control — Transportation of natural and other gas by pipeline : minimum federal safety standards
- AS 2832.4 — Australian Standard for Cathodic Protection
- ASME B31Q 0001-0191
- ASTM G 8, G 42 — Evaluating Cathodic Disbondment resistance of coatings
- DNV-RP-B401 — Cathodic Protection Design-Det Norske Veritas
- EN 12068 : 1999 — Cathodic protection. External organic coatings for the corrosion protection of buried or immersed steel pipelines used in conjunction with cathodic protection. Tapes and shrinkable materials

- EN 12473 : 2000-General principles of cathodic protection in sea water
- EN 12474 : 2001-Cathodic protection for submarine pipelines
- EN 12495 : 2000-Cathodic protection for fixed steel offshore structures
- EN 12499 : 2003-Internal cathodic protection of metallic structures
- EN 12696 : 2012-Cathodic protection of steel in concrete
- EN 12954 : 2001-Cathodic protection of buried or immersed metallic structures. General principles and application for pipelines
- EN 13173 : 2001-Cathodic protection for steel offshore floating structures
- EN 13174 : 2001-Cathodic protection for harbour installations
- EN 13509 : 2003-Cathodic protection measurement techniques
- EN 13636 : 2004-Cathodic protection of buried metallic tanks and related piping
- EN 14505 : 2005-Cathodic protection of complex structures
- EN 15112 : 2006-External cathodic protection of well casing
- EN 15280-2013-Evaluation of a.c. corrosion likelihood of buried pipelines
- EN 50162 : 2004-Protection against corrosion by stray current from direct current systems
- BS 7361-1 : 1991-Cathodic Protection
- NACE SP0169 : 2007-Control of External Corrosion on Underground or Submerged Metallic Piping Systems
- NACE TM 0497-Measurement Techniques Related to Criteria for Cathodic Protection on Underground or Submerged Metallic Piping Systems

PASSIVATION (CHEMISTRY)

Passivation, in physical chemistry and engineering, refers to a material becoming "passive," that is, being less affected by environmental factors such as air and water. Passivation involves a shielding outer-layer of corrosion, which can be applied as a micro-coating, or which occurs spontaneously in nature. As a technique, passivation is the use of a light coat of a protective material, such as metal oxide, to create a shell against corrosion. Passivation can occur only in certain conditions, and is used in microelectronics to enhance silicon. The technique of passivation is used to strengthen and preserve the appearance of metallics.

When exposed to air, many metals naturally form a hard, relatively inert surface such as the tarnishing of silver. In metals such as steel, uniform corrosion produces a somewhat rough surface by removing a substantial amount of metal, which either dissolves in the environment or reacts with it to produce a loosely adherent, porous coating of corrosion products. The reduction of the rate of corrosion will vary, depending on the metal and its environment, and is notably slowed at room-temperature air for aluminium, chromium, zinc, titanium, and silicon (a metalloid); the shell inhibits deeper corrosion, and so is the key factor of passivation. The inert surface layer, termed the "native oxide layer", is usually an oxide or a nitride, with a thickness of a monolayer (1-3Å) for a noble metal like platinum, about 15 Å for silicon and nearer to 50Å for aluminium after several years.

Mechanisms

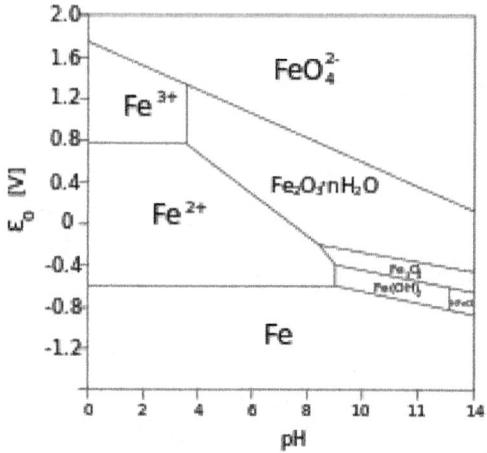

Fig. : Pourbaix diagram of iron.

There has been much interest in determining the mechanisms which describe how the thickness of the oxide layer on a material increases with time. Some of the important issues include : the relative volume of the oxide compared to the parent metal, the mechanism by which oxygen diffuses through the metal oxide to the metal-oxide interface and the relative chemical potential for the oxide to form. Boundaries between micro grains, if the oxide layer is crystalline, form an important pathway for oxygen to reach the unoxidized metal below. For this reason, vitreous oxide coatings–which lack grain boundaries–can retard oxidation. The conditions necessary (but not sufficient) for passivation are recorded in Pourbaix diagrams. Some corrosion inhibitors help the formation of a passivation layer on the surface of the metals to which they are applied. Some compounds, dissolving in solutions (chromates, molybdates) form non-reactive and low solubility films on metal surfaces.

Specific Materials

Silicon

In the area of microelectronics, the formation of a strongly adhering passivating oxide is important to the performance of silicon.

In the area of photovoltaics, a passivating surface layer such as silicon nitride, silicon dioxide or titanium dioxide can reduce surface recombination-a significant loss mechanism in solar cells.

Aluminium

Pure aluminium naturally forms a thin surface layer of aluminium oxide on contact with oxygen in the atmosphere through a process called oxidation, which creates a physical barrier to corrosion or further oxidation in most environments. Aluminium alloys, however, offer little protection against corrosion. There are three main ways to passivate these alloys: *alclading, chromate conversion coating* and *anodizing*. Alclading is the process of metallurgically bonding a thin layer of pure aluminium to the aluminium alloy. Chromate conversion coating is a common way of passivating not only aluminum, but also zinc, cadmium, copper, silver, magnesium, and tin alloys. Anodizing forms a thick oxide coating. This finish is more robust than the other processes and also provides good electrical insulation, which the other two processes do not.

For example, prior to storing hydrogen peroxide in an aluminium container, the container can be passivated by rinsing it with a dilute solution of nitric acid and peroxide alternating with deionized water. The nitric acid and peroxide oxidizes and dissolves any impurities on the inner surface of the container, and the deionized water rinses away the acid and oxidized impurities.

Ferrous Materials

Ferrous materials, including steel, may be somewhat protected by promoting oxidation ("rust") and then converting the oxidation to a metalophosphate by using phosphoric acid and further protected by surface coating. As the uncoated surface is water-soluble, a preferred method is to form manganese or zinc compounds by a process commonly known as Parkerizing or phosphate conversion. Older, less-effective but chemically-similar electro-chemical conversion coatings included black oxiding, historically known as bluing or browning. Ordinary steel forms a passivating layer in alkali environments, as reinforcing bar does in concrete.

Stainless Steel

Stainless steels are corrosion-resistant by nature, which might suggest that passivating them would be unnecessary. However, stainless steels are not completely impervious to rusting. One common mode of corrosion in corrosion-resistant steels is when small spots on the surface begin to rust because grain boundaries or embedded bits of foreign matter (such as grinding swarf) allow water molecules to oxidize some of the iron in those spots despite the alloying chromium. This is called rouging. Some grades of stainless steel are especially resistant to rouging; parts made from them may therefore, forgo any passivation step, depending on engineering decisions.

Passivation processes are generally controlled by industry standards, the most prevalent among them today being ASTM A 967 and AMS 2700. These industry standards will generally list several typical "types" of passivation processes that can be used, with the specific method to be decided between the customer and vendor. The "Method" refers to either the use of a nitric acid-based passivating bath, or a citric acid based bath (Method 2.) The various 'Types' found listed under each method refer to differences in acid bath temperature and concentration. Sodium dichromate is often required as an additive to promote oxidation in certain 'types' of nitric-based acid baths.

Common among all of the different specifications and types are the following steps : Prior to passivation, the parts must be cleaned of any contaminants and generally must undergo a validating test to prove that the surface is 'clean.' The part is then placed in an acidic passivating bath that meets the temperature and chemistry requirements of the Method and Type specified between customer and vendor. (Temperatures can range from ambient to 140 degrees Fahrenheit, while minimum passivation times are generally around 20 to 30 minutes). The parts are then neutralized using a bath of aqueous sodium hydroxide and then rinsed with clean water, dried, and the passive surface is validated using exposure to humidity, elevated temperature, a rusting agent (salt spray), or some combination of the three. However, proprietary passivation processes exist for martensitic stainless steel, which is difficult to passivate, as microscopic discontinuities can form in the surface of a machined part during passivation in a typical nitric acid bath. The passivation process removes exogenous iron, creates/restores a passive oxide layer that prevents further oxidation (rust), and cleans the parts of dirt, scale, or other welding-generated compounds (*e.g.* oxides).

It is not uncommon for some aerospace manufacturers to have additional guidelines and regulations when passivating their product that exceed the requirements in a national standard. Often, these requirements will be flowed down using NADCAP or some other accreditation system. Various testing methods are available to determine the passivation (or passive state) of stainless steel. The most common methods for validating the passivity of a part is some combination of high humidity and heat for a period of time, intended to induce rusting. Electro-chemical testers can also be utilized to commercially verify passivation.

Nickel

Nickel can be used for handling elemental fluorine, owing to the formation of a passivation layer of nickel fluoride. This fact is useful in water treatment and sewage treatment applications.

Chapter 4

KINETIC OF ELECTRODE PROCESS

CURRENT DENSITY

In electro-magnetism, and related fields in solid state physics, condensed matter physics etc. **current density** is the electric current per unit area of cross-section. It is defined as a vector whose magnitude is the electric current per cross-sectional area at a given point in space (*i.e.* it's a vector field). In SI units, the electric current density is measured in amperes per square metre.

Definition

Electric current density J is simply the electric current I (SI unit : A) per unit area A (SI unit : m^2). Its magnitude is given by the limit :

$$J = \lim_{A \to 0} \frac{I(A)}{A}$$

For current density as a vector **J**, the surface integral over a surface S, followed by an integral over the time duration t_1 to t_2, gives the total amount of charge flowing through the surface in that time $(t_2 - t_1)$:

$$q = \int_{t_1}^{t_2} \iint_S J \cdot \hat{n} d \, A dt$$

The area required to calculate the flux is real or imaginary, flat or curved, either as a cross-sectional area or a surface. For example, for charge carriers passing through an electrical conductor, the area is the cross-section of the conductor, at the section considered.

The vector area is a combination of the magnitude of the area through which the mass passes through, A, and a unit vector normal to the area, $\hat{n}$. The relation is $A = A\hat{n}$.

If the current density J passes through the area at an angle θ to the area normal $\hat{n}$, then

$$J \cdot \hat{n} = J \cos \theta$$

where $\cdot$ is the dot product of the unit vectors. This is, the component of current density passing through the surface (*i.e.* normal to it) is $J \cos \theta$, while the component of current density passing tangential to the area is $J \sin \theta$, but there is *no* current density actually passing *through* the area in the tangential direction. The *only* component of current density passing normal to the area is the cosine component.

Importance

Current density is important to the design of electrical and electronic systems.

Circuit performance depends strongly upon the designed current level, and the current density then is determined by the dimensions of the conducting elements. For example, as integrated circuits are reduced in size, despite the lower current demanded by smaller devices, there is trend toward higher current densities to achieve higher device numbers in ever smaller chip areas.

At high frequencies, current density can increase because the conducting region in a wire becomes confined near its surface, the so-called skin effect.

High current densities have undesirable consequences. Most electrical conductors have a finite, positive resistance, making them dissipate power in the form of heat. The current density must be kept sufficiently low to prevent the conductor from melting or burning up, the insulating material failing, or the desired electrical properties changing. At high current densities the material forming the interconnections actually moves, a phenomenon called *electro-migration*. In super-conductors excessive current density may generate a strong enough magnetic field to cause spontaneous loss of the super-conductive property.

The analysis and observation of current density also is used to probe the physics underlying the nature of solids, including not only metals, but also semiconductors and insulators. An elaborate theoretical formalism has developed to explain many fundamental observations.

The current density is an important parameter in Ampère's circuital law (one of Maxwell's equations), which relates current density to magnetic field.

In special relativity theory, charge and current are combined into a 4-vector.

Calculation of Current Densities in Matter

Free currents

Electric current is a coarse, average quantity that tells what is happening in an entire wire. At position **r** at time t, the *distribution* of charge flowing is described by the current density :

$$J(r, t) = \rho(r, t)\, v_d(r, t)$$

where $J(r, t)$ is the current density vector, $v_d(r, t)$ is the particles' average drift velocity (SI unit : m·s^{-1}), and

$\rho(r, t) = qn(r, t)$

is the charge density (SI unit : coulombs per cubic metre), in which $n(r,t)$ is the number of particles per unit volume ("number density") (SI unit : m^{-3}), q is the charge of the individual particles with density n (SI unit : coulombs).

A common approximation to the current density assumes the current simply is proportional to the electric field, as expressed by :

$J = \sigma E$

where E is the electric field and σ is the electrical conductivity.

Conductivity σ is the reciprocal (inverse) of electrical resistivity and has the SI units of siemens per metre ($S\ m^{-1}$), and E has the SI units of newtons per coulomb ($N\ C^{-1}$) or, equivalently, volts per metre ($V\ m^{-1}$).

A more fundamental approach to calculation of current density is based upon :

$$J(r, t) = \int_{-\infty}^{t} dt' \int d^3 r' \sigma(r - r', t - t') E(r', t')$$

indicating the lag in response by the time dependence of σ, and the non-local nature of response to the field by the spatial dependence of σ, both calculated in principle from an underlying microscopic analysis, for example, in the case of small enough fields, the linear response function for the conductive behaviour in the material.

The above conductivity and its associated current density reflect the fundamental mechanisms underlying charge transport in the medium, both in time and over distance.

A Fourier transform in space and time then results in :

$J(k, \omega) = \sigma(k, \omega) E(k, \omega)$

where $\sigma(k,\omega)$ is now a complex function.

In many materials, for example, in crystalline materials, the conductivity is a tensor, and the current is not necessarily in the same direction as the applied field. Aside from the material properties themselves, the application of magnetic fields can alter conductive behaviour.

Polarization and Magnetization Currents

Currents arise in materials when there is a non-uniform distribution of charge.

In dielectric materials, there is a current density corresponding to the net movement of electric dipole moments per unit volume, *i.e.* the polarization P :

$$J_P = \frac{\partial P}{\partial t}$$

Similarly with magnetic materials, circulations of the magnetic dipole moments per unit volume, *i.e.* the magnetization M lead to volume magnetization currents :

$$J_M = \nabla \times M$$

Together, these terms form add up to the bound current density in the material (resultant current due to movements of electric and magnetic dipole moments per unit volume) :

$$J_b = J_P + J_M$$

Total Current in Materials

The total current is simply the sum of the free and bound currents :

$$J = J_f + J_b$$

Displacement Current

There is also a displacement current corresponding to the time-varying electric displacement field D :

$$J_D = \frac{\partial D}{\partial t}$$

which is an important term in Ampere's circuital law, one of Maxwell's equations, since absence of this term would not predict electromagnetic waves to propagate, or the time evolution of electric fields in general.

Continuity Equation

Since charge is conserved, current density must satisfy a continuity equation. Here is a derivation from first principles.

The net flow out of some volume V (which can have an arbitrary shape but fixed for the calculation) must equal the net change in charge held inside the volume :

$$\int_S J \cdot dA = -\frac{d}{dt}\int_V \rho \, dV = -\int_V \frac{\partial \rho}{\partial t} dV$$

where ρ is the charge density, and dA is a surface element of the surface S enclosing the volume V. The surface integral on the left expresses the current *outflow* from the volume, and the negatively signed volume integral on the right expresses the *decrease* in the total charge inside the volume. From the divergence theorem :

$$\int_S J \cdot dA = \int_V \nabla \cdot J \, dV$$

Hence :

$$\int_V \nabla \cdot J \, dV = -\int_V \frac{\partial \rho}{\partial t} dV$$

This relation is valid for any volume, independent of size or location, which implies that :

$$\nabla \cdot J = -\frac{\partial \rho}{\partial t}$$

and this relation is called the continuity equation.

In Practice

In electrical wiring, the maximum current density can vary from $4A \cdot mm^{-2}$ for a wire with no air circulation around it, to $6A \cdot mm^{-2}$ for a wire in free air. Regulations for building wiring list the maximum allowed current of each size of cable in differing conditions. For compact designs, such as windings of SMPS transformers, the value might be as low as $2A \cdot mm^{-2}$. If the wire is carrying high frequency currents, the skin effect may affect the distribution of the current across the section by concentrating the current on the surface of the conductor. In transformers designed for high frequencies, loss is reduced if Litz wire is used for the windings. This is made of multiple isolated wires in parallel with a diameter twice the skin depth. The isolated strands are twisted together to increase the total skin area and to reduce the resistance due to skin effects.

For the top and bottom layers of printed circuit boards, the maximum current density can be as high as $35A \cdot mm^{-2}$ with a copper thickness of 35 μm. Inner layers cannot dissipate as much heat as outer layers; designers of circuit boards avoid putting high-current traces on inner layers.

In semi-conductors, the maximum current density is given by the manufacturer. A common average is $1mA \cdot \mu m^{-2}$ at 25°C for 180 nm technology. Above the maximum current density, apart from the joule effect, some other effects like electro-migration appear in the micrometer scale.

In biological organisms, ion channels regulate the flow of ions (for example, sodium, calcium, potassium) across the membrane in all cells. Current density is measured in $pA \cdot pF^{-1}$ (picoamperes per picofarad), that is, current divided by capacitance, a de facto measure of membrane area.

In gas discharge lamps, such as flashlamps, current density plays an important role in the output spectrum produced. Low current densities produce spectral line-emission and tend to favour longer wavelengths. High current densities produce continuum emission and tend to favour shorter wavelengths. Low current densities for flash lamps are generally around $1000A \cdot cm^{-2}$. High current densities can be more than $4000A \cdot cm^{-2}$.

OVERPOTENTIAL

In electro-chemistry, **overpotential** is the potential difference (voltage) between a half-reaction's thermodynamically determined reduction potential and the potential at which the redox event is experimentally observed. The term is directly related to a cell's *voltage efficiency*. In an electrolytic cell the overpotential requires more energy than thermodynamically expected to drive a reaction. In a galvanic cell overpotential means less energy is recovered than thermodynamics would predict. In each case the extra or missing energy is lost as heat. Overpotential is specific to each cell design and will vary between cells and operational conditions even for the same reaction. Practically, it is also useful to define

Thermodynamics

The four possible polarities of overpotentials are listed below :

- An electrolytic cell's anode is more positive using more energy than thermodynamics require.
- An electrolytic cell's cathode is more negative using more energy than thermodynamics require.
- A galvanic cell's anode is less negative supplying less energy than thermodynamically possible.
- A galvanic cell's cathode is less positive supplying less energy than thermodynamically possible.

The overpotential increases with increasing current density (or rate), as described by the Tafel equation. An electro-chemical reaction is a combination of two half-cells and multiple elementary steps. Each of these electro-chemical steps is associated with multiple forms of overpotential. The overall overpotential is the summation of many individual losses. *Voltage efficiency* describes the energy lost through overpotential. For an *electrolytic* cell this is the ratio of a cell's thermodynamic potential divided by the cell's experimental potential converted to a percentile. For a *galvanic* cell it is the ratio of a cell's experimental potential divided by the cell's thermodynamic potential converted to a percentile. Voltage efficiency should not be confused with Faraday efficiency. Each term refers to a mode through which electro-chemical systems can lose energy. Energy can be expressed as the product of potential, current and time (joules = volts × amps × seconds). Losses in the potential term through overpotentials are described by voltage efficiency. Losses in the current term through misdirected electrons are described by Faradaic efficiency.

Varieties of Overpotential

Overpotential can be divided into many different sub-categories that are not always well defined. For example "polarization overpotential" can refer to the electrode polarization and the hysteresis found in forward and reverse peaks of cyclic voltammetry. A likely reason for the lack of strict definitions is that it's difficult to determine how much of a measured overpotential is derived from a specific source. There is precedent for grouping overpotentials into three categories: activation, concentration, and resistance.

Activation Overpotential

Table : Activation overpotential for the evolution of selected gases on various electrode materials at 25 °C.

Material of the electrode	Hydrogen	Oxygen	Chlorine
Platinum (platinized)	−0.07 V	+0.77 V	+0.08 V
Palladium	−0.07 V	+0.93 V	
Gold	−0.09 V	+1.02 V	
Iron	−0.15 V	+0.75 V	
Platinum (shiny)	−0.16 V	+0.95 V	+0.10 V
Silver	−0.22 V	+0.91 V	
Nickel	−0.28 V	+0.56 V	
Graphite	−0.62 V	+0.95 V	+0.12 V
Lead	−0.71 V	+0.81 V	
Zinc	−0.77 V		
Mercury	−0.85 V		

The activation potential is the potential difference above the equilibrium value required to produce a current which depends on the activation energy of the redox event. While ambiguous, "activation overpotential" often refers exclusively to the activation energy necessary to transfer an electron from an electrode to an analyte. This sort of overpotential can also be called "electron transfer overpotential" and is a component of "polarization overpotential", a phenomenon observed in cyclic voltammetry and partially described by the Cottrell equation.

Reaction Overpotential

Reaction overpotential is an activation overpotential that specifically relates to chemical reactions that precede electron transfer. Reaction overpotential can be reduced or eliminated with the use of homogeneous or heterogeneous electrocatalysts. The electro-chemical reaction rate and related current density is dictated by the kinetics of the electrocatalyst and substrate concentration.

The platinum electrode common to much of electro-chemistry is electrocatalytically involved in many reactions. For example, hydrogen is oxidized and protons are reduced readily at the platinum surface of a standard hydrogen electrode in aqueous solution. Substituting an electro-catalytically inert glassy carbon electrode for the platinum electrode produces irreversible reduction and oxidation peaks with large overpotentials.

Concentration Overpotential

Concentration overpotential spans a variety of phenomena but all involve the depletion of charge-carriers at the electrode surface. Bubble overpotential is a specific form of concentration overpotential in which the concentration of charge-carriers is depleted by the physical formation of a bubble. The confusing "diffusion overpotential" can refer to a concentration overpotential created by slow diffusion rates as well as "polarization overpotential" whose overpotential is derived mostly from activation overpotential but peak current is limited by diffusion of analyte.

The potential difference is caused by differences in concentration of the charge-carriers between bulk solution and on the electrode surface. It occurs when electro-chemical reaction is sufficiently rapid to lower the surface concentration of the charge-carriers below that of bulk solution. The rate of reaction is then dependent on the ability of the charge-carriers to reach the electrode surface.

Bubble Overpotential

Bubble overpotential is a specific form of concentration overpotential and is due to the evolution of gas at either the anode or cathode. This reduces the effective area for current and increases the local current density. An example would be the electrolysis of an aqueous sodium chloride solution — although oxygen should be produced at the anode based on its potential, bubble overpotential causes chlorine to be produced instead, which allows the easy industrial production of chlorine and sodium hydroxide by electrolysis.

Resistance Overpotential

Resistance overpotentials are all the overpotentials tied to a cell design. These include "junction overpotentials" which occur at electrode surfaces and interfaces like electrolyte membranes. They can also include aspects of electrolyte diffusion, surface polarization (capacitance), and other sources of counter electromotive forces.

MASS TRANSFER COEFFICIENT

In engineering, the **mass transfer coefficient** is a diffusion rate constant that relates the mass transfer rate, mass transfer area, and concentration gradient as driving force :

$$k_c = \frac{\dot{n}_A}{A \Delta c_A}$$

Where :

- k_c is the mass transfer coefficient [mol/(s·m²)/(mol/m³)], or m/s
- $\dot{n}_A$ is the mass transfer rate [mol/s]
- A is the effective mass transfer area [m²]
- ΔC_A is the driving force concentration difference [mol/m³].

This can be used to quantify the mass transfer between phases, immiscible and partially misciblefluid mixtures (or between a fluid and a porous solid). Quantifying mass transfer allows for design and manufacture of separation process equipment that can meet specified requirements, estimate what will happen in real life situations (chemical spill), etc.

Mass transfer coefficients can be estimated from many different theoretical equations, correlations, and analogies that are functions of material properties, intensive properties and flow regime (laminar or turbulent flow). Selection of the most applicable model is dependent on the materials and the system, or environment, being studied.

Mass Transfer Coefficient Units

- $(mol/s)/(m^2 \cdot mol/m^3) = m/s$

COULOMETRY

Coulometry is the name given to a group of techniques in analytical chemistry that determine the amount of matter transformed during an electrolysis reaction by measuring the amount of electricity (in coulombs) consumed or produced. It is named after Charles-Augustin de Coulomb.

There are two basic categories of coulometric techniques. *Potentiostatic coulometry* involves holding the electric potential constant during the reaction using a potentiostat. The other, called *coulometric titration* or *amperostatic coulometry*, keeps the current (measured in amperes) constant using an amperostat.

Potentiostatic Coulometry

Potentiostatic coulometry is a technique most commonly referred to as "bulk electrolysis". The working electrode is kept at a constant potential and the current that flows through the circuit is measured. This constant potential is applied long enough to fully reduce or oxidize all of the electro-active species in a given solution. As the electro-active molecules are consumed, the current also decreases, approaching zero when the conversion is complete. The sample mass, molecular mass, number of electrons in the electrode reaction, and number of electrons passed during the experiment are all related by Faraday's laws. It follows that, if three of the values are known, then the fourth can be calculated.

Bulk electrolysis is often used to unambiguously assign the number of electrons consumed in a reaction observed through voltammetry. It also has the added benefit of producing a solution of a species (oxidation state) which may not be accessible through chemical routes. This species can then be isolated or further characterized while in solution.

The rate of such reactions is not determined by the concentration of the solution, but rather the mass transfer of the electro-active species in the solution to the electrode surface. Rates will increase when the volume of the solution is decreased, the solution is stirred more rapidly, or the area of the working electrode is increased. Since mass transfer is so important the solution is stirred during a bulk electrolysis. However, this technique is generally not considered a hydrodynamic technique, since a laminar flow of solution against the electrode is neither the objective nor outcome of the stirring.

The extent to which a reaction goes to completion is also related to how much greater the applied potential is than the reduction potential of interest. In the case where multiple reduction potentials are of interest, it is often difficult to set an electrolysis potential a "safe" distance (such as 200 mV) past a redox event. The result is incomplete conversion of the substrate, or else conversion of some of the substrate to the more reduced form. This factor must be considered when analyzing the current passed and when attempting to do further analysis/isolation/experiments with the substrate solution.

An advantage to this kind of analysis over electrogravimetry is that it does not require that the product of the reaction be weighed. This is useful for reactions where the product does not deposit as a solid, such as the determination of the amount of arsenic in a sample from the electrolysis of arsenous acid (H_3AsO_3) to arsenic acid (H_3AsO_4).

Coulometric Titration

Coulometric titrations use a constant current system to accurately quantify the concentration of a species. In this experiment, the applied current is equivalent to a titrant. Current is applied to the unknown solution until all of the unknown species is either oxidized or reduced to a new state, at which point the potential of the working electrode shifts dramatically. This potential shift indicates the endpoint. The magnitude of the current (in amperes) and the duration of the current (seconds) can be used to determine the moles of the unknown species in solution. When the volume of the solution is known, then the molarity of the unknown species can be determined.

Applications

Karl Fischer Titration

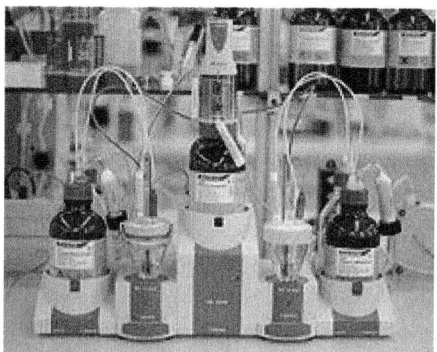

Fig. : A Karl Fischer Titrator.

Karl Fischer titration is a classic titration method in analytical chemistry that uses coulometric or volumetric titration to determine trace amounts of water in a sample. It was invented in 1935 by the German chemist Karl Fischer.

Coulometric Titration

The main compartment of the titration cell contains the anode solution plus the analyte. The anode solution consists of an alcohol (ROH), a base (B), SO_2 and I_2. A typical alcohol that may be used is methanol or diethylene glycol monoethyl ether, and a common base is imidazole.

The titration cell also consists of a smaller compartment with a cathode immersed in the anode solution of the main compartment. The two compartments are separated by an ion-permeable membrane.

The Pt anode generates I_2 when current is provided through the electric circuit. The net reaction as shown below is oxidation of SO_2 by I_2. One mole of I_2 is consumed for each mole of H_2O. In other words, 2 moles of electrons are consumed per mole of water.

$$B \cdot I_2 + B \; SO_2 + B + H_2O \rightarrow 2BH^+I^- + BSO_3$$
$$BSO_3 + ROH \rightarrow BH^+ROSO_3^-$$

The end point is detected most commonly by a bipotentiometric method. A second pair of Pt electrodes are immersed in the anode solution. The detector circuit maintains a constant current between the two detector electrodes during titration. Prior to the equivalence point, the solution contains I^- but little I_2. At the equivalence point, excess I_2 appears and an abrupt voltage drop marks the end point. The amount of current needed to generate I_2 and reach the end point can then be used to calculate the amount of water in the original sample.

Volumetric Titration

The volumetric titration is based on the same principles as the coulometric titration except that the anode solution above now is used as the titrant solution. The titrant consists of an alcohol (ROH), base (B), SO_2 and a known concentration of I_2.

One mole of I_2 is consumed for each mole of H_2O. The titration reaction proceeds as above, and the end point may be detected by a bipotentiometric method as described above.

Advantage of Analysis

The popularity of the Karl Fischer titration is due in large part to several practical advantages that it holds over other methods of moisture determination, including:

- High accuracy and precision-typically within 1% of available water, *i.e.* 3.00% appears as 2.97–3.03%.
- Selectivity for water
- Small sample quantities required
- Easy sample preparation
- Short analysis duration
- Nearly unlimited measuring range (1ppm to 100%)

- Suitability for analyzing :
  - Solids
  - Liquids
  - Gases.
- Independence of presence of other volatiles
- Suitability for automation
- Linearity-single-point calibration, no calibration curves necessary.

  In contrast, loss on drying will detect the loss of *any* volatile substance.

The major disadvantage is that the water has to be accessible and easily brought into methanol solution. Many common substances, especially foods such as chocolate, release water slowly and with difficulty, and require additional efforts to reliably bring the total water content into contact with the Karl Fischer reagents.

Determination of Film Thickness

Coulometry can be used in the determination of the thickness of metallic coatings. This is performed by measuring the quantity of electricity needed to dissolve a well-defined area of the coating. The film thickness Δ is proportional to the constant current i, the molecular weight M of the metal, the density ρ of the metal, and the surface area A :

$$\Delta \, \alpha \, \frac{iM}{A\rho}$$

The electrodes for this reaction are often platinum electrode and an electrode that relates to the reaction. For tin coating on a copper wire, a tin electrode is used, while a sodium chloride-zinc sulfate electrode would be used to determine the zinc film on a piece of steel. Special cells have been created to adhere to the surface of the metal to measure its thickness. These are basically columns with the internal electrodes with magnets or weights to attach to the surface. The results obtained by this coulometric method are similar to those achieved by other chemical and metallurgic techniques.

Coulometers

Electronic Coulometer

The electronic coulometer is based on the application of the operational amplifier in the "integrator"-type circuit. The current passed through the resistor R_1 makes a potential drop which is integrated by operational amplifier on the capacitor plates; the higher current, the larger the potential drop. The current need not be constant. In such scheme V_{out} is proportional of the passed charge ($i*t$). Sensitivity of the coulometer can be changed by choosing of the appropriate value of R_1.

Electro-chemical Coulometers

There are three common types of coulometers based on electro-chemical processes :

- Copper coulometer
- Mercury coulometer
- Hofmann voltameter.

   "Voltameter" is a synonym for "coulometer".

Membrane Potential

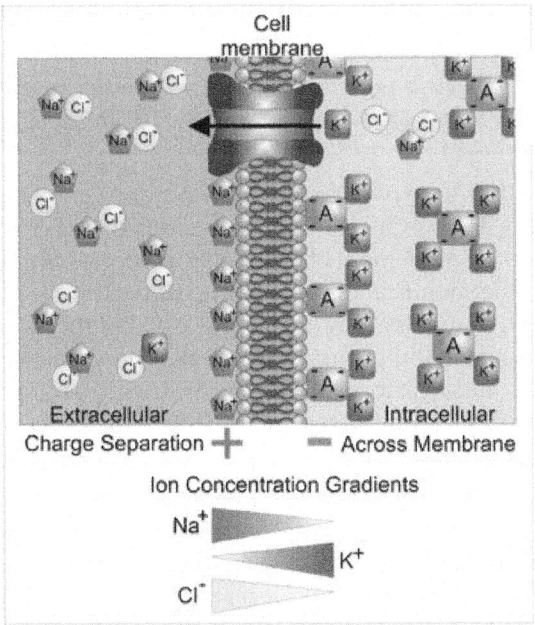

Differences in concentration of ions on opposite sides of a cellular membrane lead to a voltage called the membrane potential. Typical values of membrane potential are in the range–40 mV to–80 mV. Many ions have a concentration gradient across the membrane, including potassium (K^+), which is at a high inside and a low concentration outside the membrane. Sodium (Na^+) and chloride (Cl^-) ions are at high concentrations in the extra-cellular region, and low concentrations in the intra-cellular regions. These concentration gradients provide the potential energy to drive the formation of the membrane potential. This voltage is established when the membrane has permeability to one or more ions. In the simplest case, illustrated here, if the membrane is selectively permeable to potassium, these positively charged ions can diffuse down the concentration gradient to the outside of the cell, leaving behind uncompensated negative charges. This separation of charges is what causes the membrane potential. Note that the system as a whole is electro-neutral. The "uncompensated" positive charges outside the cell, and the uncompensated negative charges inside the cell, physically line up on the membrane surface and attract each other across membrane. Thus, the membrane potential is physically located only in the immediate vicinity of the membrane. It is the separation of these charges across the membrane that is the basis of the membrane voltage. Note also that this diagram is only an approximation of the ionic contributions to the membrane potential. Other ions including sodium, chloride, calcium and others play a more minor role, even though they have strong concentration gradients, because they have more limited permeability than potassium. Key : Blue pentagons-sodium ions; Purple squares-potassium ions; Yellow circles-Choloride ions; Orange rectangles-Anions (these arise from a variety of sources including proteins). The large purple structure with an arrow represents a transmembrane potassium channel and the direction of net potassium movement.

Membrane potential (also **transmembrane potential** or **membrane voltage**) is the difference in electric potential between the interior and the exterior of a biological cell. With respect to the exterior of the cell, typical values of membrane potential range from–40 mV to–80 mV.

All animal cells are surrounded by a membrane composed of a lipid bilayer with proteins embedded in it. The membrane serves as both an insulator and a diffusion barrier to the movement of ions. Ion transporter/pump proteins actively push ions across the membrane to establish concentration gradients across the membrane, and ion channels allow ions to move across the membrane down those concentration gradients. Ion pumps and ion channels are electrically equivalent to a set of batteries and resistors inserted in the membrane, and therefore, create a voltage difference between the two sides of the membrane.

Virtually all eukaryotic cells (including cells from animals, plants, and fungi) maintain a non-zero transmembrane potential, usually with a negative voltage in the cell interior as compared to the cell exterior ranging from–40 mV to–80 mV. The membrane potential has two basic functions. First, it allows a cell to function as a battery, providing power to operate a variety of "molecular devices" embedded in the membrane. Second, in electrically excitable cells such as neurons and muscle cells, it is used for transmitting signals between different parts of a cell. Signals are generated by opening or closing of ion channels at one point in the membrane, producing a local change in the membrane potential. This change in the electric field can quickly be detected by either adjacent or more distant ion channels in the membrane. Those ion channels can then open or close in response to the potential change, reproducing the signal.

In non-excitable cells, and in excitable cells in their baseline states, the membrane potential is held at a relatively stable value, called the resting potential. For neurons, typical values of the resting potential range from–70 to–80 millivolts; that is, the interior of a cell has a negative baseline voltage of a bit less than one-tenth of a volt. The opening and closing of ion channels can induce a departure from the resting potential. This is called a depolarization if the interior voltage becomes less negative (say from–70 mV to–60 mV), or a hyperpolarization if the interior voltage becomes more negative (say from–70 mV to–80 mV). In excitable cells, a sufficiently large depolarization can evoke an action potential, in which the membrane potential changes rapidly and significantly for a short time (on the order of 1 to 100 milliseconds), often reversing its polarity. Action potentials are generated by the activation of certain voltage-gated ion channels.

In neurons, the factors that influence the membrane potential are diverse. They include numerous types of ion channels, some of which are chemically gated and some of which are voltage-gated. Because voltage-gated ion channels are controlled by the membrane potential, while the membrane potential itself is influenced by these same ion channels, feedback loops that allow for complex temporal dynamics arise, including oscillations and regenerative events such as action potentials.

Physical Basis

The membrane potential in a cell derives ultimately from two factors : electrical force and diffusion. Electrical force arises from the mutual attraction between particles with opposite electrical charges (positive and negative) and the mutual repulsion between particles with the same type of charge (both positive or both negative). Diffusion arises from the statistical tendency of particles to redistribute from regions where they are highly concentrated to regions where the concentration is low (due to thermal energy).

Voltage

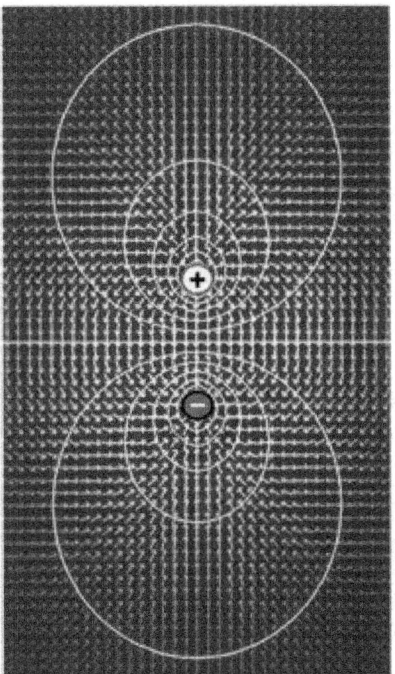

Fig. : Electric field (arrows) and contours of constant voltage created by a pair of oppositely charged objects. The electric field is at right angles to the voltage contours, and the field is strongest where the spacing between contours is the smallest.

Voltage, which is synonymous with *difference in electrical potential*, is the ability to drive an electric current across a resistance. Indeed, the simplest definition of a voltage is given by Ohm's law : $V = IR$, where V is voltage, I is current and R

is resistance. If a voltage source such as a battery is placed in an electrical circuit, the higher the voltage of the source the greater the amount of current that it will drive across the available resistance. The functional significance of voltage lies only in potential *differences* between two points in a circuit. The idea of a voltage at a single point is meaningless. It is conventional in electronics to assign a voltage of zero to some arbitrarily chosen element of the circuit, and then assign voltages for other elements measured relative to that zero point. There is no significance in which element is chosen as the zero point — the function of a circuit depends only on the differences not on voltages *per se*. However, in most cases and by convention, the zero level is most often assigned to the portion of a circuit that is in contact with ground.

The same principle applies to voltage in cell biology. In electrically active tissue, the potential difference between any two points can be measured by inserting an electrode at each point, for example one inside and one outside the cell, and connecting both electrodes to the leads of what is in essence a specialized voltmeter. By convention, the zero potential value is assigned to the outside of the cell and the sign of the potential difference between the outside and the inside is determined by the potential of the inside relative to the outside zero.

In mathematical terms, the definition of voltage begins with the concept of an electric field E, a vector field assigning a magnitude and direction to each point in space. In many situations, the electric field is a conservative field, which means that it can be expressed as the gradient of a scalar function V, that is, $E = -\nabla V$. This scalar field V is referred to as the voltage distribution. Note that the definition allows for an arbitrary constant of integration — this is why absolute values of voltage are not meaningful. In general, electric fields can be treated as conservative only if magnetic fields do not significantly influence them, but this condition usually applies well to biological tissue.

Because the electric field is the gradient of the voltage distribution, rapid changes in voltage within a small region imply a strong electric field; on the converse, if the voltage remains approximately the same over a large region, the electric fields in that region must be weak. A strong electric field, equivalent to a strong voltage gradient, implies that a strong force is exerted on any charged particles that lie within the region.

Ions and the Forces Driving their Motion

Electrical signals within biological organisms are, in general, driven by ions. The most important cations for the action potential are sodium (Na^+) and potassium (K^+). Both of these are *monovalent* cations that carry a single positive charge. Action potentials can also involve calcium (Ca^{2+}),[4] which is a *divalent* cation that carries a double positive charge. The chloride anion (Cl^-) plays a major role in the action potentials of some algae, but plays a negligible role in the action potentials of most animals.

Ions cross the cell membrane under two influences : diffusion and electric fields. A simple example wherein two solutions — A and B — are separated by a porous barrier illustrates that diffusion will ensure that they will eventually mix into equal solutions. This mixing occurs because of the difference in their concentrations. The region with high concentration will diffuse out toward the region with low concentration. To extend the example, let solution A have 30 sodium ions and 30 chloride ions. Also, let solution B have only 20 sodium ions and 20 chloride ions. Assuming the barrier allows both types of ions to travel through it, then a steady state will be reached whereby both solutions have 25 sodium ions and 25 chloride ions. If, however, the porous barrier is selective to which ions are let through, then diffusion alone will not determine the resulting solution. Returning to the previous example, let's now construct a barrier that is permeable only to sodium ions. Now, only sodium is allowed to diffuse cross the barrier from its higher concentration in solution A to the lower concentration in solution B. This will result in a greater accumulation of sodium ions than chloride ions in solution B and a lesser number of sodium ions than chloride ions in solution A.

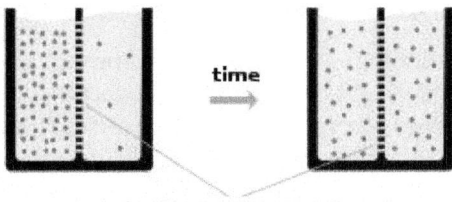

semipermeable membrane

Fig. : Ions (pink circles) will flow across a membrane from the higher concentration to the lower concentration (down a concentration gradient), causing a current. However, this creates a voltage across the membrane that opposes the ions' motion. When this voltage reaches the equilibrium value, the two balance and the flow of ions stops.

This means that there is a net positive charge in solution B from the higher concentration of positively charged sodium ions than negatively charged chloride ions. Likewise, there is a net negative charge in solution A from the greater concentration of negative chloride ions than positive sodium ions. Since opposite charges attract and like charges repel, the ions are now also influenced by electrical fields as well as forces of diffusion. Therefore, positive sodium ions will be less likely to travel to the now-more-positive B solution and remain in the now-more-negative A solution. The point at which the forces of the electric fields completely counteract the force due to diffusion is called the equilibrium potential. At this point, the net flow of the specific ion (in this case sodium) is zero.

Plasma Membranes

Every animal cell is enclosed in a plasma membrane, which has the structure of a lipid bilayer with many types of large molecules embedded in it. Because it is made of lipid molecules, the plasma membrane intrinsically has a high electrical resistivity, in other words a low intrinsic permeability to ions. However, some of the molecules embedded in the membrane are capable either of actively transporting ions from one side of the membrane to the other or of providing channels through which they can move.

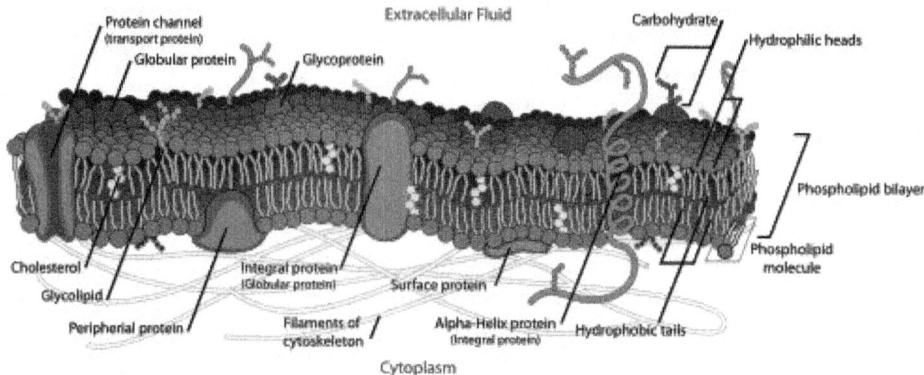

Fig. : The cell membrane, also called the plasma membrane or plasmalemma, is a semipermeable lipid bilayer common to all living cells. It contains a variety of biological molecules, primarily proteins and lipids, which are involved in a vast array of cellular processes.

In electrical terminology, the plasma membrane functions as a combined resistor and capacitor. Resistance arises from the fact that the membrane impedes the movement of charges across it. Capacitance arises from the fact that the lipid bilayer is so thin that an accumulation of charged particles on one side gives rise to an electrical force that pulls oppositely charged particles toward the other side. The capacitance of the membrane is relatively unaffected by the molecules that are embedded in it, so it has a more or less invariant value estimated at about 2 μF/ cm^2 (the total capacitance of a patch of membrane is proportional to its area). The conductance of a pure lipid bilayer is so low, on the other hand, that in biological situations it is always dominated by the conductance of alternative pathways provided by embedded molecules. Thus, the capacitance of the membrane is more or less fixed, but the resistance is highly variable.

The thickness of a plasma membrane is estimated to be about 7–8 nanometers. Because the membrane is so thin, it does not take a very large transmembrane voltage to create a strong electric field within it. Typical membrane potentials in animal cells are on the order of 100 millivolts (that is, one tenth of a volt), but calculations show that this generates an electric field close to the maximum that the membrane can sustain — it has been calculated that a voltage difference much larger than 200 millivolts could cause dielectric breakdown, that is, arcing across the membrane.

Facilitated Diffusion and Transport

The resistance of a pure lipid bilayer to the passage of ions across it is very high, but structures embedded in the membrane can greatly enhance ion movement, either actively or passively, *via* mechanisms called facilitated transport and facilitated diffusion. The two types of structure that play the largest roles are ion channels and ion pumps, both usually formed from assemblages of protein molecules. Ion channels provide passageways through which ions can move. In most cases, an ion channel is permeable only to specific types of ions (for example, sodium and potassium but not chloride or calcium), and sometimes the permeability varies depending on the direction of ion movement. Ion pumps, also known as ion transporters or carrier proteins, actively transport specific types of ions from one side of the membrane to the other, sometimes using energy derived from metabolic processes to do so.

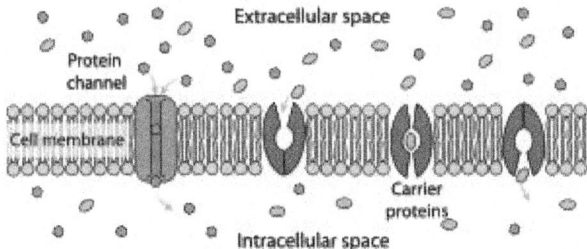

Fig. : Facilitated diffusion in cell membranes, showing ion channels and carrier proteins

Ion Pumps

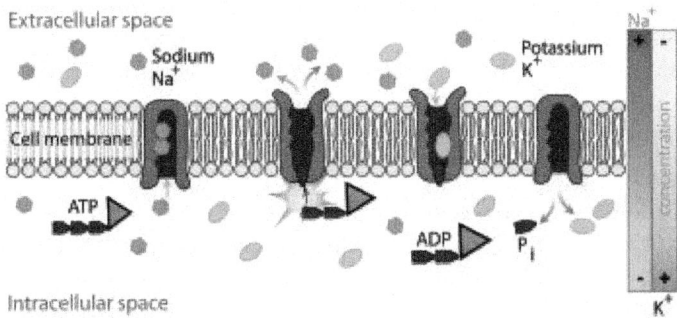

Fig. : The sodium-potassium pump uses energy derived from ATP to exchange sodium for potassium ions across the membrane.

Ion pumps are integral membrane proteins that carry out active transport, *i.e.*, use cellular energy (ATP) to "pump" the ions against their concentration gradient. Such ion pumps take in ions from one side of the membrane (decreasing its concentration there) and release them on the other side (increasing its concentration there). The ion pump most relevant to the action potential is the sodium–potassium pump, which transports three sodium ions out of the cell and two potassium ions in. As a consequence, the concentration of potassium ions K^+ inside the neuron is roughly 20-fold larger than the outside concentration, whereas the sodium concentration outside is roughly nine-fold larger than inside. In a similar manner, other ions have different concentrations inside and outside the neuron, such as calcium, chloride and magnesium.

Ion pumps influence the action potential only by establishing the relative ratio of intra-cellular and extra-cellular ion concentrations. The action potential involves mainly the opening and closing of ion channels not ion pumps. If the ion pumps are turned off by removing their energy source, or by adding an inhibitor such as ouabain, the axon can still fire hundreds of thousands of action potentials before their amplitudes begin to decay significantly. In particular, ion pumps play no significant role in the repolarization of the membrane after an action potential.

A major contribution to establishing the membrane potential is made by the sodium-potassium pump. This is a complex of proteins embedded in the membrane that derives energy from ATP in order to transport sodium and potassium ions across the membrane. On each cycle, the pump exchanges three Na^+ ions from the intra-cellular space for two K^+ ions from the extra-cellular space. If the numbers of each type of ion were equal, the pump would be electrically neutral, but, because of the three-for-two exchange, it gives a net movement of one positive charge from intra-cellular to extra-cellular for each cycle, thereby contributing to a positive voltage difference. The pump has three effects : (1) it makes the sodium concentration high in the extra-cellular space and low in the intra-cellular space; (2) it makes the potassium concentration high in the intra-cellular space and low in the extra-cellular space; (3) it gives the intra-cellular space a negative voltage with respect to the extra-cellular space.

The sodium-potassium exchange pump is relatively slow in operation. If a cell were initialized with equal concentrations of sodium and potassium everywhere, it would take hours for the pump to establish equilibrium. The pump operates constantly, but becomes progressively less efficient as the concentrations of sodium and potassium available for pumping are reduced.

Another functionally important ion pump is the sodium-calcium exchanger. This pump operates in a conceptually similar way to the sodium-potassium pump, except that in each cycle it exchanges three Na^+ from the extra-cellular space for one Ca^{++} from the intra-cellular space. Because the net flow of charge is inward, this pump runs "downhill", in effect, and therefore does not require any energy source except the membrane voltage. Its most important effect is to pump calcium outward — it also allows an inward flow of sodium, thereby counteracting the sodium-potassium pump, but, because overall sodium and potassium concentrations are much higher than calcium concentrations, this effect is relatively unimportant. The net result of the sodium-calcium exchanger is that in the resting state, intra-cellular calcium concentrations become very low.

Ion Channels

Ion channels are integral membrane proteins with a pore through which ions can travel between extra-cellular space and cell interior. Most channels are specific (selective) for one ion; for example, most potassium channels are characterized by 1000 : 1 selectivity ratio for potassium over sodium, though potassium and sodium ions have the same charge and differ only slightly in their radius. The channel pore is typically so small that ions must pass through it in single-file order. Channel pores can be either open or closed for ion passage, although a number of channels demonstrate various sub-conductance levels. When a channel is open, ions permeate through the channel pore down the transmembrane concentration gradient for that particular ion. Rate of ionic flow through the channel, *i.e.* single-channel current amplitude, is determined by the maximum channel conductance and electro-chemical driving force for that ion, which is the difference between the instantaneous value of the membrane potential and the value of the reversal potential.

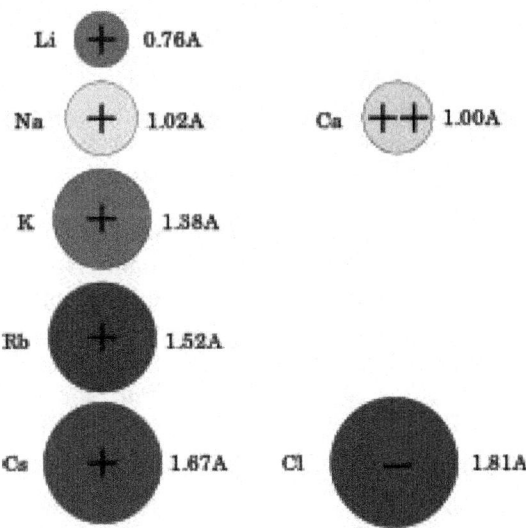

Fig.: Despite the small differences in their radii, ions rarely go through the "wrong" channel. For example, sodium or calcium ions rarely pass through a potassium channel.

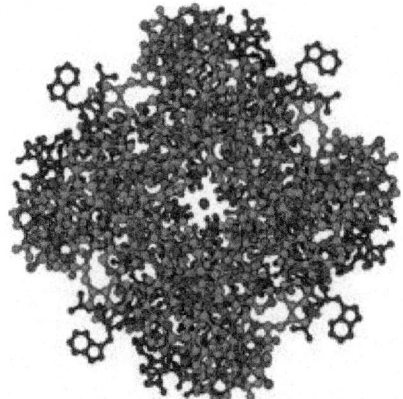

Fig.: Depiction of the open potassium channel, with the potassium ion shown in purple in the middle, and hydrogen atoms omitted. When the channel is closed, the passage is blocked.

A channel may have several different states (corresponding to different conformations of the protein), but each such state is either open or closed. In general, closed states correspond either to a contraction of the pore — making it impassable to the ion — or to a separate part of the protein, stoppering the pore. For example, the voltage-dependent sodium channel undergoes *inactivation*, in which a portion of the protein swings into the pore, sealing it. This inactivation shuts off the sodium current and plays a critical role in the action potential.

Ion channels can be classified by how they respond to their environment. For example, the ion channels involved in the action potential are *voltage-sensitive channels*; they open and close in response to the voltage across the membrane. *Ligand-gated channels* form another important class; these ion channels open and close in response to the binding of a ligand molecule, such as a neurotransmitter. Other ion channels open and close with mechanical forces. Still other ion channels — such as those of sensory neurons — open and close in response to other stimuli, such as light, temperature or pressure.

Leakage Channels

Leakage channels are the simplest type of ion channel, in that their permeability is more or less constant. The types of leakage channels that have the greatest significance in neurons are potassium and chloride channels. It should be noted that even these are not perfectly constant in their properties : First, most of them are voltage-dependent in the sense that they conduct better in one direction than the other (in other words, they are rectifiers); second, some of them are capable of being shut off by chemical ligands even though they do not require ligands in order to operate.

Ligand-gated Channels

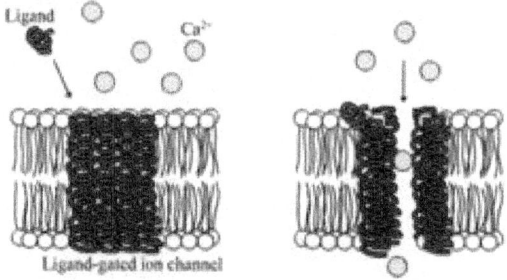

Fig. : Ligand-gated calcium channel in closed and open states.

Ligand-gated ion channels are channels whose permeability is greatly increased when some type of chemical ligand binds to the protein structure. Animal cells contain hundreds, if not thousands, of types of these. A large sub-set function as neurotransmitter receptors — they occur at post-synaptic sites, and the chemical ligand that gates them is released by the pre-synaptic axon terminal. One example of this type is the AMPA receptor, a receptor for the neurotransmitter glutamate that when activated allows passage of sodium and potassium ions. Another example is the $GABA_A$ receptor, a receptor for the neurotransmitter GABA that when activated allows passage of chloride ions.

Neurotransmitter receptors are activated by ligands that appear in the extra-cellular area, but there are other types of ligand-gated channels that are controlled by interactions on the intra-cellular side.

Voltage-dependent Channels

Voltage-gated ion channels, also known as *voltage dependent ion channels*, are channels whose permeability is influenced by the membrane potential. They form another very large group, with each member having a particular ion selectivity and a particular voltage dependence. Many are also time-dependent—in other words, they do not respond immediately to a voltage change but only after a delay.

One of the most important members of this group is a type of voltage-gated sodium channel that underlies action potentials—these are sometimes called *Hodgkin-Huxley sodium channels* because they were initially characterized by Alan Lloyd Hodgkin and Andrew Huxley in their Nobel Prize-winning studies of the physiology of the action potential. The channel is closed at the resting voltage level, but opens abruptly when the voltage exceeds a certain threshold, allowing a large influx of sodium ions that produces a very rapid change in the membrane potential. Recovery from an action potential is partly dependent on a type of voltage-gated potassium channel that is closed at the resting voltage level but opens as a consequence of the large voltage change produced during the action potential.

Reversal Potential

The reversal potential (or *equilibrium potential*) of an ion is the value of transmembrane voltage at which diffusive and electrical forces counterbalance, so that there is no net ion flow across the membrane. This means that the transmembrane voltage exactly opposes the force of diffusion of the ion, such that the net current of the ion across the membrane is zero and unchanging. The reversal potential is important because it gives the voltage that acts on channels permeable to that ion—in other words, it gives the voltage that the ion concentration gradient generates when it acts as a battery.

The equilibrium potential of a particular ion is usually designated by the notation E_{ion}. The equilibrium potential for any ion can be calculated using the Nernst equation. For example, reversal potential for potassium ions will be as follows :

$$E_{eq,k+} = \frac{RT}{zF} \ln \frac{[K^+]_o}{[K^+]_i},$$

where :

- $E_{eq,K}{}^+$ is the equilibrium potential for potassium, measured in volts
- R is the universal gas constant, equal to 8.314 joules $\cdot K^{-1} \cdot mol^{-1}$
- T is the absolute temperature, measured in kelvins (= K = degrees Celsius + 273.15)
- z is the number of elementary charges of the ion in question involved in the reaction
- F is the Faraday constant, equal to 96,485 coulombs $\cdot mol^{-1}$ or $J \cdot V^{-1} \cdot mol^{-1}$
- $[K^+]_o$ is the extra-cellular concentration of potassium, measured in mol $\cdot m^{-3}$ or m mol $\cdot l^{-1}$
- $[K^+]_i$ is the intra-cellular concentration of potassium.

Even if two different ions have the same charge (*i.e.*, K^+ and Na^+), they can still have very different equilibrium potentials, provided their outside and/or inside concentrations differ. Take, for example, the equilibrium potentials of potassium and sodium in neurons. The potassium equilibrium potential E_K is −84 mV with 5 mM potassium outside and 140 mM inside. On the other hand, the sodium equilibrium potential, E_{Na}, is approximately +40 mV with approximately 12 mM sodium inside and 140 mM outside.

Equivalent Circuit

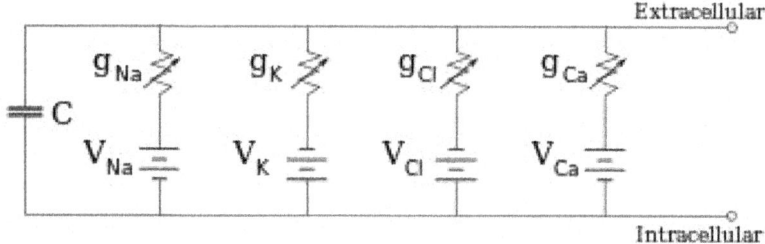

Fig. : Equivalent circuit for a patch of membrane, consisting of a fixed capacitance in parallel with four pathways each containing a battery in series with a variable conductance.

Electro-physiologists model the effects of ionic concentration differences, ion channels, and membrane capacitance in terms of an equivalent circuit, which is intended to represent the electrical properties of a small patch of membrane. The equivalent circuit consists of a capacitor in parallel with four pathways each consisting of a battery in series with a variable conductance. The capacitance is determined by the properties of the lipid bilayer, and is taken to be fixed. Each of the four parallel pathways comes from one of the principal ions, sodium, potassium, chloride, and calcium. The voltage of each ionic pathway is determined by the concentrations of the ion on each side of the membrane. The conductance of each ionic pathway at any point in time is determined by the states of all the ion channels that are potentially permeable to that ion, including leakage channels, ligand-gated channels, and voltage-gated ion channels.

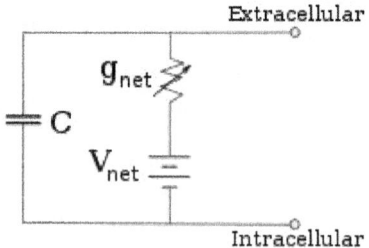

Fig. : Reduced circuit obtained by combining the ion-specific pathways using the Goldman equation.

For fixed ion concentrations and fixed values of ion channel conductance, the equivalent circuit can be further reduced, using the Goldman equation as described below, to a circuit containing a capacitance in parallel with a battery and conductance. In electrical terms, this is a type of RC circuit (resistance-capacitance circuit), and its electrical properties are very simple. Starting from any initial state, the current flowing across either the conductance or the capacitance decays with an exponential time course, with a time constant of $\tau = RC$, where C is the capacitance of the membrane patch, and $R = 1/g_{net}$ is the net resistance. For realistic situations, the time constant usually lies in the $1-100$ millisecond range. In most cases, changes in the conductance of ion channels occur on a faster time scale, so an RC circuit is not a good approximation; however, the differential equation used to model a membrane patch is commonly a modified version of the RC circuit equation.

Resting Potential

When the membrane potential of a cell can go for a long period of time without changing significantly, it is referred to as a resting potential or resting voltage. This term is used for the membrane potential of non-excitable cells, but also for the membrane potential of excitable cells in the absence of excitation. In excitable cells, the other possible states are graded membrane potentials (of variable amplitude), and action potentials, which are large, all-or-nothing rises in membrane potential that usually follow a fixed time course. Excitable cells include neurons, muscle cells, and some secretory cells in glands. Even in other types of cells, however, the membrane voltage can undergo changes in response to environmental or intra-cellular stimuli. For example, depolarization of the plasma membrane appears to be an important step in programmed cell death.

The interactions that generate the resting potential are modeled by the Goldman equation. This is similar in form to the Nernst equation shown above, in that it is based on the charges of the ions in question, as well as the difference between their inside and outside concentrations. However, it also takes into consideration the relative permeability of the plasma membrane to each ion in question.

$$E_m = \frac{RT}{F}\ln\left(\frac{P_K[K^+]_{out} + P_{Na}[Na^+]_{out} + P_{Cl}[Cl^-]_{in}}{P_K[K^+]_{in} + P_{Na}[Na^+]_{in} + P_{Cl}[Cl^-]_{out}}\right)$$

The three ions that appear in this equation are potassium (K), sodium (Na^+), and chloride (Cl^-). Calcium is omitted, but can be added to deal with situations in which it plays a significant role. Being an anion, the chloride terms are treated differently from the cation terms; the intra-cellular concentration is in the numerator, and the extra-cellular concentration in the denominator, which is reversed from the cation terms. P_i stands for the relative permeability of the ion type i.

In essence, the Goldman formula expresses the membrane potential as a weighted average of the reversal potentials for the individual ion types, weighted by permeability. In most animal cells, the permeability to potassium is much higher in the resting state than the permeability to sodium. As a consequence, the resting potential is usually close to the potassium reversal potential. The permeability to chloride can be high enough to be significant, but, unlike the other ions, chloride is not actively pumped, and therefore equilibrates at a reversal potential very close to the resting potential determined by the other ions.

Values of resting membrane potential in most animal cells usually vary between the potassium reversal potential (usually around-80 mV) and around-40 mV. The resting potential in excitable cells (capable of producing action potentials) is usually near-60 mV — more depolarized voltages would lead to spontaneous generation of action potentials. Immature or undifferentiated cells show highly variable values of resting voltage, usually significantly more positive than in differentiated cells. In such cells, the resting potential value correlates with the degree of differentiation : undifferentiated cells in some cases may not show any transmembrane voltage difference at all.

Maintenance of the resting potential can be metabolically costly for a cell because of its requirement for active pumping of ions to counteract losses due to leakage channels. The cost is highest when the cell function requires an especially depolarized value of membrane voltage. For example, the resting potential in daylight-adapted blowfly (*Calliphora vicina*) photoreceptors can be as high as-30 mV. This elevated membrane potential allows the cells to respond very rapidly to visual inputs; the cost is that maintenance of the resting potential may consume more than 20% of overall cellular ATP.

On the other hand, the high resting potential in undifferentiated cells can be a metabolic advantage. This apparent paradox is resolved by examination of the origin of that resting potential. Little-differentiated cells are characterized by extremely high input resistance, which implies that few leakage channels are present at this stage of cell life. As an apparent result, potassium permeability becomes similar to that for sodium ions, which places resting potential in-between the reversal potentials for sodium and potassium as discussed above. The reduced leakage currents also mean there is little need for active pumping in order to compensate, therefore low metabolic cost.

Graded Potentials

As explained above, the potential at any point in a cell's membrane is determined by the ion concentration differences between the intra-cellular and extra-cellular areas, and by the permeability of the membrane to each type of ion. The ion concentrations do not normally change very quickly (with the exception of Ca^{2+}, where the baseline intra-cellular concentration is so low that even a small influx may increase it by orders of magnitude), but the permeabilities of the ions can change in a fraction of a millisecond, as a result of activation of ligand-gated ion channels. The change in membrane potential can be either large or small, depending on how many ion channels are activated and what type they are, and can be either long or short, depending on the lengths of time that the channels remain open. Changes of this type are referred to as **graded potentials**, in contrast to action potentials, which have a fixed amplitude and time course.

As can be derived from the Goldman equation shown above, the effect of increasing the permeability of a membrane to a particular type of ion shifts the membrane potential toward the reversal potential for that ion. Thus, opening Na^+ channels pulls the membrane potential toward the Na^+ reversal potential, which is usually around +100 mV. Likewise, opening K^+ channels pulls the membrane potential toward about–90 mV, and opening Cl^-channels pulls it toward about–70 mV (resting potential of most membranes). Because–90 to +100 mV is the full operating range of membrane potential, the effect is that Na^+ channels always pull the membrane potential up, K^+ channels pull it down, and Cl^-channels pull it toward the resting potential.

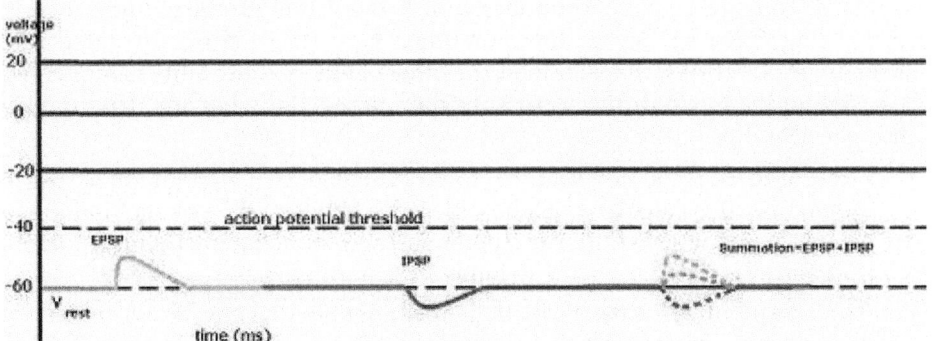

Fig. : Graph displaying an EPSP, an IPSP, and the summation of an EPSP and an IPSP.

Graded membrane potentials are particularly important in neurons, where they are produced by synapses — a temporary change in membrane potential produced by activation of a synapse by a single graded or action potential is called a post-synaptic potential. Neurotransmitters that act to open Na^+ channels typically cause the membrane potential to become more positive, while neurotransmitters that act on K^+ channels typically cause it to become more negative.

Whether a post-synaptic potential is considered excitatory or inhibitory depends on the reversal potential for the ions of that current, and the threshold for a cell to fire an action potential (around–50mV). A post-synaptic potential with a reversal potential above threshold, such a typical Na^+ current, is considered excitatory. A potential with a reversal potential below threshold, such as a typical K^+ or Cl^-current, is considered inhibitory. Even if a current depolarizes a cell, it will inhibit the cell if its reversal potential is below threshold. This is due to the fact that multiple post-synaptic potentials do not have an added effect but average, so a current with a reversal potential above the resting potential, but below threshold, will not contribute to reaching threshold. Thus, neurotransmitters that act to open Na^+ channels produce excitatory post-synaptic potentials, or EPSPs, whereas neurotransmitters that act to open K^+ or Cl^-channels produce inhibitory post-synaptic potentials, or IPSPs. When multiple types of channels are open within the same time period, their post-synaptic potentials summate (add) non-linearly.

Other Values

From the viewpoint of biophysics, the *resting* membrane potential is merely the membrane potential that results from the membrane permeabilities that predominate when the cell is resting. The above equation of weighted averages always applies, but the following approach may be more easily visualized. At any given moment, there are two factors for an ion that determine how much influence that ion will have over the membrane potential of a cell :

1. That ion's driving force
2. That ion's permeability.

This seems to be easy to understand. If the driving force is high, then the ion is being "pushed" across the membrane hard (more correctly stated : It is diffusing in one direction faster than the other). If the permeability is high, it will be easier for the ion to diffuse across the membrane. But what are 'driving force' and 'permeability'?

*   **Driving force** is the net electrical force available to move that ion across the membrane. It is calculated as the difference between the voltage that the ion "wants" to be at (its equilibrium potential) and the actual membrane potential (E_m). So, in formal terms, the driving force for an ion = E_m-E_{ion}
*   For example, at our earlier calculated resting potential of −73 mV, the driving force on potassium is 7 mV : (−73 mV) − (−80 mV) = 7 mV. The driving force on sodium would be (−73 mV) − (60 mV) = −133 mV.
*   **Permeability** is a measure of how easily an ion can cross the membrane. It is normally measured as the (electrical) conductance and the unit, siemens, corresponds to 1 C·s^{-1}·V^{-1}, that is one coulomb per second per volt of potential.

So, in a resting membrane, while the driving force for potassium is low, its permeability is very high. Sodium has a huge driving force but almost no resting permeability. In this case, potassium carries about 20 times more current than sodium, and thus has 20 times more influence over E_m than does sodium.

However, consider another case — the peak of the action potential. Here, permeability to Na is high and K permeability is relatively low. Thus, the membrane moves to near E_{Na} and far from E_K.

The more ions are permeant the more complicated it becomes to predict the membrane potential. However, this can be done using the Goldman-Hodgkin-Katz equation or the weighted means equation. By plugging in the concentration gradients and the permeabilities of the ions at any instant in time, one can determine the membrane potential at that moment. What the GHK equations means is that, at any time, the value of the membrane potential will be a weighted average of the equilibrium potentials of all permeant ions. The "weighting" is the ions relative permeability across the membrane.

Effects and Implications

While cells expend energy to transport ions and establish a transmembrane potential, they use this potential in turn to transport other ions and metabolites such as sugar. The transmembrane potential of the mitochondria drives the production of ATP, which is the common currency of biological energy.

Cells may draw on the energy they store in the resting potential to drive action potentials or other forms of excitation. These changes in the membrane potential enable communication with other cells (as with action potentials) or initiate changes inside the cell, which happens in an egg when it is fertilized by a sperm.

In neuronal cells, an action potential begins with a rush of sodium ions into the cell through sodium channels, resulting in depolarization, while recovery involves an outward rush of potassium through potassium channels. Both these fluxes occur by passive diffusion.

Ion Selective Electrode

An **ion-selective electrode (ISE)**, also known as a **specific ion electrode (SIE)**, is a transducer (or sensor) that converts the activity of a specific ion dissolved in a solution into an electrical potential, which can be measured by a voltmeter or pH meter. The voltage is theoretically dependent on the logarithm of the ionic activity, according to the Nernst equation. The sensing part of the electrode is usually made as an ion-specific membrane, along with a reference electrode. Ion-selective electrodes are used in bio-chemical and bio-physical research, where measurements of ionicconcentration in an aqueous solution are required, usually on a real time basis.

Types of Ion-selective Membrane

There are four main types of ion-selective membrane used in ion-selective electrodes (ISEs) : glass, solid state, liquid based, and compound electrode.

Glass Membranes

Glass membranes are made from an ion-exchange type of glass (silicate or chalcogenide). This type of ISE has good selectivity, but only for several single-charged cations; mainly H^+, Na^+, and Ag^+. Chalcogenide glass also has selectivity for double-charged metal ions, such as Pb^{2+}, and Cd^{2+}. The glass membrane has excellent chemical durability and can work in very aggressive media. A very common example of this type of electrode is the pH glass electrode.

Crystalline Membranes

Crystalline membranes are made from mono-or polycrystallites of a single substance. They have good selectivity, because only ions which can introduce themselves into the crystal structure can interfere with the electrode response. Selectivity of crystalline membranes can be for both cation and anion of the membrane-forming substance. An example is the fluoride selective electrode based on LaF_3 crystals.

Ion-exchange Resin Membranes

Ion-exchange resins are based on special organic polymer membranes which contain a specific ion-exchange substance (resin). This is the most widespread type of ion-specific electrode. Usage of specific resins allows preparation of selective electrodes for tens of different ions, both single-atom or multi-atom. They are also the most widespread electrodes with anionic selectivity. However, such electrodes have low chemical and physical durability as well as "survival time". An example is the potassium selective electrode, based on valinomycin as an ion-exchange agent.

Construction

These electrodes are prepared from glass capillary tubing approximately 2 millimeters in diameter, a large batch at a time. Polyvinyl chloride is dissolved in a solvent and plasticizers (typically phthalates) added, in the standard fashion used when making something out of vinyl. In order to provide the ionic specificity, a specific ion channel or carrier is added to the solution; this allows the ion to pass through the vinyl, which prevents the passage of other ions and water.

One end of a piece of capillary tubing about an inch or two long is dipped into this solution and removed to let the vinyl solidify into a plug at that end of the tube. Using a syringe and needle, the tube is filled with salt solution from the other end, and may be stored in a bath of the salt solution for an indeterminate period. For convenience in use, the open end of the tubing is fitted through a tight o-ring into a somewhat larger diameter tubing containing the same salt solution, with a silver or platinum electrode wire inserted. New electrode tips can thus be changed very quickly by simply removing the older electrode and replacing it with a new one.

Applications

In use, the electrode wire is connected to one terminal of a galvanometer or pH meter, the other terminal of which is connected to a reference electrode, and both electrodes are immersed in the solution to be tested. The passage of the ion through the vinyl *via* the carrier or channel creates an electrical current, which registers on the galvanometer; by calibrating against standard solutions of varying concentration, the ionic concentration in the tested solution can be estimated from the galvanometer reading.

In practice there are several issues which affect this measurement, and different electrodes from the same batch will differ in their properties. Leakage between the vinyl and the wall of the capillary, thereby allowing passage of any ions, will cause the meter reading to show little or no change between the various calibration solutions, and requires that that electrode be discarded. Similarly, with use the ion-sensitive channels in the vinyl appear to gradually become blocked or otherwise inactivated, causing the electrode to lose sensitivity. The response of the electrode and galvanometer is temperature sensitive, and also 'drifts' over time, requiring recalibration frequently during a series of measurements, ideally at least one calibration sample before and after each test sample. On the other hand, after immersion in the solution there is a 'settling time' which can be five minutes or even longer, before the electrode and galvanometer equilibrate to a new reading; so that timing of the reading is critical in order to find the most accurate 'window' after the response has settled, but before it has drifted appreciably.

Enzyme Electrodes

Enzyme electrodes definitely are not true **ion**-selective electrodes but usually are considered within the ion-specific electrode topic. Such an electrode has a "double reaction" mechanism-an enzyme reacts with a specific substance, and the product of this reaction (usually H^+ or OH^-) is detected by a true ion-selective electrode, such as a pH-selective electrodes. All these reactions occur inside a special membrane which covers the true ion-selective electrode, which is why enzyme electrodes sometimes are considered as ion-selective. An example is glucose selective electrodes.

Interferences

The most serious problem limiting use of ion-selective electrodes is interference from other, undesired, ions. No ion-selective electrodes are completely ion-specific; all are sensitive to other ions having similar physical properties, to an extent which depends on the degree of similarity. Most of these interferences are weak enough to be ignored, but in some cases the electrode may actually be much more sensitive to the interfering ion than to the desired ion, requiring that the interfering ion be present only in relatively very low concentrations, or entirely absent. In practice, the relative sensitivities of each type of ion-specific electrode to various interfering ions is generally known and should be checked for each case; however the precise degree of interference depends on many factors, preventing precise correction of readings. Instead, the calculation of relative degree of interference from the concentration of interfering ions can only be used as a guide to determine whether the approximate extent of the interference will allow reliable measurements, or whether the experiment will need to be redesigned so as to reduce the effect of interfering ions. The nitrate electrode has various ionic interferences, *i.e.* perchlorate, iodide, chloride, and sulfate. These interferences vary markedly in the extent to which they interfere. Thus, perchlorate gives a response which is about 50,000x as great as an equal amount of nitrate, while 1000x as much sulfate produces about a 10% error in the reading. Chloride causes a 10% error when present at about 30x the nitrate level, but can be removed by the addition of silver sulfate. Alternatively, nitrate can be determined by using an ammonia gas sensing electrode. This technique allows the user to determine both ammonium and nitrate ions sequentially. The procedure makes use of the reducing ability of titanium chloride. Trivalent titanium reduces any nitrate ion, up to 20 ppm, to ammonium ion (*i.e.*, reverse nitrification). At pH 12-13, any ammonium ion in the sample is converted to ammonia gas and is ultimately detected by the electrode.

Microbial Metabolism

Microbial metabolism is the means by which a microbe obtains the energy and nutrients (*e.g.* carbon) it needs to live and reproduce. Microbes use many different types of metabolic strategies and species can often be differentiated from each other based on metabolic characteristics. The specific metabolic properties of a microbe are the major factors in determining that microbe's ecological niche, and often allow for that microbe to be useful in industrial processes or responsible for biogeochemical cycles.

Types of Microbial Metabolism

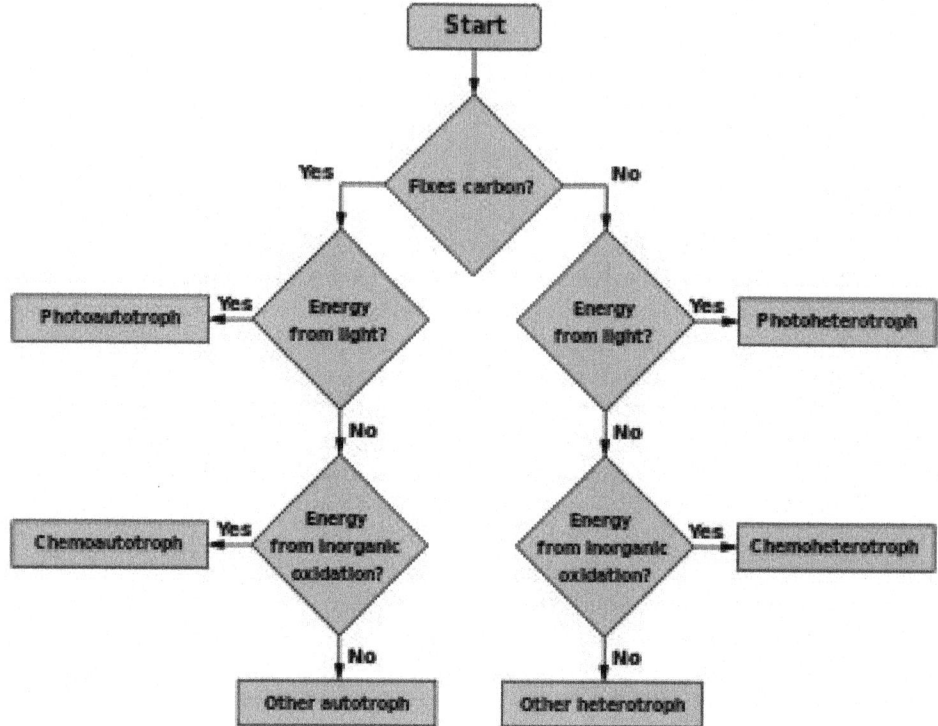

Fig. : Flow chart to determine the metabolic characteristics of micro-organisms.

All microbial metabolisms can be arranged according to three principles :

1. How the organism obtains carbon for synthesising cell mass :
   - *Autotrophic :* Carbon is obtained from carbon dioxide (CO_2)
   - *Heterotrophic :* Carbon is obtained from organic compounds
   - *Mixotrophic :* Carbon is obtained from both organic compounds and by fixing carbon dioxide.

2. How the organism obtains reducing equivalents used either in energy conservation or in bio-synthetic reactions :
   - *Lithotrophic :* Reducing equivalents are obtained from inorganic compounds
   - *Organotrophic :* Reducing equivalents are obtained from organic compounds.

3. How the organism obtains energy for living and growing :
   - *Chemotrophic :* Energy is obtained from external chemical compounds
   - *Phototrophic :* Energy is obtained from light.

In practice, these terms are almost freely combined. Typical examples are as follows :

- **Chemolithoautotrophs** obtain energy from the oxidation of inorganic compounds and carbon from the fixation of carbon dioxide. Examples: Nitrifying bacteria, Sulfur-oxidizing bacteria, Iron-oxidizing bacteria, Knallgas-bacteria.

- **Photolithoautotrophs** obtain energy from light and carbon from the fixation of carbon dioxide, using reducing equivalents from inorganic compounds. Examples : Cyanobacteria (water (H_2O) as reducing equivalent donor), Chlorobiaceae, Chromatiaceae (hydrogen sulfide (H_2S) as reducing equivalent donor), *Chloroflexus* (hydrogen (H_2) as reducing equivalent donor).

- **Chemolithoheterotrophs** obtain energy from the oxidation of inorganic compounds, but cannot fix carbon dioxide (CO_2). Examples : some *Thiobacilus*, some *Beggiatoa*, some *Nitrobacter* spp., *Wolinella* (with H_2 as reducing equivalent donor), some Knallgas-bacteria, some sulfate-reducing bacteria.

- **chemo-organoheterotrophs** obtain energy, carbon, and reducing equivalents for bio-synthetic reactions from organic compounds. Examples : most bacteria, e. g. *Escherichia coli*, *Bacillus* spp., *Actinobacteria.*

- **photo-organoheterotrophs** obtain energy from light, carbon and reducing equivalents for bio-synthetic reactions from organic compounds. Some species are strictly heterotrophic, many others can also fix carbon dioxide and are mixotrophic. Examples : *Rhodobacter*, *Rhodopseudomonas*, *Rhodospirillum*, *Rhodomicrobium*, *Rhodocyclus*, *Heliobacterium*, *Chloroflexus* (alternatively to photolithoautotrophy with hydrogen)

Heterotrophic Microbial Metabolism

Most microbes are heterotrophic (more precisely chemo-organoheterotrophic), using organic compounds as both carbon and energy sources. Heterotrophic microbes live off of nutrients that they scavenge from living hosts (as commensals or parasites) or find in dead organic matter of all kind (saprophages). Microbial metabolism is the main contribution for the bodily decay of all organisms after death. Many eukaryotic micro-organisms are heterotrophic by predation or parasitism, properties also found in some bacteria such as *Bdellovibrio* (an intra-cellular parasite of other bacteria, causing death of its victims) and Myxobacteria such as *Myxococcus* (predators of other bacteria which are killed and lysed by co-operating swarms of many single cells of Myxobacteria). Most pathogenic bacteria can be viewed as heterotrophic parasites of humans or the other eukaryotic species they affect. Heterotrophic microbes are extremely abundant in nature and are responsible for the breakdown of large organic polymers such as cellulose, chitin or lignin which are generally indigestible to larger animals. Generally, the breakdown of large polymers to carbon dioxide (mineralization) requires several different organisms, with one breaking down the polymer into its constituent monomers, one able to use the monomers and excreting simpler waste compounds as by-products, and one able to use the excreted wastes. There are many variations on this theme, as different organisms are able to degrade different polymers and secrete different waste products. Some organisms are even able to degrade more recalcitrant compounds such as petroleum compounds or pesticides, making them useful in bio-remediation.

Bio-chemically, prokaryotic heterotrophic metabolism is much more versatile than that of eukaryotic organisms, although many prokaryotes share the most basic metabolic models with eukaryotes, e. g. using glycolysis (also called EMP pathway) for sugar metabolism and the citric acid cycle to degrade acetate, producing energy in the form of ATP and reducing power in the form of NADH or quinols. These basic pathways are well conserved because they are also involved in bio-synthesis of many conserved building blocks needed for cell growth (sometimes in reverse direction). However, many bacteria and archaea utilize alternative metabolic pathways other than glycolysis and the citric acid cycle. A well-studied example is sugar metabolism *via* the keto-deoxy-phosphogluconate pathway (also called ED pathway) in *Pseudomonas*. Moreover, there is a third alternative sugar-catabolic pathway used by some bacteria, the pentose phosphate pathway. The metabolic diversity and ability of prokaryotes to use a large variety of organic compounds arises from the much deeper evolutionary history and diversity of prokaryotes, as compared to eukaryotes. It is also noteworthy that the mitochondrion, the small membrane-bound intra-cellular organelle that is the site of eukaryotic energy metabolism, arose from the endosymbiosis of a bacterium related to obligate intra-cellular *Rickettsia*, and also to plant-associated *Rhizobium* or *Agrobacterium*. Therefore it is not surprising that all mitrochondriate eukaryotes share metabolic properties with these Proteobacteria. Most microbes respire (use an electron transport chain), although oxygen is not the only terminal electron acceptor that may be used. As discussed below, the use of terminal electron acceptors other than oxygen has important biogeochemical consequences.

Fermentation

Fermentation is a specific type of heterotrophic metabolism that uses organic carbon instead of oxygen as a terminal electron acceptor. This means that these organisms do not use an electron transport chain to oxidize NADH to NAD+ and therefore must have an alternative method of using this reducing power and maintaining a supply of NAD+ for the proper functioning of normal metabolic pathways (*e.g.* glycolysis). As oxygen is not required, fermentative organisms are anaerobic. Many organisms can use fermentation under anaerobic conditions and aerobic respiration when oxygen is present. These organisms are facultative anaerobes. To avoid the over-production of NADH, obligately fermentative organisms usually do not have a complete citric acid cycle. Instead of using an ATP synthase as in respiration, ATP in fermentative organisms is produced by substrate-level phosphorylation where a phosphate group is transferred from a high-energy organic compound to ADP to form ATP. As a result of the need to produce high energy phosphate-containing organic compounds (generally in the form of Coenzyme A-esters) fermentative organisms use NADH and other cofactors to produce many different reduced metabolic by-products, often including hydrogen gas (H2). These reduced organic compounds are generally small organic acids and alcohols derived from pyruvate, the end product of glycolysis. Examples include ethanol, acetate, lactate, and butyrate. Fermentative organisms are very important industrially and are used to make many different types of food products. The different metabolic end products produced by each specific bacterial species are responsible for the different tastes and properties of each food.

Not all fermentative organisms use substrate-level phosphorylation. Instead, some organisms are able to couple the oxidation of low-energy organic compounds directly to the formation of a proton (or sodium) motive force and therefore, ATP synthesis. Examples of these unusual forms of fermentation include succinate fermentation by *Propionigenium modestum* and oxalate fermentation by *Oxalobacter formigenes*. These reactions are extremely low-energy yielding. Humans and other higher animals also use fermentation to produce lactate from excess NADH, although this is not the major form of metabolism as it is in fermentative micro-organisms.

Special Metabolic Properties

Methylotrophy

Methylotrophy refers to the ability of an organism to use C1-compounds as energy sources. These compounds include methanol, methyl amines, formaldehyde, and formate. Several other less common substrates may also be used for metabolism, all of which lack carbon-carbon bonds. Examples of methylotrophs include the bacteria *Methylomonas* and *Methylobacter*. Methanotrophs are a specific type of methylotroph that are also able to use methane (CH_4) as a carbon source by oxidizing it sequentially to methanol (CH_3OH), formaldehyde (CH_2O), formate ($HCOO-$), and carbon dioxide CO_2 initially using the enzyme methane monooxygenase. As oxygen is required for this process, all (conventional) methanotrophs are obligate aerobes. Reducing power in the form of quinones and NADH is produced during these oxidations to produce a proton motive force and therefore ATP generation. Methylotrophs and methanotrophs are not considered as autotrophic, because they are able to incorporate some of the oxidized methane (or other metabolites) into cellular carbon before it is completely oxidized to CO_2 (at the level of formaldehyde), using either the serine pathway (*Methylosinus, Methylocystis*) or the ribulose monophosphate pathway (*Methylococcus*), depending on the species of methylotroph.

In addition to aerobic methylotrophy, methane can also be oxidized anaerobically. This occurs by a consortium of sulfate-reducing bacteria and relatives of methanogenicArchaea working syntrophically. Little is currently known about the bio-chemistry and ecology of this process.

Methanogenesis is the biological production of methane. It is carried out by methanogens, strictly anaerobic Archaea such as *Methanococcus, Methanocaldococcus, Methanobacterium, Methanothermus, Methanosarcina, Methanosaeta* and *Metha-*

nopyrus. The bio-chemistry of methanogenesis is unique in nature in its use of a number of unusual cofactors to sequentially reduce methanogenic substrates to methane, such as coenzyme M and methanofuran. These cofactors are responsible (among other things) for the establishment of a proton gradient across the outer membrane thereby driving ATP synthesis. Several types of methanogenesis occur, differing in the starting compounds oxidized. Some methanogens reduce carbon dioxide (CO_2) to methane (CH_4) using electrons (most often) from hydrogen gas (H_2) chemolithoautotrophically. These methanogens can often be found in environments containing fermentative organisms. The tight association of methanogens and fermentative bacteria can be considered to be syntrophic because the methanogens, which rely on the fermentors for hydrogen, relieve feedback inhibition of the fermentors by the build-up of excess hydrogen that would otherwise inhibit their growth. This type of syntrophic relationship is specifically known as interspecies hydrogen transfer. A second group of methanogens use methanol (CH_3OH) as a substrate for methanogenesis. These are chemo-organotrophic, but still autotrophic in using CO_2 as only carbon source. The bio-chemistry of this process is quite different from that of the carbon dioxide-reducing methanogens. Lastly, a third group of methanogens produce both methane and carbon dioxide from acetate (CH_3COO-) with the acetate being split between the two carbons. These acetate-cleaving organisms are the only chemoorganoheterotrophic methanogens. All autotrophic methanogens use a variation of the acetyl-CoA pathway to fix CO_2 and obtain cellular carbon.

Syntrophy

Syntrophy, in the context of microbial metabolism, refers to the pairing of multiple species to achieve a chemical reaction that, on its own, would be energetically unfavourable. The best studied example of this process is the oxidation of fermentative end products (such as acetate, ethanol and butyrate) by organisms such as *Syntrophomonas*. Alone, the oxidation of butyrate to acetate and hydrogen gas is energetically unfavourable. However, when a hydrogenotrophic (hydrogen-using) methanogen is present the use of the hydrogen gas will significantly lower the concentration of hydrogen (down to 10^{-5} atm) and thereby shift the equilibrium of the butyrate oxidation reaction under standard conditions ($\Delta G^{o\prime}$) to non-standard conditions ($\Delta G'$). Because the concentration of one product is lowered, the reaction is "pulled" towards the products and shifted towards net energetically favourable conditions (for butyrate oxidation : $\Delta G^{o\prime}$= +48.2 kJ/mol, but $\Delta G'$ =-8.9 kJ/mol at 10^{-5} atm hydrogen and even lower if also the initially produced acetate is further metabolized by methanogens). Conversely, the available free energy from methanogenesis is lowered from $\Delta G^{o\prime}$=-131 kJ/mol under standard conditions to $\Delta G'$ =-17 kJ/mol at 10^{-5} atm hydrogen. This is an example of intraspecies hydrogen transfer. In this way, low energy-yielding carbon sources can be used by a consortium of organisms to achieve further degradation and eventual mineralization of these compounds. These reactions help prevent the excess sequestration of carbon over geologic time scales, releasing it back to the biosphere in usable forms such as methane and CO_2.

Anaerobic Respiration

While aerobic organisms during respiration use oxygen as a terminal electron acceptor, anaerobic organisms use other electron acceptors. These inorganic compounds have a lower reduction potential than oxygen, meaning that respiration is less efficient in these organisms and leads to slower growth rates than aerobes. Many facultative anaerobes can use either oxygen or alternative terminal electron acceptors for respiration depending on the environmental conditions.

Most respiring anaerobes are heterotrophs, although some do live autotrophically. All of the processes described below are dissimilative, meaning that they are used during energy production and not to provide nutrients for the cell (assimilative). Assimilative pathways for many forms of anaerobic respiration are also known.

Denitrification—Nitrate as Electron Acceptor

Denitrification is the utilization of nitrate (NO_{-3}) as a terminal electron acceptor. It is a widespread process that is used by many members of the Proteobacteria. Many facultative anaerobes use denitrification because nitrate, like oxygen, has a high reduction potential. Many denitrifying bacteria can also use ferric iron (Fe_{3+}) and some organic electron acceptors. Denitrification involves the stepwise reduction of nitrate to nitrite (NO_{-2}), nitric oxide (NO), nitrous oxide (N_2O), and dinitrogen (N_2) by the enzymes nitrate reductase, nitrite reductase, nitric oxide reductase, and nitrous oxide reductase, respectively. Protons are transported across the membrane by the initial NADH reductase, quinones, and nitrous oxide reductase to produce the electro-chemical gradient critical for respiration. Some organisms (*e.g. E. coli*) only produce nitrate reductase and therefore can accomplish only the first reduction leading to the accumulation of nitrite. Others (*e.g. Paracoccus denitrificans* or *Pseudomonas stutzeri*) reduce nitrate completely. Complete denitrification is an environmentally significant process because some intermediates of denitrification (nitric oxide and nitrous oxide) are important greenhouse gases that react with sunlight and ozone to produce nitric acid, a component of acid rain. Denitrification is also important in biological wastewater treatment where it is used to reduce the amount of nitrogen released into the environment thereby reducing eutrophication.

Sulfate Reduction—Sulfate as Electron Acceptor

Sulfate reduction is a relatively energetically poor process used by many Gram negative bacteria found within the δ-Proteobacteria, Gram-positive organisms relating to *Desulfotomaculum* or the archaeon *Archaeoglobus*. Hydrogen sulfide (H_2S) is produced as a metabolic end product. For sulfate reduction electron donors and energy are needed.

Electron Donors

Many sulfate reducers are organotrophic, using carbon compounds such as lactate and pyruvate (among many others) as electron donors, while others are lithotrophic, using hydrogen gas (H_2) as an electron donor. Some unusual autotrophic sulfate-reducing bacteria (*e.g. Desulfotignum phosphitoxidans*) can use phosphite (HPO_{-3}) as an electron donor whereas others (*e.g. Desulfovibrio sulfodismutans, Desulfocapsa thiozymogenes, Desulfocapsa sulfoexigens*) are capable of sulfur disproportionation (splitting one compound into two different compounds, in this case an electron donor and an electron acceptor) using elemental sulfur (S0), sulfite (SO_{2-3}), and thiosulfate (S_2O_{2-3}) to produce both hydrogen sulfide (H_2S) and sulfate (SO_{2-4}).

Energy for Reduction

All sulfate-reducing organisms are strict anaerobes. Because sulfate is energetically stable, before it can be metabolized it must first be activated by adenylation to form APS (adenosine 5'-phosphosulfate) thereby consuming ATP. The APS is then reduced by the enzyme APS reductase to form sulfite (SO_{2-3}) and AMP. In organisms that use carbon compounds as electron donors, the ATP consumed is accounted for by fermentation of the carbon substrate. The hydrogen produced during fermentation is actually what drives respiration during sulfate reduction.

Acetogenesis—Carbon Dioxide as Electron Acceptor

Acetogenesis is a type of microbial metabolism that uses hydrogen (H_2) as an electron donor and carbon dioxide (CO_2) as an electron acceptor to produce acetate, the same electron donors and acceptors used in methanogenesis. Bacteria that can autotrophically synthesize acetate are called homoacetogens. Carbon dioxide reduction in all homoacetogens occurs by the acetyl-CoA pathway. This pathway is also used for carbon fixation by autotrophic sulfate-reducing bacteria and hydrogenotrophic methanogens. Often homoacetogens can also be fermentative, using the hydrogen and carbon dioxide produced as a result of fermentation to produce acetate, which is secreted as an end product.

Other Inorganic Electron Acceptors

Ferric iron (Fe_{3+}) is a widespread anaerobic terminal electron acceptor both for autotrophic and heterotrophic organisms. Electron flow in these organisms is similar to those in electron transport, ending in oxygen or nitrate, except that in ferric iron-reducing organisms the final enzyme in this system is a ferric iron reductase. Model organisms include *Shewanella putrefaciens* and *Geobacter metallireducens*. Since some ferric iron-reducing bacteria (*e.g. G. metallireducens*) can use toxic hydrocarbons such as toluene as a carbon source, there is significant interest in using these organisms as bio-remediation agents in ferric iron-rich contaminated aquifers.

Although ferric iron is the most prevalent inorganic electron acceptor, a number of organisms (including the iron-reducing bacteria mentioned above) can use other inorganic ions in anaerobic respiration. While these processes may often be less significant ecologically, they are of considerable interest for bio-remediation, especially when heavy metals or radionuclides are used as electron acceptors. Examples include :

*   Manganic ion (Mn4+) reduction to manganous ion (Mn2+)

- Selenate (SeO_{2-4}) reduction to selenite (SeO_{2-3}) and selenite reduction to inorganic selenium (Se0)
- Arsenate (AsO_{3-4}) reduction to arsenite (AsO_{3-3})
- Uranyl ion ion (UO_{2+2}) reduction to uranium dioxide (UO_2)

Organic Terminal Electron Acceptors

A number of organisms, instead of using inorganic compounds as terminal electron acceptors, are able to use organic compounds to accept electrons from respiration. Examples include :

- Fumarate reduction to succinate
- Trimethylamine N-oxide (TMAO) reduction to trimethylamine (TMA)
- Dimethyl sulfoxide (DMSO) reduction to Dimethyl sulfide (DMS)
- Reductive dechlorination.

TMAO is a chemical commonly produced by fish, and when reduced to TMA produces a strong odour. DMSO is a common marine and freshwater chemical which is also odiferous when reduced to DMS. Reductive dechlorination is the process by which chlorinated organic compounds are reduced to form their non-chlorinated endproducts. As chlorinated organic compounds are often important (and difficult to degrade) environmental pollutants, reductive dechlorination is an important process in bio-remediation.

Chemolithotrophy

Chemolithotrophy is a type of metabolism where energy is obtained from the oxidation of inorganic compounds. Most chemolithotrophic organisms are also autotrophic. There are two major objectives to chemolithotrophy : the generation of energy (ATP) and the generation of reducing power (NADH).

Hydrogen Oxidation

Many organisms are capable of using hydrogen (H_2) as a source of energy. While several mechanisms of anaerobic hydrogen oxidation have been mentioned previously (*e.g.* sulfate reducing-and acetogenic bacteria), hydrogen can also be used as an energy source aerobically. In these organisms, hydrogen is oxidized by a membrane-bound hydrogenase causing proton pumping *via* electron transfer to various quinones and cytochromes. In many organisms, a second cytoplasmic hydrogenase is used to generate reducing power in the form of NADH, which is subsequently used to fix carbon dioxide *via* the Calvin cycle. Hydrogen-oxidizing organisms, such as *Cupriavidus necator* (formerly *Ralstonia eutropha*), often inhabit oxic-anoxic interfaces in nature to take advantage of the hydrogen produced by anaerobic fermentative organisms while still maintaining a supply of oxygen.

Sulfur Oxidation

Sulfur oxidation involves the oxidation of reduced sulfur compounds (such as sulfide H_2S), inorganic sulfur (S0), and thiosulfate (S_2O_{2-3}) to form sulfuric acid (H_2SO_4). A classic example of a sulfur-oxidizing bacterium is *Beggiatoa*, a microbe originally described by Sergei Winogradsky, one of the founders of environmental microbiology. Another example is *Paracoccus*. Generally, the oxidation of sulfide occurs in stages, with inorganic sulfur being stored either inside or outside of the cell until needed. This two step process occurs because energetically sulfide is a better electron donor than inorganic sulfur or thiosulfate, allowing for a greater number of protons to be translocated across the membrane. Sulfur-oxidizing organisms generate reducing power for carbon dioxide fixation *via* the Calvin cycle using reverse electron flow, an energy-requiring process that pushes the electrons against their thermodynamic gradient to produce NADH. Bio-chemically, reduced sulfur compounds are converted to sulfite (SO_{2-3}) and subsequently converted to sulfate (SO_{2-4}) by the enzyme sulfite oxidase. Some organisms, however, accomplish the same oxidation using a reversal of the APS reductase system used by sulfate-reducing bacteria. In all cases the energy liberated is transferred to the electron transport chain for ATP and NADH production. In addition to aerobic sulfur oxidation, some organisms (*e.g. Thiobacillus denitrificans*) use nitrate (NO_{-3}) as a terminal electron acceptor and therefore grow anaerobically.

Ferrous Iron (Fe2+) Oxidation

Ferrous iron is a soluble form of iron that is stable at extremely low pHs or under anaerobic conditions. Under aerobic, moderate pH conditions ferrous iron is oxidized spontaneously to the ferric (Fe3+) form and is hydrolyzed abiotically to insoluble ferric hydroxide ($Fe(OH)_3$). There are three distinct types of ferrous iron-oxidizing microbes. The first are acidophiles, such as the bacteria *Acidithiobacillus ferrooxidans* and *Leptospirillum ferrooxidans*, as well as the archaeon *Ferroplasma*. These microbes oxidize iron in environments that have a very low pH and are important in acid mine drainage. The second type of microbes oxidize ferrous iron at cirum-neutral pH. These micro-organisms (for example *Gallionella ferruginea*, *Leptothrix ochracea*, or *Mariprofundus ferrooxydans*) live at the oxic-anoxic interfaces and are micro-aerophiles. The third type of iron-oxidizing microbes are anaerobic photo-synthetic bacteria such as Rhodopseudomonas, which use ferrous iron to produce NADH for autotrophic carbon dioxide fixation. Bio-chemically, aerobic iron oxidation is a very energetically poor process which therefore requires large amounts of iron to be oxidized by the enzyme rusticyanin to facilitate the formation of proton motive force. Like sulfur oxidation, reverse electron flow must be used to form the NADH used for carbon dioxide fixation *via* the Calvin cycle.

Nitrification

Nitrification is the process by which ammonia (NH_3) is converted to nitrate (NO_{-3}). Nitrification is actually the net result of two distinct processes : oxidation of ammonia to nitrite (NO_{-2}) by nitrosifying bacteria (*e.g. Nitrosomonas*) and oxidation of nitrite to nitrate by the nitrite-oxidizing bacteria (*e.g. Nitrobacter*). Both of these processes are extremely energetically poor leading to very slow growth rates for both types of organisms. Bio-chemically, ammonia oxidation occurs by the step-wise oxidation of ammonia to hydroxylamine (NH_2OH) by the enzyme ammonia mono-oxygenase in the cytoplasm, followed by the oxidation of hydroxylamine to nitrite by the enzyme hydroxylamine oxidoreductase in the periplasm.

Electron and proton cycling are very complex but as a net result only one proton is translocated across the membrane per molecule of ammonia oxidized. Nitrite reduction is much simpler, with nitrite being oxidized by the enzyme nitrite oxidoreductase coupled to proton translocation by a very short electron transport chain, again leading to very low growth rates for these organisms. Oxygen is required in both ammonia and nitrite oxidation, meaning that both nitrosifying and nitrite-oxidizing bacteria are aerobes. As in sulfur and iron oxidation, NADH for carbon dioxide fixation using the Calvin cycle is generated by reverse electron flow, thereby placing a further metabolic burden on an already energy-poor process.

Anammox

Anammox stands for anaerobic ammonia oxidation and the organisms responsible were relatively recently discovered, in the late 1990s. This form of metabolism occurs in members of the Planctomycetes (*e.g.* Candidatus *Brocadia anammoxidans*) and involves the coupling of ammonia oxidation to nitrite reduction. As oxygen is not required for this process these organisms are strict anaerobes. Amazingly, hydrazine (N2H4 – rocket fuel) is produced as an intermediate during anammox metabolism. To deal with the high toxicity of hydrazine, anammox bacteria contain a hydrazine-containing intra-cellular organelle called the anammoxasome, surrounded by highly compact (and unusual) ladderane lipid membrane. These lipids are unique in nature, as is the use of hydrazine as a metabolic intermediate. Anammox organisms are autotrophs although the mechanism for carbon dioxide fixation is unclear. Because of this property, these organisms could be used in industry to remove nitrogen in wastewater treatment processes. Anammox has also been shown have widespread occurrence in anaerobic aquatic systems and has been speculated to account for approximately 50% of nitrogen gas production in the ocean.

Phototrophy

Many microbes (phototrophs) are capable of using light as a source of energy to produce ATP and organic compounds such as carbohydrates, lipids, and proteins. Of these, algae are particularly significant because they are oxygenic, using water as an electron donor for electron transfer during photo-synthesis. Phototrophic bacteria are found in the phyla Cyanobacteria, Chlorobi, Proteobacteria, Chloroflexi, and Firmicutes. Along with plants these microbes are responsible for all biological generation of oxygen gas on Earth. Because chloroplasts were derived from a lineage of the Cyanobacteria, the general principles of metabolism in these endosymbionts can also be applied to chloroplasts. In addition to oxygenic photo-synthesis, many bacteria can also photo-synthesize anaerobically, typically using sulfide (H_2S) as an electron donor to produce sulfate. Inorganic sulfur (S0), thiosulfate (S_2O_{2-3}) and ferrous iron (Fe_{2+}) can also be used by some organisms. Phylogenetically, all oxygenic photo-synthetic bacteria are Cyanobacteria, while anoxygenic photo-synthetic bacteria belong to the purple bacteria (Proteobacteria), Green sulfur bacteria (*e.g.* Chlorobium), Green non-sulfur bacteria (*e.g.* Chloroflexus), or the heliobacteria (Low%G+C Gram positives). In addition to these organisms, some microbes (*e.g.* the Archaeon *Halobacterium* or the bacterium *Roseobacter*, among others) can utilize light to produce energy using the enzyme bacteriorhodopsin, a light-driven proton pump. However, there are no known Archaea that carry out photo-synthesis.

As befits the large diversity of photo-synthetic bacteria, there are many different mechanisms by which light is converted into energy for metabolism. All photo-synthetic organisms locate their photo-synthetic reaction centers within a membrane, which may be invaginations of the cytoplasmic membrane (Proteobacteria), thylakoid membranes (Cyanobacteria), specialized antenna structures called chlorosomes (Green sulfur and non-sulfur bacteria), or the cytoplasmic membrane itself (heliobacteria). Different photo-synthetic bacteria also contain different photo-synthetic pigments, such as chlorophylls and carotenoids, allowing them to take advantage of different portions of the electromagnetic spectrum and thereby inhabit different niches. Some groups of organisms contain more specialized light-harvesting structures (*e.g.* phycobilisomes in Cyanobacteria and chlorosomes in Green sulfur and non-sulfur bacteria), allowing for increased efficiency in light utilization.

Bio-chemically, anoxygenic photo-synthesis is very different from oxygenic photo-synthesis. Cyanobacteria (and by extension, chloroplasts) use the Z scheme of electron flow in which electrons eventually are used to form NADH. Two different reaction centers (photo-systems) are used and proton motive force is generated both by using cyclic electron flow and the quinone pool. In anoxygenic photo-synthetic bacteria, electron flow is cyclic, with all electrons used in photo-synthesis eventually being transferred back to the single reaction center. A proton motive force is generated using only the quinone pool. In heliobacteria, Green sulfur, and Green non-sulfur bacteria, NADH is formed using the protein ferredoxin, an energetically favourable reaction. In purple bacteria, NADH is formed by reverse electron flow due to the lower chemical potential of this reaction center. In all cases, however, a proton motive force is generated and used to drive ATP production *via* an ATPase.

Most photo-synthetic microbes are autotrophic, fixing carbon dioxide *via* the Calvin cycle. Some photo-synthetic bacteria (*e.g.* Chloroflexus) are photo-heterotrophs, meaning that they use organic carbon compounds as a carbon source for growth. Some photo-synthetic organisms also fix nitrogen.

Nitrogen Fixation

Nitrogen is an element required for growth by all biological systems. While extremely common (80% by volume) in the atmosphere, dinitrogen gas (N_2) is generally biologically inaccessible due to its high activation energy. Throughout all of nature, only specialized bacteria and Archaea are capable of nitrogen fixation, converting dinitrogen gas into ammonia (NH_3), which is easily assimilated by all organisms. These prokaryotes, therefore, are very important ecologically and are often essential for the survival of entire ecosystems. This is especially true in the ocean, where nitrogen-fixing cyanobacteria are often the only sources of fixed nitrogen, and in soils, where specialized symbioses exist between legumes and their nitrogen-fixing partners to provide the nitrogen needed by these plants for growth.

Nitrogen fixation can be found distributed throughout nearly all bacterial lineages and physiological classes but is not a universal property. Because the enzyme nitrogenase, responsible for nitrogen fixation, is very sensitive to oxygen which will inhibit it irreversibly, all nitrogen-fixing organisms must possess some mechanism to keep the concentration of oxygen low. Examples include :

- Heterocyst formation (cyanobacteria *e.g. Anabaena*) where one cell does not photo-synthesize but instead fixes nitrogen for its neighbors which in turn provide it with energy
- Root nodule symbioses (*e.g. Rhizobium*) with plants that supply oxygen to the bacteria bound to molecules of leghaemoglobin
- Anaerobic lifestyle (*e.g. Clostridium pasteurianum*)
- Very fast metabolism (*e.g. Azotobacter vinelandii*)

The production and activity of nitrogenases is very highly regulated, both because nitrogen fixation is an extremely energetically expensive process (16–24 ATP are used per N_2 fixed) and due to the extreme sensitivity of the nitrogenase to oxygen.

Electrophoretic Deposition

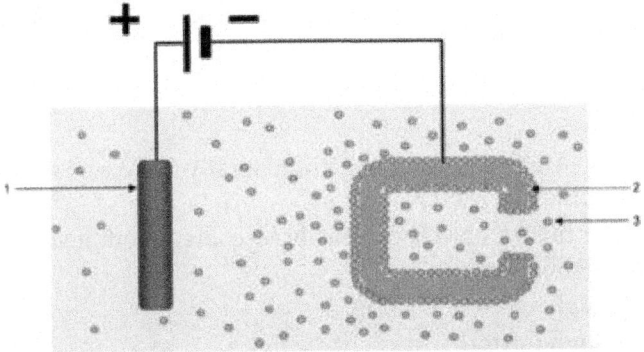

Fig. : Electrophoretic Deposition Process.

Electrophoretic deposition (EPD), is a term for a broad range of industrial processes which includes electrocoating, e-coating, cathodic electrodeposition, anodic electrodeposition, and electrophoretic coating, or electrophoretic painting. A characteristic feature of this process is that colloidal particles suspended in a liquid medium migrate under the influence of an electric field (electrophoresis) and are deposited onto an electrode. All colloidal particles that can be used to form stable suspensions and that can carry a charge can be used in electrophoretic deposition. This includes materials such as polymers, pigments, dyes, ceramics and metals.

The process is useful for applying materials to any electrically conductive surface. The materials which are being deposited are the major determining factor in the actual processing conditions and equipment which may be used.

Due to the wide utilization of electrophoretic painting processes in many industries, aqueous EPD is the most common commercially used EPD process. However, non-aqueous electrophoretic deposition applications are known. Applications of non-aqueous EPD are currently being explored for use in the fabrication of electronic components and the production of ceramic coatings. Non-aqueous processes have the advantage of avoiding the electrolysis of water and the oxygen evolution which accompanies electrolysis.

Uses of EPD

This process is industrially used for applying coatings to metal fabricated products. It has been widely used to coat automobile bodies and parts, tractors and heavy equipment, electrical switch gear, appliances, metal furniture, beverage containers, fasteners, and many other industrial products.

EPD processes are often applied for the fabrication of supported titanium dioxide (TiO_2) photocatalysts for water purification applications, using precursor powders which can be immobilised using EPD methods onto various support materials. Thick films produced this way allow cheaper and more rapid synthesis relative to sol-gel thin-films, along with higher levels of photocatalyst surface area.

In the fabrication of Solid Oxide Fuel Cells EPD techniques are widely employed for the fabrication of porous ZrO_2 anodes from powder precursors onto conductive substrates.

EPD processed have a number of advantages which have made such methods widely used :

1. The process applies coatings which generally have a very uniform coating thickness without porosity.
2. Complex fabricated objects can easily be coated, both inside cavities as well as on the outside surfaces.
3. Relatively high speed of coating.
4. Relatively high purity.
5. Applicability to wide range of materials (metals, ceramics, polymers, etc.)
6. Easy control of the coating composition.
7. The process is normally automated and requires less human labour than other coating processes.
8. Highly efficient utilization of the coating materials result in lower costs relative to other processes.
9. The aqueous process which is commonly used has less risk of fire relative to the solvent-borne coatings that they have replaced.
10. Modern electrophoretic paint products are significantly more environmentally friendly than many other painting technologies.

Thick, complex ceramic pieces have been made in several research laboratories. Furthermore, EPD has been used to produce customized micro-structures, such as functional gradients and laminates, through suspension control during processing.

History of Electrophoretic Painting

The first patent for the use of electrophoretic painting was awarded in 1917 to Davey and General Electric. Since the 1920s, the process has been used for the deposition of rubber latex. In the 1930s the first patents were issued which described base neutralized, water dispersible resins specifically designed for EPD.

Electrophoretic coating began to take its current shape in the late 1950s, when Dr. George E. F. Brewer and the Ford Motor Company team began working on developing the process for the coating of automobiles. The first commercial anodic automotive system began operations in 1963.

The first patent for a cathodic EPD product was issued in 1965 and assigned to BASF AG. PPG Industries, Inc. was the first to introduce commercially cathodic EPD in 1970. The first cathodic EPD use in the automotive industry was in 1975. Today, around 70% of the volume of EPD in use in the world today is the cathodic EPD type, largely due to the high usage of the technology in the automotive industry. It is probably the best system ever developed and has resulted in great extension of body life in the automotive industry

There are thousands of patents which have been issued relating to various EPD compositions, EPD processes, and articles coated with EPD. Although patents have been issued by various government patent offices, virtually all of the significant developments can be followed by reviewing the patents issued by the U.S. Patent and Trademark Office.

Process of Electrophoretic Painting

The overall industrial process of electrophoretic deposition consists of several sub-processes :

1. The object to be coated needs to be prepared for coating. This normally consists of some kind of cleaning process and may include the application of a conversion coating, typically an inorganic phosphate coating.

2. The coating process itself. This normally involves submerging the part into a container or vessel which holds the coating bath or solution and applying direct current electricity through the EPD bath using electrodes. Typically voltages of 25-400 volts DC are used in electrocoating or electrophoretic painting applications. The object to be coated is one of the electrodes, and a set of "counter-electrodes" are used to complete the circuit.

3. After deposition, the object is normally rinsed to remove the undeposited bath. The rinsing process may utilize an ultra-filter to dewater a portion of the bath from the coating vessel to be used as rinse material. If an ultra-filter is used, all of the rinsed off materials can be returned to the coating vessel, allowing

for high utilization efficiency of the coating materials, as well as reducing the amount of waste discharged into the environment.

4.　A baking or curing process is normally used following the rinse. This will cross-link the polymer and allows the coating, which will be porous due to the evolution of gas during the deposition process, to flow out and become smooth and continuous.

During the EPD process itself, direct current is applied to a solution of polymers with ionizable groups or a colloidal suspension of polymers with ionizable groups which may also incorporate solid materials such a pigments and fillers. The ionizable groups incorporated into the polymer are formed by the reaction of an acid and a base to form a salt. The particular charge, positive or negative, which is imparted to the polymer depends on the chemical nature of the ionizable group. If the ionizable groups on the polymer are acids, the polymer will carry a negative charge when salted with a base. If the ionizable groups on the polymer are bases, the polymer will carry a positive charge when salted with an acid.

There are two types of EPD processes, anodic and cathodic. In the anodic process, negatively charged material is deposited on the positively charged electrode, or anode. In the cathodic process, positively charged material is deposited on the negatively charged electrode, or cathode.

When an electric field is applied, all of the charged species migrate by the process of electrophoresis towards the electrode with the opposite charge. There are several mechanisms by which material can be deposited on the electrode :

1.　Charge destruction and the resultant decrease in solubility.
2.　Concentration coagulation.
3.　Salting out.

The primary electro-chemical process which occurs during aqueous electro-deposition is the electrolysis of water. This can be shown by the following two half reactions which occur at the two electrodes :

Anode : $2H_2O ---> O_2(gas) + 4H(+) + 4e(-)$

Cathode : $4H_2O + 4e(-) ---> 4OH(-) + 2H_2(gas)$.

In anodic deposition, the material being deposited will have salts of an acid as the charge bearing group. These negatively charged anions react with the positively charged hydrogen ions (protons) which are being produced at the anode by the electrolysis of water to reform the original acid. The fully protonated acid carries no charge (charge destruction) and is less soluble in water, and may precipitate out of the water onto the anode.

The analogous situation occurs in cathodic deposition except that the material being deposited will have salts of a base as the charge bearing group. If the salt of the base has been formed by protonation of the base, the protonated base will react with the hydroxyl ions being formed by electrolysis of water to yield the neutral charged base (again charge destruction) and water. The uncharged polymer is less soluble in water than it was when was charged, and precipitation onto the cathode occurs.

Onium salts, which have been used in the cathodic process, are not protonated bases and do not deposit by the mechanism of charge destruction. These type of materials can be deposited on the cathode by concentration coagulation and salting out. As the colloidal particles reach the solid object to be coated, they become squeezed together, and the water in the interstices is forced out. As the individual micelles are squeezed, they collapse to form increasingly larger micelles. Colloidal stability is inversely proportional to the size of the micelle, so as the micelles get bigger, they become less and less stable until they precipitate from solution onto the object to be coated. As more and more charged groups are concentrated into a smaller volume, this increases the ionic strength of the medium, which also assists in precipitating the materials out of solution. Both of these processes are occurring simultaneously and both contribute to the deposition of material.

Factors Affecting Electrophoretic Painting

During the aqueous deposition process, gas is being formed at both electrodes. Hydrogen gas is being formed at the cathode, and oxygen gas at the anode. It should be noted that for a given amount of charge transfer, exactly twice as much hydrogen is generated compared to oxygen on a molecular basis.

This has some significant effects on the coating process. The most obvious is in the appearance of the deposited film prior to the baking process. The cathodic process results in considerably more gas being trapped within the film than the anodic process. Since the gas has a higher electrical resistance than either depositing film or the bath itself, the amount of gas has a significant effect on the current at a given applied voltage. This is why cathodic processes are often able to be operated at significantly higher voltages than the corresponding anodic processes.

The deposited coating has significantly higher resistance than the object which is being coated. As the deposited film precipitates, the resistance increases. The increase in resistance is proportional to the thickness of the deposited film, and thus, at a given voltage, the electric current decreases as the film gets thicker until it finally reaches a point where deposition has slowed or stopped occurring (self-limiting). Thus the applied voltage is the primary control for the amount of film applied.

The ability for the EPD coating to coat interior recesses of a part is called the "throwpower". In many applications, it is desirable to use coating materials with a high throwpower. The throwpower of a coating is dependent on a number of variables, but generally it can be stated that the higher the coating voltage, the further a given coating will "throw" into recesses. High throwpower electrophoretic paints typically use application voltages in excess of 300 volts DC.

The coating temperature is also an important variable affecting the EPD process. The coating temperature has an effect on the bath conductivity and deposited film conductivity, which increases as temperature increases. Temperature also has an effect on the viscosity of the deposited film, which in turn affects the ability of the deposited film to release the gas bubbles being formed.

The coalescence temperature of the coating system is also an important variable for the coating designer. It can be determined by plotting the film build of a given system versus coating temperature keeping the coating time and voltage application profile constant. At temperatures below the coalescence temperature, film growth behaviour and rupturing behaviour is quite different from the usual practice as a result of porous deposition.

The coating time also is an important variable in determining the film thickness, the quality of the deposited film, and the throwpower. Depending on the type of object being coated, coating times of several seconds up to several minutes may be appropriate.

The maximum voltage which can be utilized depends on the type of coating system and a number of other factors. As already stated, film thickness and throwpower are dependent on the application voltage. However, at excessively high voltages, a phenomenon called "rupture" can occur. The voltage where this phenomenon occurs is called the "rupture voltage". The result of rupture is a film that is usually very thick and porous. Normally this is not an acceptable film cosmetically or functionally. The causes and mechanisms for rupturing are not completely understood, however the following is known :

1. Commercially available anodic EPD coating chemistries typically exhibit rupturing at voltages significantly lower than their commercially available cathodic counterparts.

2. For a given EPD chemistry, the higher the bath conductivity, the lower the rupture voltage.

3. For a given EPD chemistry, the rupture voltages normally decrease as the temperature is increased (for temperatures above the coalescence temperature).

4. Additions to a given bath composition of organic solvents and plasticizers which reduce the deposited film's viscosity will often produce higher film thicknesses at a given voltage, but will generally also reduce the throwpower and the rupture voltage.

5. The type and preparation of the substrate (material used to make the object being coated) can also have a significant effect on rupturing phenomenon.

Types of EPD Chemistries for Electrophoretic Painting

There are two major categories of EPD chemistries : anodic and cathodic. Both continue to be used commercially, although the anodic process has been in use industrially for a longer period of time and is thus considered to be the older of the two processes. There are advantages and disadvantages for both types of processes, and different experts may have different perspectives on some of the pros and cons of each.

The major advantages that are normally touted for the anodic process are :

1. Lower costs compared to cathodic process.

2. Simpler and less complex control requirements.

3. Fewer problems with inhibition of cure of subsequent topcoating layers.

4. Less sensitivity to variations in substrate quality.

5. The substrate is not subjected to highly alkaline conditions, which may dissolve phosphate and other conversion coatings.

6. Certain metals, such as zinc, may become imbrittled from the hydrogen gas which is evolved at the cathode. The anodic process avoids this effect since oxygen is being generated at the anode.

The major advantages that are normally touted for the cathodic processes are :

1. Higher levels of corrosion protection are possible. (While many people believe that cathodic technologies have higher corrosion protection capability, other experts argue that this probably has more to do with the coating polymer and cross-linking chemistry rather than on which electrode the film is deposited.)

2. Higher throwpower can be designed into the product. (While this may be true with the currently commercially available technologies today, high throwpower anodic systems are known and have been used commercially in the past.)

3. Oxidation only occurs at the anode, and thus staining and other problems which may result from the oxidation of the electrode substrate itself is avoided in the cathodic process.

A significant and real difference which is not often mentioned is the fact that acid catalyzed cross-linking technologies are more appropriate to the anodic process. Such cross-linkers are widely used in all types of coating applications. These include such popular and relatively inexpensive cross-linkers such as melamine-formaldehyde, phenol-formaldehyde, urea-formaldehyde, and acrylamide-formaldehyde cross-linkers.

Melamine-formaldehyde type cross-linkers in particular are widely used in anodic electrocoatings. These types cross-linkers are relatively inexpensive and provide a wide range of cure and performance characteristics which allow the coating designer to tailor the product for the desired end use. Coatings formulated with this type of cross-linker can have acceptable UV light resistance. Many of them are relatively low viscosity materials and can act as a reactive plasticizer, replacing some of the organic solvent that otherwise might be necessary. The amount of free formaldehyde, as well as formaldehyde which may be released during the baking process is of concern as these are considered to be hazardous air pollutants.

The deposited film in cathodic systems is quite alkaline, and acid catalyzed cross-linking technologies have not been preferred in cathodic products in general, although there have been some exceptions. The most common type of cross-linking chemistry in use today with cathodic products are based on urethane and urea chemistries.

The aromatic polyurethane and urea type cross-linker is one of the significant reasons why many cathodic electrocoats show high levels of protection against corrosion. Of course, it is not the only reason, but if one compares electrocoating compositions with aromatic urethane cross-linkers to analogous systems containing aliphatic urethane cross-linkers, consistently systems with aromatic urethane cross-linkers perform significantly better. However, coatings containing aromatic urethane cross-linkers generally do not perform well in terms of UV light resistance. If the resulting coating contains aromatic urea cross-links, the UV resistance will be considerably worse than if only urethane cross-links can occur. A disadvantage of aromatic urethanes is that they can also cause yellowing of the coating itself as well as cause yellowing in subsequent topcoat layers. A significant undesired side reaction which occurs during the baking process produces aromatic polyamines. Urethane cross-linkers based on toluene diisocyanate (TDI) can be expected to produce toluene diamine as a side reaction, whereas those based on Methylene diphenyl diisocyanate produce diaminodiphenylmethane and higher order aromatic polyamines. The undesired aromatic polyamines can inhibit the cure of subsequent acid catalysed topcoat layers, and can cause delamination of the subsequent topcoat layers after exposure to sunlight. Although the industry has never acknowledged this problem, many of these undesired aromatic polyamines are known or suspected carcinogens.

Besides the two major categories of anodic and cathodic, EPD products can also be described by the base polymer chemistry which is utilized. The are several polymer types that have been used commercially. Many of the earlier anodic types were based on maleinized oils of various types, tall oil and linseed oil being two of the more common. Today, epoxy and the acrylic types predominate. The description and the generally touted advantages are as follows :

1. *Epoxy :* Although aliphatic epoxy materials have been used, the majority of EPD epoxy types are based on aromatic epoxy polymers, most commonly based on polymerization of diglycidal ethers of bis phenol A. The polymer backbone may be modified with other types of chemistries to achieve the desired performance characteristics. Generally, this type of chemistry is used in primer applications where the coating will receive a topcoat, particularly if the coated object needs to withstand sunlight. This chemistry generally does not have good resistance to UV light. However, this chemistry is often used where high corrosion resistance is required.

2. *Acrylic :* These polymers are based on free radical initiated polymers containing monomers based on acrylic acid and methacrylic acid and their many esters which are available. Such polymers often also include styrene as a monomer. Generally, this type of chemistry is utilized when UV resistance is desirable. These polymers also have the advantage of allowing a wider colour palette since the polymer is less prone to yellowing when compared to epoxies.

Non-aqueous Electrophoretic Deposition

In certain applications, such as the deposition of ceramic materials, voltages above 3-4V cannot be applied in aqueous EPD if it is necessary to avoid the electrolysis of water. However, higher application voltages may be desirable in order to achieve higher coating thicknesses or to increase the rate of deposition. In such applications, organic solvents are used instead of water as the liquid medium.

The organic solvents used are generally polar solvents such as alcohols and ketones. Ethanol, acetone, and methyl ethyl ketone are examples of solvents which have been reported as suitable candidates for use in electrophoretic deposition.

Electro-chemical Regeneration

The **electro-chemical regeneration** of activated carbon based adsorbents involves the removal of molecules adsorbed onto the surface of the adsorbent with the use of an electric current in an electro-chemical cell restoring the carbon's adsorptive capacity. Electro-chemical regeneration represents an alternative to thermal regeneration commonly used in waste water treatment applications. Common adsorbents include powdered activated carbon (PAC), granular activated carbon (GAC) and activated carbon fibre.

Regeneration for Adsorbent Re-use

In waste water treatment, the most commonly used adsorbent is granular activated carbon (GAC), often used as to treat both liquid and gas phase volatile organic compounds and organic pollutants. Activated carbon beds vary in lifetime depending on the concentration of the pollutant(s) being removed, their associated adsorption isotherms, inlet flow rates and required discharge consents. Life-times of these beds can range between hours and months. Activated carbon is often landfilled at the end of its useful life but sometimes it is possible to regenerate it restoring its adsorptive capacity allowing it to be re-used. Thermal regeneration is the most prolific regeneration technique but has drawbacks in terms of high energy and commercial costs and a significant carbon footprint. These drawbacks have encouraged research into alternative regeneration techniques such as electro-chemical regeneration.

Electro-chemically Regenerating Activated Carbons

Once the adsorptive capacity of the activated carbon bed has been exhausted by the adsorption of pollutant molecules, the carbon is transferred to an electro-chemical cell (to either the anode or the cathode) in which electro-chemical regeneration can occur.

Principles

There are several mechanisms by which passing a current through the electro-chemical cell can encourage pollutant desorption. Ions generated at the electrodes can change local pH conditions in the divided cell which affect the adsorption equilibrium and have been shown to promote desorption of organic pollutants such as phenols from the carbon surface. Other mechanisms include reactions between the ions generated and the adsorbed pollutants resulting in the formation of a species with a lower adsorptive affinity for activated carbon that subsequently desorb, or the oxidative destruction of the organics on the carbon surface. It is agreed that the main mechanisms are based on desorption induced regeneration as electro-chemical effects are confined to the surface of the porous carbons so cannot be responsible for bulk regeneration. The performance of different regeneration methods can be directly compared using the regeneration efficiency. This is defined as :

$$\frac{\text{adsorptive capacity be fore adsorption}}{\text{adsorptive capacity after adsorption and electrochemical regeneration}} \times 100$$

Cathodic Regeneration

The cathode is the reducing electrode and generates OH⁻ions which increases local pH conditions. An increase in pH can have the effect of promoting the desorption of pollutants into solution where they can migrate to the anode and undergo oxidation hence destruction. Studies on cathodic regeneration have shown regeneration efficiencies for adsorbed organic pollutants such as phenols of the order of 85% based on regeneration times of 4 hours with applied currents between 10-100 mA. However, due to mass transfer limitations between the cathode and anode, there is often residual pollutant left in the cathode unless large currents or long regeneration times are employed.

Anodic Regeneration

The anode is the oxidising electrode and as a result has a lower localised pH during electrolysis which also promotes desorption of some organic pollutants. Regeneration efficiencies of activated carbon in the anodic compartment are lower than that achievable in the cathodic compartment by between 5-20% for the same regeneration times and currents, however there is no observed residual organic due to the strong oxidising nature of the anode.

Repeated Adsorption-Regeneration

For the bulk of carbonaceous adsorbents regeneration efficiency decreases over subsequent cycles as a result of pore blockages and damage to adsorption sites by the applied current. Decreases in regeneration efficiency are typically a further 2% per cycle. Current leading edge research focuses on developing adsorbents able to regenerate 100% of their adsorptive capacity through electro-chemical regeneration.

Commercial Systems

Currently there are a very limited number of commercially available carbon based adsorption-electro-chemical regeneration systems. One system that does exist uses a carbon adsorbent called Nyex in a continuous adsorption-regeneration system that uses electro-chemical regeneration to adsorb and destroy organic pollutants.

Chapter 5

DIRECT ELECTROCHEMISTRY OF HEMOGLOBIN IMMOBILIZED ON A FUNCTIONALIZED MULTI-WALLED CARBON NANOTUBES AND GOLD NANOPARTICLES NANOCOMPLEX-MODIFIED GLASSY CARBON ELECTRODE

Jun Hong[1,*], Ying-Xue Zhao[1], Bao-Lin Xiao[1], Ali Akbar Moosavi-Movahedi[2,*], Hedayatollah Ghourchian[2] and Nader Sheibani[3]

[1] School of Life Sciences, Henan University, JinMing Road, Kaifeng 475000, China; E-Mails: 66yingxue@163.com (Y.-X.Z.); arixxl@163.com (B.-L.X.)
[2] Institute of Biochemistry and Biophysics, University of Tehran, Enquelab Avenue, P.O. Box 13145-1384, Tehran, Iran; E-Mail: hadi@ibb.ut.ac.ir
[3] Department of Ophthalmology and Visual Sciences, University of Wisconsin, 600 Highland Avenue, K6/456 CSC, Madison, WI 53792-4673, USA; E-Mail: nsheibanikar@wisc.edu

* Author to whom correspondence should be addressed; E-Mails: hongjun@henu.edu.cn (J.H.); moosavi@ibb.ut.ac.ir (A.A.M.-M.); Tel.: +86-137-8116-1597 (J.H.); Fax: +86-378-388-6258 (J.H.); Tel.: +98-21-640-3957 (A.A.M.-M.); Fax: +98-21-640-4680 (A.A.M.-M.).

ABSTRACT

Direct electron transfer of hemoglobin (Hb) was realized by immobilizing Hb on a carboxyl functionalized multi-walled carbon nanotubes (FMWCNTs) and gold nanoparticles (AuNPs) nanocomplex-modified glassy carbon electrode. The ultraviolet-visible absorption spectrometry (UV-Vis), transmission electron microscopy (TEM) and Fourier transform infrared (FTIR) methods were utilized for additional characterization of the AuNPs and FMWCNTs. The cyclic voltammogram of the modified electrode has a pair of well-defined quasi-reversible redox peaks with a formal potential of -0.270 ± 0.002 V (vs. Ag/AgCl) at a scan rate of 0.05 V/s. The heterogeneous electron transfer constant (ks) was evaluated to be 4.0 ± 0.2 s^{-1}. The average surface concentration of electro-active Hb on the surface of the modified glassy carbon electrode was calculated to be $6.8 \pm 0.3 \times 10^{-10}$ mol cm^{-2}. The cathodic peak current of the modified electrode increased linearly with increasing concentration of hydrogen peroxide (from 0.05 nM to 1 nM) with a detection limit of 0.05 ± 0.01 nM. The apparent Michaelis-Menten constant (K_m^{app}) was calculated to be 0.85 ± 0.1 nM. Thus, the modified electrode could be applied as a third generation biosensor with high sensitivity, long-term stability and low detection limit.

Keywords

Hemoglobin; direct electrochemistry; functionalized multi-walled carbon nanotubes; gold nanoparticles; nanocomplex

1. INTRODUCTION

Direct electrochemistry of redox proteins immobilized on different electrodes has recently attracted great attention. These methods can provide a suitable model for understanding the electron transfer mechanisms in biological systems and to establish a foundation for fabrication of electrochemical biosensors and devices [1–4]. Successful approaches have included cast films of redox proteins with different materials and membranes [5–23]. Achieving direct electron transfer between redox proteins or enzymes and electrodes simplifies these devices, which has a great significance in preparing the third generation of biosensors [24].

Hemoglobin (Hb) is an oxygen carrier and as a pro-oxidant takes part in complex redox processes in the blood. This protein consists of two alpha and two beta subunits with a molecular weight of about 67,000 and each subunit has one peptide chain and one protoheme [25–27]. Hb structure is similar to the peroxidases and can catalyze the reduction of hydrogen peroxide. [28,29]. Direct electron transfer of Hb immobilized on various electrode materials has become popular. In addition, the structure of Hb is known and can be used as a model to explain the relationship between the protein structure and function and to construct new functional biosensors without mediators [13,30–32].

Carbon nanotubes are widely used for the fabrication of electrochemical biosensors, because of their special structure and properties such as high surface area, which makes them a suitable material to transfer electrons [33–35]. Gold nanoparticles are one of the most stable metal nanoparticles, and have been widely applied in analytical chemistry and electrochemistry, because of their novel optical, electrical and catalytic properties and favorable biocompatibility [36,37]. Each type of nano-material has its own physical and chemical characteristics, which makes the design and preparation of biosensors based on only one type of nanomaterial laborious. Thus, the use of composites from several type of materials could be preferable [38]. Recently, several nano materials, including carbon nanotubes and nanoparticles of Au, Ag, TiO_2, Fe_3O_4 or MnO_2, have been applied in electrochemical studies of hemoglobin and other redox proteins [39]. Immoblization of redox proteins on nanocomplexes is a new way to realize their direct electrochemistry.

In our previous study, a nanocomplex consisting of carboxylic acid functionalized multi-walled carbon nanotubes (FMWCNTs) and gold nanoparticles (AuPNs) was modified on a glassy carbon (GC) electrode, and applied to analysis the electrochemical properties of catalase [40] and heme-containing artificial peroxidase [41].

In the present study, direct electron transfer of Hb was realized when it was immobilized on the nanocomplex-modified glassy carbon (GC) electrode. Thus, this electrode could be used as a high sensitivity hydrogen peroxide (H_2O_2) biosensor.

2. EXPERIMENTAL SECTION

2.1. Chemicals

Hb from bovine erothrocyte, L-cysteine (Cys), NF (5%), $HAuCl_4$ and sodium citrate were from Sigma (Saint Louis, MO, USA) and used without further purification. Multi-wall carbon nanotubes (MWCNTs), prepared by chemical vapor deposition, were purchased from Shenzhen Nanotech Port Ltd. Co. (Shenzhen, China). Hydrogen peroxide, sodium dihydrogen phosphate (NaH_2PO_4) and disodium hydrogen phosphate (Na_2HPO_4) were obtained from Shanghai Chemicals Company (Shanghai, China). All solutions were prepared in double-distilled deionized water. The stock solutions of hydrogen peroxide (H_2O_2) were prepared by appropriate dilutions of 30% (v/v) H_2O_2 in deionized water. All other chemicals were of analytical grade and used without further purification.

2.2. Preparation of Gold Nanoparticles (AuNPs)

AuNPs were prepared as previously reported in the literature [42-44]. Briefly, 0.01% $HAuCl_4$ solution was heated, then 0.02 M sodium citrate solution was dropped quickly into the hot $HAuCl_4$ solution while agitating vigorously. At first, the color of solution changed from light yellow to grey, then to black and gradually to wine red color without further change. At this stage, the color no longer changed, the heating was stopped, and the solution mixture was continuously agitated until it was cooled to room temperature. The prepared gold colloidal nanoparticles (AuNPs) were stored in dark at 4 °C.

2.3. Preparation of Functional Multi-Walled Carbon Nanotubes (FMWCNTs)

MWCNTs were functionalized according to published methods [23]. Briefly, purified MWCNTs were treated with a concentrated mixture of H_2SO_4 and HNO_3 (v/v = 1/3) under supersonic bath condition (KQ-100B Supersonic Cleaner, Kunshan Shumei, Kunshan, China) for 4 h at 80 °C to introduce carboxyl groups on their surface. The solution pH was then adjusted to 7 with 1 M NaOH solution, centrifuged and washed with water three times. The obtained MWCNTs-COOH (FMWCNTs) were dried at room temperature.

2.4. Preparation of Functional Membrane Modified Glassy Carbon (GC) Electrode

The preparation of the GC electrode was as previously described [45-49]. Prior to coating, the GC electrode was mechanically polished twice with alumina (particle sizes 1.00, 0.30 and 0.05 µm, respectively) to a mirror finish. The electrode was then treated electrochemically in 0.2 M sulfuric acid, cycling between -1.0 and +0.5 V *(vs.* Ag/AgCl) at a sweep rate of 0.1 V/s for approximately 10 min. Thereafter, the working electrode was placed in a 50 mM PBS (pH 7.0), and an anodic potential of 1.70 V *(vs.* Ag/AgCl) was applied for 3-5 min. After the electrode was washed, the AuNPs (negative charged) were electro-deposited on a cleaned bare GC electrode in the range of 0.0 to1.1 V 25 cycles at a scan rate of 0.1 V/s [50]. The GC electrode was then dipped in 1.0 mM L-cysteine (Cys) for 30 min, washed with water, 3 µL of FMWCNTs (2 mg/mL) was dropped onto the surface of the electrode, and dried at room temperature. The electrode was dipped in a Hb solution (80 µM) for 24 h at 4 °C, and for protection, 2 µL Nafion (NF, 5%) was dropped on the electrode surface. The preparation process of functional membrane modified glassy carbon (GC) electrode was also shown in Figure 1.

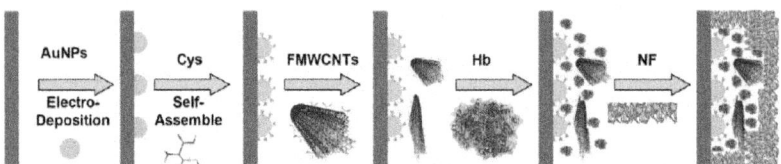

Figure 1. Preparation process of functional membrane modified glassy carbon (GC) electrode.

2.5. Apparatus and Measurements

Electrochemical studies were carried out in a conventional three-electrode cell powered by an electrochemical system comprising of CHI650C (CHI Instruments, Austin, TX, USA). An Ag/AgCl-saturated KCl, a Pt wire and a GC electrode of 3 mm diameter (CHI Instruments) were used as the reference, counter and working electrodes, respectively. All of the potentials in this article were with respect to Ag/AgCl. The electrochemical measurements were carried out in N_2-saturated 0.05 M sodium phosphate buffer solution (PBS) at pH 7.0, 20 °C. Electron microscopic images (TEM) of FMWCNT and AuNPs were obtained using a JEM-1400 (JEOL, Musashino, Japan). Fourier transform infrared (FTIR) spectra of FMWCNTs by KBr pellets were cellected in the range of 1,000-3,500 cm^{-1} on a FTIR 4300 (Shimadzu, city, Japan) spectrometer at room temperature. UV-vis absorption spectra of the AuNPs were collected using a TU-1901 spectrophotometer (Beijing Purkinje General Instrument Company, Beijing, China), with 1 cm path length cells equipped with a thermostat holder and an external temperature controller (Shanghai Hengping Instrument Company, Shanghai, China) at 25 ± 0.1 °C.

3. RESULTS AND DISCUSSION

3.1. Characteristics of the Nanomaterials

The AuPNs and FMWCNTs were characterized by TEM. 3 µL of AuPNs (0.254 mM) or FMWCNTs (3 mg/mL) was dropped onto the surface of fomvar/carbon coated grids (300 mesh), dried and then viewed by TEM operating at 80 kV [40], respectively. UV-vis spectroscopy of the prepared AuNPs exhibited a maximum absorption at 522 nm. The mean size of the AuNPs was then determined to be 18.4 ± 1.1 nm [40,51]. The FTIR spectrum of FMWCNTs shows the characteristic peaks at 1,709, 1,172, and 3,402 cm^1 correspond to the C=O, C- O and O-H stretching vibration of the carboxyl group [52-54], respectively, which indicates that carboxyl groups were modified on the MWCNTs (data not shown).

3.2. Electrochemical Studies

Figure 2(A) presents the cyclic voltammograms (CVs) of: (a) bare electrode; (b) NF/Hb/GC electrode; (c) NF/FMWCNTs/Cys/AuNPs /GC electrode; (d) NF/Hb/FMWCNTs/GC electrode; and (e) NF/Hb/FMWCNTs/Cys/AuNPs/ GC electrode at a scan rate of 0.05 V/s. It can be seen that either (d) or (e) show a well-defined redox wave. The electrode (e) had a stronger redox peak current than that of (d), and the AuPNs could help to significantly increase the redox peaks. Moreover, cathodic and anodic peak potentials were -0.309 V and -0.231 V (vs. Ag/AgCl), respectively. Thus, the formal potential ($E^{o'} = (E_{pa} + Ep_c)/2$) of the electrode (e) was -0.270 ± 0.002 V (vs. Ag/AgCl). This value is consistent with the $E^{o'}$ obtained for an Hb/PLGA/ILs/GC electrode (-0.318 V vs. SCE) [21], an Hb/BMS/CS/GC electrode (-0.32 V vs. Ag/AgCl) [55], or an Hb/Chit-[bmim] PF$_6$-TiO$_2$-Gr/GC electrode (-0.206 V vs. SCE) [17], which is the characteristics of the Hb heme Fe$^{(III)}$/Fe$^{(II)}$ redox couple. It is notable that the positive potential for the electrode facilitated the electrode reaction (Equation (1)) and led to a more efficient bio-catalytic reduction [46,47,49]:

$$k_s = A e^{-\Delta G / RT} e^{-\alpha n FE}. \tag{1}$$

The separation between the anodic and cathodic peak potential ($E_p = Ep_a - E_{pc} = 78$ mV), and the current ratio of the anodic peak current to the cathodic one ($I_{pa}/I_{pc} \approx 1$), indicate that the electrochemical process of the modified GC electrode (e) is quasi-reversible [46,47]. Thus, AuNPs can greatly improve the redox current-modified electrode and facilitate a fast direct electron transfer of Hb entrapped in the composite film. In addition, it is worth noting that there was a pair of unstable redox peaks for curves at site 1 and 2, which should be due to Cys-modified Au, and this redox peak would disappear gradually while the electrode was working. CVs of NF/Hb/FMWCNTs/Cys/AuNPs/GC electrode in 50 mM PBS (pH 7.0) at various scan rates are shown in Figure 2(B). The peak currents increased with increasing the scan rate (v) and were linearly proportional to v (Figure 2C) (not $v^{1/2}$). The linear regression equations for cathodic (I_{pc}) and anodic peak (I_{pa}) currents are: $I_{pc} = 12.65v + 0.14$ and $I_{pa} = -12.45v - 0.13$, with the correlation coefficients of 0.9981 and 0.9977, respectively. The CVs remained essentially unchanged on consecutive potential cycling, indicating that modified electrode is stably confined on the glassy carbon electrode. Figure 2(D) shows the relationship between the peak potential (Ep) and the natural logarithm of scan rate (lnv) for the modified electrode. In the range from 1.2 to 4 V/s, the cathodic peak potential (E_{pc}) changed linearly $vs.$ lnv with a linear regression equation of $E_{pc} = -0.0302$ ln(v) - 0.324, $r = 0.991$. According to Laviron's Equation [56]:

$$E_p = E^{\alpha} + \frac{RT}{\alpha nF} - \frac{RT}{\alpha nF} \ln \upsilon \qquad (2)$$

where α is the cathodic electron transfer coefficient, n is the number of electrons, T is the temperature (293 K here), R the gas constant (8.314 JK^{-1}mol^{-1}) and F the Faraday constant (96,485 C mol^{-1}), respectively. RT/α nF was 0.0302 here, then, αn could be calculated to be 0.55 It could be concluded that $n = 1$ and $α = 0.55 \pm 0.05$ [57]. The $n = 1$ can be also obtained from the width of the peak at mid-height with a low scan rate. Thus, the redox reaction between Hb and the glassy carbon electrode is a single electron transfer process.

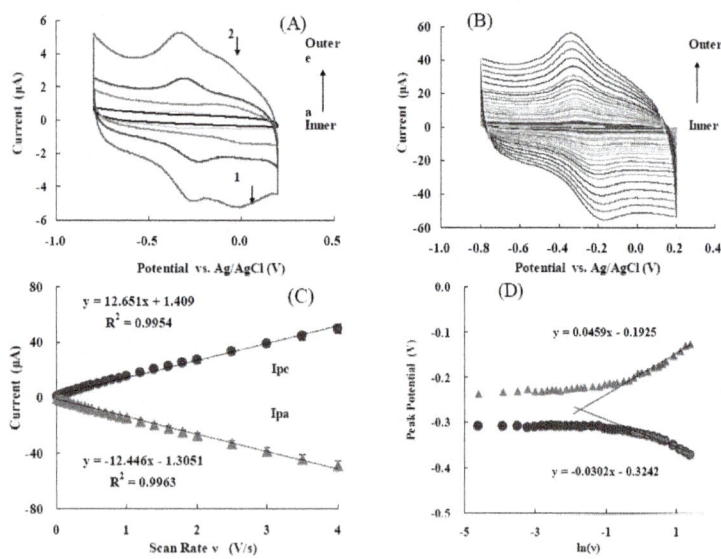

Figure 2. (A) CVs of different modified electrodes (from inner to outer): (a) Bare GC electrode; (b)NF/Hb/GC; (c) NF/FMWCNTs/Cys/AuNPs/GC; (d) NF/Hb/FMWCNTs/GC; (e) NF/Hb/FMWCNTs/Cys/AuNPs/GC. The experiments were carried out in 0.05 M PBS (pH7.0) at a scan rate of 0.05 V/s. (B) CVs of NF/Hb/FMWCNTs/Cys/AuNPs/GC electrode in 0.05 M PBS (pH 7.0) at various scan rates (from inner to outer): 0.02, 0.04, 0.06, 0.08, 0.1, 0.12, 0.14, 0.16, 0.18... 4 V/s, respectively; (C) Plot of peck current I_p $vs.$ scan rate v; (D) Plot of peak potential E_p $vs.$ ln (v).

The value of the apparent heterogeneous electron transfer rate constant k_s could be calculated using the following equation based on Laviron's Equation [58]:

$$\ln k_s = \alpha \ln(1-\alpha) + (1-\alpha)\ln \alpha - \ln(\frac{RT}{nFv}) - \alpha(1-\alpha)\frac{nF\Delta E_p}{RT}. \tag{3}$$

Then, k_s was calculated to be 4.0 ± 0.2 s^{-1}. This value was higher than the most reported k_s values of Hb immobilized on GC electrodes [5,9,10,17,23], due to the high specific surface area and good biocompatibility of the nano-complex.

The average surface concentration (Γ) of electro-active sites (heme groups) of Hb on the surface of glassy carbon electrode could be estimated based on the slope of I_p vs. v (Equation (4)) [59]:

$$I_p = \frac{n^2 F^2 A \Gamma \upsilon}{4RT}. \tag{4}$$

The value of Γ was calculated to be $6.8 \pm 0.3 \times 10^{-10}$ mol $\cdot$ cm^{-2}, which is higher than the Γ value of monolayer of Hb 1.89×10^{-11} molc $\cdot$m^{-2} [60,61]. This high surface concentration can be attributed to the AuNPs and FMWCNTs nanocomplex. The larger surface area and good biocompatibility of the nanocomplex may be helpful for more efficient Hb entrapped in the nanocomplex membrane and provides more activity sites to take part in the electron transfer process (see also Figure 1).

3.3. pH Effect

Figure 3(A) represents CVs of the modified electrode in 50 mM PBS at various pH values. An increase in pH of the solution from 5.0 to 8.0 led to a negative

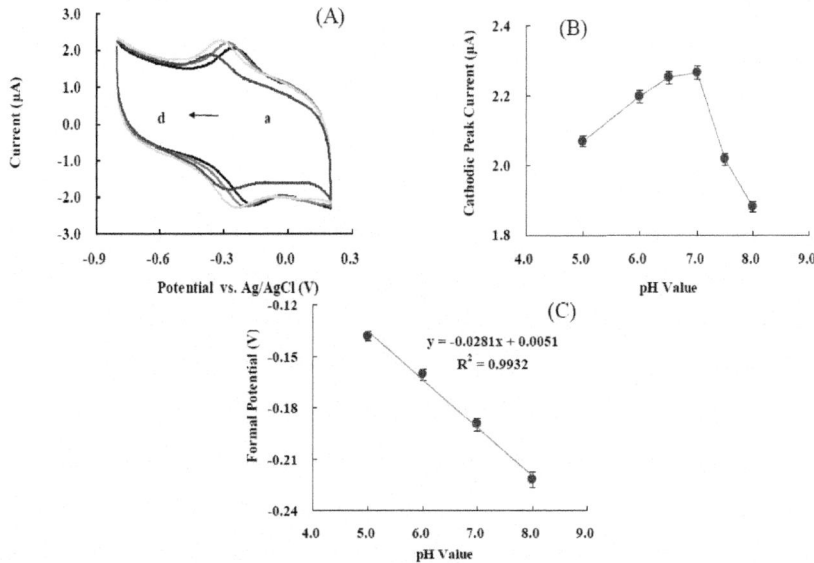

Figure 3. (A) CVs of NF/Hb/ FMWCNTs/Cys/AuNPs/GC electrode in 0.05 M PBS at different pH values: (a) 5.0, (b) 6.0, (c) 7.0, and (d) 8.0, respectively; (B) plot of I_{pc} vs. pH value; (C) Plot of $E^{o\prime}$ vs. pH value.

shift in both reduction and oxidation peak potentials. Figure 3(B) shows that the cathodic peak's current increased with pH changes from 5.0 to 7.0, and then decreased when the pH was greater than 7.0. The maximum cathodic current was obtained at pH 7.0. Figure 3(C) shows that the formal potential of the electrode is pH dependent. These results indicated that the slope was 28.1 ± 0.4 mV/pH over a pH range of 5.0 to 8.0. This value was much smaller than the ideal Nernst's value of 59.2 mV/pH for a one electron and one proton process [59]. The reason might be the biocompatible micro-environment provided by NF. This makes the electrode more stable to pH changes [45-48], influences the protonation state of *trans* ligands to the heme iron and amino acids around the heme, or the protonation of the water molecules coordinated to the central iron [62].

3.4. Optimum Monitoring Potential

Figure 4(A) shows the linear sweep voltammograms (LSVs) of NF/Hb/ FMWCNTs/Cys/AuNPs/GC electrode in the absence (a) and presence of H_2O_2 and with 0.1, 0.13, 0.16, 0.2 mM H_2O_2 (b-e). The cathodic peak current increased with increased concentration of H_2O_2. Thus, the electrode is sensitive to the tested concentrations of the chosen substrate and used as a biosensor of H_2O_2. In addition, the ΔI reached a maximum value for each concentration of H_2O_2 at about -350 mV. Hence, the potential of -350 mV (*vs.* Ag/AgCl) was selected as the optimized monitoring potential throughout this study.

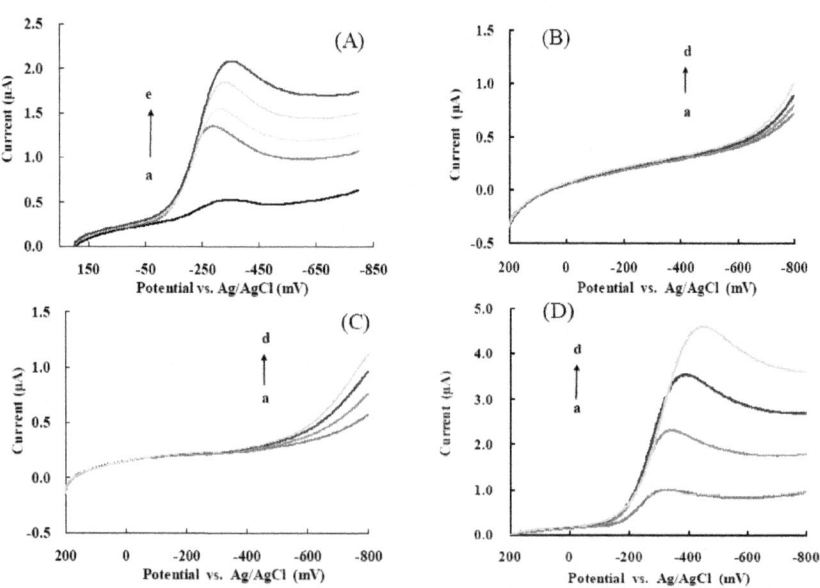

Figure 4. LSVs of (A) NF/Hb/FMWCNTs/Cys/AuNPs/GC electrode in the absence or presence of different concentrations of H_2O_2 (from curve a to curve e): 0, 0.1, 0.13, 0.16, 0.20 mM, respectively. LSVs of (B) Cys/AuNPs/GC; (C) NF/Cys/AuNPs/GC and (D) NF/ Hb/Cys/AuNPs/GC electrodes, respectively in the presence of different concentrations of H_2O_2(from a to curve d): 0.1, 0.3, 0.5 and 0.7 mM, respectively.

It is worth noting that either Cys/AuNPs/GC (Figure 4(B)) or NF/Cys/ AuNPs/GC (Figure 4(C)) electrodes exhibited an amperometric response to H_2O_2 at potentials higher than -0.6 V (*vs.* Ag/AgCl). These results were similar to those reported in the literature [63,64]. However, Figure 4(D) shows that the amperometric response of the NF/Hb/AuNPs/GC electrode toward H_2O_2 occurred at an even lower potential (-400 mV) with stronger current intensity. Thus, these results could be attributed to the bioelectro-catalytic behavior of Hb.

3.5. Electro-Catalytic Behavior of Modified Electrode and Detection Limit

Figure 5 presents the current responses of the modified electrode to successive additions of 5 µL of 1 nM (a), 10 nM (b) or 20 nM (c) of H_2O_2 in 5 mL of 0.05 M PBS (pH 7.0) at the applied potential of -0.35 V (*vs.* Ag/AgCl). The inset shows the typical current response for each addition process. Current at state 1 is a steady state (I_{sa}) with no addition, when H_2O_2 is added the current increases rapidly and reaches a maximum, state 2 (I_{ma}). The current then reduces gradually to a steady state, state 3 (I_{sb}), before the next addition and next maximum current response (I_{mb}), state 4. Though the amount of maximum response current was great, this state was unsuitable for use, as the change of the maximum current response ($\Delta I = I_{mb} - I_{ma}$) was not stable for each addition, and was affected by the mainly addition position and diffusion rate of H_2O_2. The steady state current response ($\Delta I = I_{sb} - I_{sa}$) increased when the addition concentration of added H_2O_2 was over 10 nM with the final concentration of H_2O_2 (0.01-50 nM) and a linear range from 0.05 to 1 nM (see also Figure 6(A)).

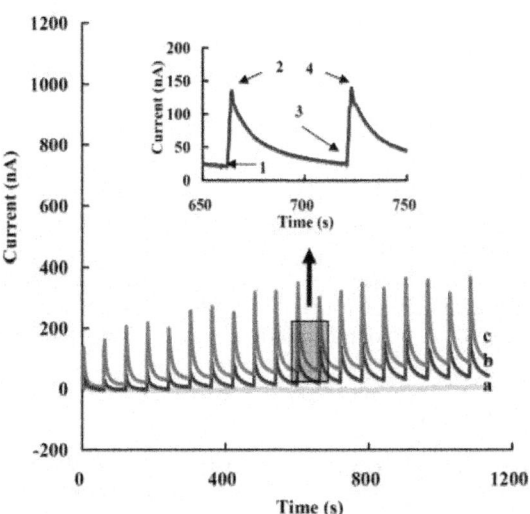

Figure 5. Amperometirc response of the modified electrode to successive additions of 5 µL of 1 nM (a), 10 nM(b) or 20 nM(c) H_2O_2 in 5 mL of 0.05 M PBS, pH 7.0, at the applied potential of -0.35 V (*vs.* Ag/AgCl). Inset shows the typical current response for each addition process: (1) previous steady state current; (2) maximum response current; (3) steady state current; (4). next maximum response current.

To determine the detection limit (minimum concentration of H_2O_2 that could be detected by this method), the steady current (I_s) of each addition was measured while the final concentration of H_2O_2 ([H_2O_2]) was increased gradually (Figure 6(B)). The detection limit was determined to be 0.05 ± 0.01 nM from the cross point of the lines fitted to the linear segments of the I_s vs. [H_2O_2] [65,66], which was lower than the most reported values, e.g., an Hb/GNPs/Hb/MWNT/GC electrode (80 nM) [67] and a CAT/[bmim][PF$_6$]/MWCNTs/GC electrode (0.25 nM) [68].

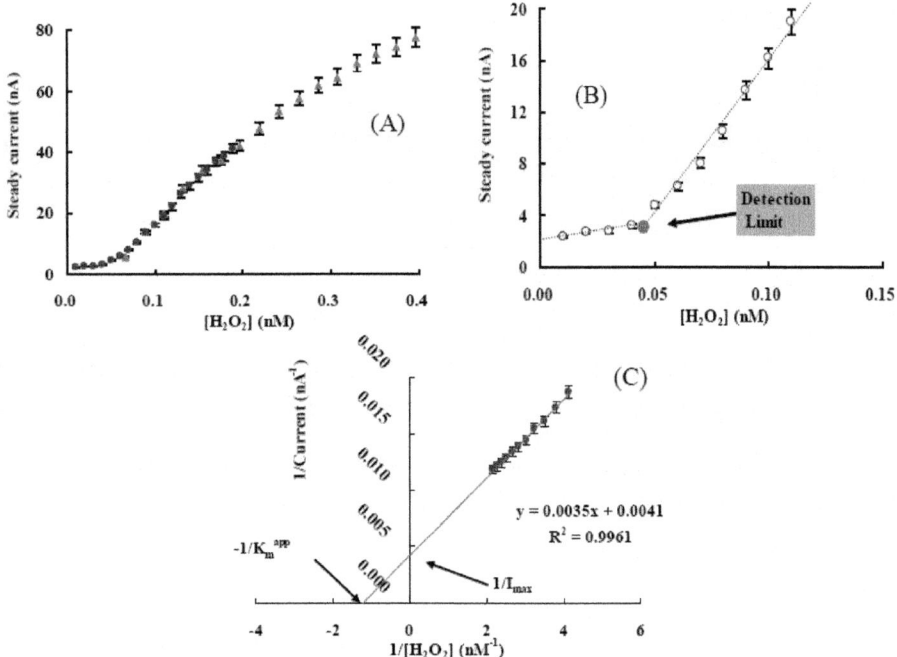

Figure 6. (A) The typical steady current vs. [H_2O_2] in the process of successive additions of 5 µL of 10 nM (•) and 20 nM (▲) H_2O_2 in 5 mL of 50 mM PBS (pH 7.0) at the applied potential of -0.35 V (vs. Ag/AgCl). **(B)** The determination of the H_2O_2 detection limit for NF/Hb/FMWCNTs/Cys/AuNPs/GC electrode. The detection limit was determined from the cross point of the lines fitted to the linear segments of the steady current Is vs. [H_2O_2] in the process of successive additions of 5 µL of 10 nM H_2O_2 in 5 mL of 50 mM PBS (pH7.0). **(C)** Lineweaver-Burk plot for K_m^{app} determination.

3.6. Kinetic Parameters

Overall, the electrode reaction could be supposed to be carried out by two steps [40]:

Step One: $Hb(Fe^{III}) + e + H^+ = Hb(Fe^{II}-H)$

Step Two: $Hb(Fe^{II}-H) + 1/2\ H_2O_2 \rightarrow Hb(Fe^{III}) + H_2O$

where, $Hb(Fe^{III})$ and $Hb(Fe^{II}-H)$ denote the oxidized and reduced forms of the modified Hb, respectively. Initially $Hb(Fe^{III})$ undergoes the electron transfer reaction with the electrode resulting in the production of $Hb(Fe^{II}\text{-}H)$, as shown in Step One, this is an one-electron and one-proton (pH dependent) process. Alternatively, $Hb(Fe^{II}\text{-}H)$ can be oxidized by H_2O_2 in the solution to regenerate $Hb(Fe^{III})$ as shown in Step Two. In this catalytic reaction, the peak current kept rising until $Hb(Fe^{III})$ was consumed in the Step One and was compensated by its production in Step Two.

The apparent Michaelis-Menten constant K_m^{app} is a reflection of both enzyme affinity and ratio of microscopic kinetic constants. It can be obtained from the electrochemical version of the Linewearver-Burk Equation [13]:

$$\frac{1}{I_{ss}} = \frac{1}{I_{max}} + \frac{K_m^{app}}{I_{max}c} \tag{5}$$

where, c is the substrate (H_2O_2) concentration in the solution ($[H_2O_2]$ here), I_{ss} the steady-state current after the addition of substrate, and I_{max} is the maximum current measured under saturated substrate conditions. K_m^{app} value of the modified electrode was calculated to be 0.85 ± 0.1 nM (Figure 6(C)), which was much lower than the most reported values, e.g., an HRP-AQ/GC electrode (51 nM) [49], a PLGA/ILs/Hb/GC electrode (69 µM) [21], and an Hb/Chit-[bmim]PF_6-TiO_2-Gr/GC electrode (1.245 mM) [17] . A low K_m^{app} value indicates a strong substrate binding and exhibits a higher affinity of H_2O_2 for this modified electrode. FM-WCNTs and AuNPs complex system helps to reduce the bridge length between electroactive center (heme group) of Hb and GC electrode [47], and result in the high sensitivity of the modified electrode to H_2O_2 (Table 1).

Modified Electrode	$E^{o'}$ (mV)	k_s (s^{-1})	Γ (mol·cm^{-2})	K_m^{app}	Linear Range	Detection Limit	Ref
NF/Hb/FMWCNTs/Cys/AuNPs/GC	-270 ± 2 [a]	4.0 ± 0.2	$6.8 \pm 0.3 \times 10^{-10}$	0.85 nM	0.05–1 nM	0.05 nM	This work
Hb/PdNPs/GR-CS/GC	-240 [b]	0.69	1.74×10^{-10}	16 µM	2–1100 µM	660 nM	[5]
Hb/NiO/GC	-70 [a]	5.2 ± 0.5	1.73×10^{-11}	1370 µM	1–2000 µM	630 nM	[6]
NF/Hb/PAM-P123/GC	-317 [b]	-	7.64×10^{-11}	36 µM	1–30 µM	400 nM	[7]
Hb/Gel/GC	-380 [b]	-	-	-	50–1200 µM	3400 µM	[8]
Hb/NGC-SF/GC	-380 [b]	1.98	-	-	0.6–1.7 and 2–22 mM	-	[9]
Polymer–Hb–CNTs/GC	-273 [b]	0.90	1.1×10^{-10}	140 µM	8–240 µM	4 µM	[10]
Hb/CS-[bmim]PF$_6$-TiO$_2$-GR/GC	-206 [b]	0.73–3.96	3.21×10^{-10}	1245 µM	1–1170 µM	0.3 µM	[17]
Hb/PLGA/ILs/GC	-318 [b]	5.02 ± 0.16	4.74×10^{-10}	69 µM	5–8050 µM	0.237 µM	[21]
Hb/ATP/GC	-362 [b]	4.6 ± 0.65	6.7×10^{-11}	490 µM	5.4–400 µM	2.4 µM	[22]
EDC-Hb-CNTs/GC	-268 [b]	1.02 ± 0.05	4.7×10^{-9}	-	0.25–140 µM	0.18 µM	[23]
Hb/BMS/CS/GC	-320 [a]	-	9.34×10^{-11}	-	2.5–245 µM	0.83 µM	[55]
Hb/GNPs/Hb/MWNT/GC	-355 [b]	-	-	260 µM	0.21–3000 µM	80 nM	[67]

NF: Nafion; GR: graphene; PdNPs: palladium nanoparticles; CS: chitosan; NiO: nickel oxide nanoparticles; PLGA: poly lactic-co-glycolic acid; ILs: ionic liquid, 1-butyl-3-methylimidazolium tetrafluoroborate ([BMIM]BF4; BMS: bimodal mesoporous silica; PAM-P123: polyacrylamide-P123; Gel: gelatine; NGC: nanostructured gold colloid; SF: silk fibroin; EDC: 1-ethyl-3-(3-dimethylaminopropyl) carbodiimide; ATP: attapulgite; GNPs: gold colloidal nanoparticles; [a] vs. Ag/AgCl; [b] vs. SCE.

3.7. Stability

Long-term stability is an important parameter for biosensors. The operational stability of the modified electrode was determined by the CV method. The cathodic peak current was reduced by less than 5% after 50 cycles at the scan rate of 0.05 V/s, while the peak potential remained unchanged. As for the storage stability, the CVs showed minimal change after two weeks of storage in a bottle over the PBS solution at 4 °C . NF may offer a biocompatible micro-environment to confine bio-macromolecules at their ionic cluster region (30–50 nm), and this view was also consistent with our previous study [45–48]. Moreover, Nafion may be helpful to restrict, confirm and protect Hb/FMWCNTs/Cys/AuNPs system on GC electrodes.

3.8. Interference Determination

The degree of interference from interfering substances can be evaluated by the value of the cathodic current ratio which were calculated by reading the cathodic current (I_{pc1}) of the proposed biosensor in 50 mM PBS (pH 7.0) containing 0.10 mM H_2O_2 and a 0.20 mM hampering substance, and then, comparing it with the cathodic current (I_{pc0}) from the proposed biosensor in the same solution containing only 0.10 mM H_2O_2. Five interfering substances were tested here and the results are listed in Table 2. It can be observed that none of the tested interferents could cause interference to the determination of H_2O_2, which is largely attributed to the low working potential of -350 mV used in the determination of H_2O_2 and the negatively charged NF protection membrane.

Table 2. Effects of possible interferences on the hydrogen peroxide biosensor.

Possible Interferences	I_{pc1}/I_{pc0}	R.S.D (%)
Glucose	1.04 ± 0.08	3.2
Adenosine Triphosphate	1.03 ± 0.06	3.4
L-Histidine	1.01 ± 0.05	2.9
Ascorbic Acid	0.98 ± 0.06	3.1
Thiol	1.00 ± 0.04	2.5

I_{pc1}: The cathodic current for a mixture of a 0.20 mM interfering substance and 0.10 mM H_2O_2; I_{pc0}: The cathodic current for 0.10 mM H_2O_2 alone in a 50 mM PBS (pH 7.0), at -350 mV vs. Ag/AgCl. R.S.D: Relative standard deviation, obtained for nine measurements.

4. CONCLUSIONS

The direct electrochemical properties of immobilized Hb on a FMWCNTs/Cys/AuNPs-modified glassy carbon electrode were found to be due to the excellent microenvironment provided by NF, AuNPs and FMWCNTs for Hb. The small value of K_m^{app}, high sensitivity, long-term stability and low detection limit were other characteristics of the biosensor. The modified electrode showed the ability to be used as a third generation biosensor for determination of H_2O_2 at ultra-trace levels. Moreover, a redox protein on the functional nano complex modified electrode may be a new useful electrochemical tool for the analysis of relationship between the structure and function of redox proteins, especially for a heme-containing protein.

Acknowledgments

The financial supports of Henan University Science Foundation and the Research Council of University of Tehran and the Iran National Science Foundation (INSF) are gratefully acknowledged.

Conflict of Interest

The authors declare no conflict of interest.

REFERENCES

1. Armstrong, F.A.; Hill, H.A.O.; Walton, N.J. Direct electrochemistry of redox proteins. *Acc. Chem. Res.* **1988**, *21*, 407–413.

2. Armstrong, F.A.; Wilson, G.S. Recent developments in faradaic bioelectrochemistry. *Electrochim. Acta* **2000**, *45*, 2623–2645.

3. Hill, H.A.O. Direct electrochemistry of cytochrome c. *Coord. Chem. Rev.* **1996**, *151*, 115–123.

4. Thévenot, D.R.; Toth, K.; Durst, R.A.; Wilson, G.S. Electrochemical biosensors: Recommended definitions and classification. *Biosens. Bioelectron.* **2001**, *16*, 121–131.

5. Sun, A.; Sheng, Q.; Zheng, J. A hydrogen peroxide biosensor based on direct electrochemistry of hemoglobin in palladium nanoparticles/graphene–chitosan nanocomposite film. *Appl. Biochem. Biotechnol.* **2012**, *166*, 764–773.

6. Salimi, A.; Sharifi, E.; Noorbakhsh, A.; Soltanian, S. Direct voltammetry and electrocatalytic properties of hemoglobin immobilized on a glassy carbon electrode modified with nickel oxide nanoparticles. *Electrochem. Commun.* **2006**, *8*, 1499–1508.

7. Li, J.; Tang, J.; Zhou, L.; Han, X.; Liu, H. Direct electrochemistry and electrocatalysis of hemoglobin immobilized on polyacrylamide-P123 film modified glassy carbon electrode. *Bioelectrochemistry* **2012**, *86*, 60–66.

8. Yao, H.; Li, N.; Xu, J.Z.; Zhu, J.J. Direct electrochemistry and electrocatalysis of hemoglobin in gelatine film modified glassy carbon electrode. *Talanta* **2007**, *71*, 550–554.

9. Guo, H.L.; Liu, D.Y.; Yu, X.D.; Xia, X.H. Direct electrochemistry and electrocatalysis of hemoglobin on nanostructured gold colloid-silk fibroin modified glassy carbon electrode. *Sens. Actuators B Chem.* **2009**, *139*, 598–603.

10. Chen, L.; Lu, G. Direct electrochemistry and electrocatalysis of hybrid film assembled by polyelectrolyte–surfactant polymer, carbon nanotubes and hemoglobin. *J. Electroanal. Chem.* **2006**, *597*, 51–59.

11. Lee, K.P.; Gopalan, A.I.; Komathi, S. Direct electrochemistry of cytochrome c and biosensing for hydrogen peroxide on polyaniline grafted multi-walled carbon nanotube electrode. *Sens. Actuators B Chem.* **2009**, *141*, 518–525.

12. Liu, X.J.; Zhang, W.J.; Huang, Y.X.; Li, G.X. Enhanced electron-transfer reactivity of horseradish peroxidase in phosphatidylcholine films and its catalysis to nitric oxide. *J. Biotechnol.* **2004**, *108*, 145–157.

13. Liu, Y.; Han, T.; Chen, C.; Bao, N.; Yu, C.M.; Gu, H.Y. A novel platform of hemoglobin on core-shell structurally Fe3O4@Au nanoparticles and its direct electrochemistry. *Electrochim. Acta* **2011**, *56*, 3238–3247.

14. Rhieu, S.Y.; Ludwig, D.R.; Siu, V.S.; Palmore, G.T.R. Direct electrochemistry of cytochrome P450 27B1 in surfactant films. *Electrochem. Commun.* **2009**, *11*, 1857–1860.

15. Rusling, J.F.; Nassar, A.E.F. Enhanced electron transfer for myoglobin in surfactant films on electrodes. *J. Am. Chem. Soc.* **1993**, *115*, 11891–11897.

16. Shang, L.B.; Sun, Z.Y.; Wang, X.W.; Li, G.X. Enhanced peroxidase activity of hemoglobin in a DNA membrane and its application to an unmediated hydrogen peroxide biosensor. *Anal. Sci.* **2003**, *11*, 1537–1539.

17. Sun, J.Y.; Huang, K.J.; Zhao, S.F.; Fan, Y.; Wu, Z.W. Direct electrochemistry and electrocatalysis of hemoglobin on chitosan-room temperature ionic liquid-TiO$_2$-graphene nanocomposite film modified electrode. *Bioelectrochemistry* **2011**, *82*, 125–130.

18. Xu, J.S.; Zhao, G.C. A third-generation biosensor based on the enzyme-like activity of cytochrome c on a room temperature ionic liquid and gold nanoparticles composite. *Int. J. Electrochem. Sci.* **2008**, *3*, 519–527.

19. Yang, J.; Hu, N. Direct electron transfer for hemoglobin in biomembrane-like dimyristoyl phosphatidylcholine films on pyrolytic graphite electrodes. *Bioelectrochem. Bioenerg.* **1999**, *48*, 117–127.

20. Yang, N.; Hoffmann, R.; Smirnov, W.; Kriele, A.; Nebel, C.E. Direct electrochemistry of cytochrome c on nanotextured diamond surface. *Electrochem. Commun.* **2010**, *12*, 1218–1221.

21. Zhang, Y.; Sun, X.; Jia, N. Direct electrochemistry and electrocatalysis of hemoglobin immobilized into poly (lactic-co-glycolic acid)/room temperature ionic liquid composite film. *Sens. Actuators B Chem.* **2011**, *157*, 527–532.

22. Xu, J.; Li, W.; Yin, Q.; Zhong, H.; Zhu, Y.; Jin, L. Direct electron transfer and bioelectrocatalysis of hemoglobin on nano-structural attapulgite clay-modified glassy carbon electrode. *J. Colloid Interface Sci.* **2007**, *315*, 170–176.

23. Zhang, R.; Wang, X.; Shiu, K.K. Accelerated direct electrochemistry of hemoglobin based on hemoglobin–carbon nanotube (Hb–CNT) assembly. *J. Colloid Interface Sci.* **2007**, *316*, 517–522.

24. Gorton, L.; Lindgren, A.; Larsson, T.; Munteanu, F.D.; Ruzgas, T.; Gazaryan, I. Direct electron transfer between heme-containing enzymes and electrodes as basis for third generation biosensors. *Anal. Chim. Acta* **1999**, *400*, 91–108.

25. Olsson, M.G.; Allhorn, M.; Olofsson, T.; AKerstrom, B. Up-regulation of alpha1-microglobulin by hemoglobin and reactive oxygen species in hepatoma and blood cell lines. *Free Radic. Biol. Med.* **2007**, *42*, 842–851.

26. Buehler, P.W.; Haney, C.R.; Gulati, A.; Ma, L.; Hsia, C.J.C. Polynitroxyl hemoglobin: A pharmacokinetic study of covalently bound nitroxides to hemoglobin platforms. *Free Radic. Biol. Med.* **2004**, *37*, 124–135.

27. Royer, W.E., Jr.; Knapp, J.E.; Strand, K.; Heaslet, H.A. Cooperative hemoglobins: Conserved fold, diverse quaternary assemblies and allosteric mechanisms. *Trends Biochem. Sci.* **2001**, *26*, 297–304.

28. Ding, Y.; Wang, Y.; Li, B.K.; Lei, Y. Electrospun hemoglobin microbelts based biosensor for sensitive detection of hydrogen peroxide and nitrite. *Biosens. Bioelectron.* **2010**, *25*, 2009–2015.

29. He, X.Y.; Zhu, L. Direct electrochemistry of hemoglobin in cetylpyridinium bromide film: Redox thermodynamics and electrocatalysis to nitric oxide. *Electrochem. Commun.* **2006**, *8*, 615–620.

30. Ferapontova, E.E.; Gorton, L. Direct electrochemistry of heme multicofactor-containing enzymes on alkanethiol-modified gold electrodes. *Bioelectrochemistry* **2005**, *66*, 55–63.

31. Parak, F.G.; Nienhaus, G.U. Myoglobin, a paradigm in the study of protein dynamics. *Chemphyschem* **2002**, *3*, 249–254.

32. Cao, W.X.; Christian, J.F.; Champion, P.M.; Rosca, F.; Sage, J.T. Water penetration and binding to ferric myoglobin. *Biochemistry* **2001**, *40*, 5728–5737.

33. Ajayan, P.M. Nanotubes from carbon. *Chem. Rev.* **1999**, *99*, 1787–1800.

34. Ebbesen, T.W.; Ajayan, P.M. Large-Scale synthesis of carbon nanotubes. *Nature* **1992**, *358*, 220–222.

35. Iijima, S. Helical microtubules of graphitic carbon. *Nature* **1991**, *354*, 56–58.

36. Chen, H.J.; Wang, Y.L.; Wang, Y.Z.H.; Dong, S.H.J.; Wang, E.K. One-Step preparation and characterization of PDDA-protected gold nanoparticles. *Polymer* **2006**, *47*, 763–766.

37. Nada, M.D.; David, M.B. Radiolytically induced formation and optical absorption spectra of colloidal silver nanoparticles in supercritical ethane. *J. Phys. Chem. B* **2001**, *105*, 954–959.

38. Chen, G.F.; Liang, Z.Q.; Li, G.X. Progress of electrochemical biosensors fabricated with nano-materials. *Acta Biophys. Sin.* **2010**, *26*, 711–725.

39. Scheller, F.W.; Bistolas, N.; Liu, S.Q.; Jänchen, M.; Katterle, M.; Wollenberger, U. Thirty years of haemoglobin electrochemistry. *Adv. Colloid Interface Sci.* **2005**, *116*, 111–120.

40. Hong, J.; Yang, W.Y.; Zhao, Y.X.; Xiao, B.L.; Gao, Y.F.; Yang, T.; Ghourchian, H.; Moosavi-Movahedi, Z.; Sheibani, N.; Li, J.G.; *et al.* Catalase immobilized on a functionalized multi-walled carbon nanotubes–gold nanocomposite as a highly sensitive bio-sensing system for detection of hydrogen peroxide. *Electrochim. Acta* **2013**, *89*, 317–325.

41. Yang, W.Y.; Hong, H.; Zhao, Y.X.; Xiao, B.L.; Gao, Y.F.; Yang, T.; Moosavi-Movahedi, A.A.; Ghourchian, H.; Moosavi-Movahedi, Z. Electrochemical study of a nano vesicular artificial peroxidase on a functional nano complex modified glassy carbon electrode. *J. New Mat. Electrochem. Syst.* **2013**, *16*, 89–95.

42. Lin, S.Q.; Ju, H.X. Renewable reagentless hydrogen peroxide sensor based on direct electron transfer of horseradish peroxidase immobilized on colloidal gold-modified electrode. *Anal. Biochem.* **2002**, *307*, 110–116.

43. Daniel, M.C.; Astruc, D. Gold Nanoparticles: Assembly, supramolecular chemistry, quantum-Size-related properties, and applications toward biology, catalysis, and nanotechnology. *Chem. Rev.* **2004**, *104*, 293–346.

44. Xian, Y.; Hu, Y.; Liu, F.; Xian, Y.; Wang, H.; Jin, L. Glucose biosensor based on Au nanoparticles–conductive polyaniline nanocomposite. *Biosens. Bioelectron.* **2006**, *21*, 1996–2000.

45. Hong, J.; Ghourchian, H.; Moosavi-Movahedi, A.A. Direct electron transfer of redox proteins on a Nafion-cysteine modified gold electrode. *Electrochem. Commun.* **2006**, *8*, 1572–1576.

46. Hong, J.; Ghourchian, H.; Rezaei-zarchi, S.; Moosavi-Movahedi, A.A.; Ahmadian, S.; Saboury, A.A. Nafion-methylene blue functional memberane and its application in chemical/bio-sensing. *Anal. Lett.* **2007**, *40*, 483–496.

47. Hong, J.; Moosavi-Movahedi, A.A.; Ghourchian, H.; Molaei Rad, A. Direct electron transfer of horseradish peroxidase on Nafion-cysteine modified gold electrode. *Electrochim. Acta* **2007**, *52*, 6261–6267.

48. Rezaei-Zarchi, S.; Saboury, A.A.; Hong, J.; Norouzi, P.; Moghaddam, A.B.; Ghourchian, H.; Ganjali, M.R.; Moosavi-Movahedi, A.A.; Javed, A.; Mohammadian, A. Electrochemical behavior of redox proteins immobilized on Nafion-riboflavin modified gold electrode. *Bull. Korean Chem. Soc.* **2007**, *28*, 2266–2270.

49. Shourian, M.; Ghourchian, H. Biosensing improvement of horseradish peroxidase towards hydrogen peroxide upon modifying the accessible lysines. *Sens. Actuators B Chem.* **2010**, *145*, 607–612.

50. Tian, Y.; Mao, L.; Okajima, T.; Ohsaka, T. A carbon fiber microelectrode-based third-generation biosensor for superoxide anion. *Biosens. Bioelectron.* **2005**, *21*, 557–564.

51. Haiss, W.; Thanh, N.T.K.; Aveyard, J.; Fernig, D.G. Determination of size and concentration of gold nanoparticles from UV-vis spectra. *Anal. Chem.* **2007**, *79*, 4215–4221.

52. Park, M.J.; Lee, J.K.; Lee, B.S.; Lee, Y.W.; Choi, I.S.; Lee, S.G. Covalent modification of multi-walled carbon nanotubes with imidazolium-based ionic liquids: Effect of anions on solubility. *Chem. Mater.* **2006**, *18*, 1546–1551.

53. Shen, J.; Huang, W.; Wu, L.; Hu, Y.; Ye, M. Study on amino-functionalized multiwalled carbon nanotubes. *Mater. Sci. Eng. A* **2007**, *464*, 151–156.

54. Wang, J.; Fang, Z.; Gu, A.; Xu, L.; Liu, F. Effect of amino-functionalization of multi-walled carbon nanotubes on the dispersion with epoxy resin matrix. *J. Appl. Polym. Sci.* **2006**, *100*, 97–104.

55. Zhang, L.; Zhang, Q.; Li, J. Direct electrochemistry and electrocatalysis of hemoglobin immobilized in bimodal mesoporous silica and chitosan inorganic–organic hybrid film. *Electrochem. Commun.* **2007**, *9*, 1530–1535.

56. Laviron, E.J. Adsorption, autoinhibition and autocatalysis in polarography and in linear potential sweep voltammetry. *J. Electroanal. Chem.* **1974**, *52*, 355–393.

57. Ma, H.; Hu, N.; Rusling, J.F. Electroactive myoglobin films grown layer-by-layer with poly(styrenesulfonate) on pyrolytic graphite electrodes. *Langmuir* **2000**, *16*, 4969–4975.

58. Laviron, E.J. General expression of the linear potential sweep voltammogram in the case of diffusionless electrochemical systems. *J. Electroanal. Chem.* **1979**, *101*, 19–28.

59. Rahimi, P.; Rafiee-Pour, H.; Ghourchian, H.; Norouzi, P.; Ganjali, M.R. Ionic-liquid/ NH2-MWCNTs as a highly sensitive nano-composite for catalase direct electrochemistry. *Biosen. Bioelectron.* **2010**, *25*, 1301–1306.

60. Wang, S.; Chen, T.; Zhang, Z.; Shen, X.; Lu, Z.; Pang, D.; Wong, K. Direct electrochemistry and electrocatalysis of heme proteins entrapped in agarose hydrogel films in room-temperature ionic liquids. *Langmuir* **2005**, *21*, 9260–9266.

61. Wang, S.; Chen, T.; Zhang, Z.; Pang, D.; Wong, K. Effects of hydrophilic room-temperature ionic liquid 1-butyl-3-methylimidazolium tetrafluoroborate on direct electrochemistry and bioelectrocatalysis of heme proteins entrapped in agarose hydrogel films. *Electrochem. Commun.* **2007**, *9*, 1709–1714.

62. Yamazaki, I.; Araiso, T.; Hayashi, Y.; Yamada, H.; Makino, R. Analysis of acid-base properties of peroxidase and myoglobin. *Adv. Biophys.* **1978**, *11*, 249–281.

63. Wang, L.; Bo, X.; Bai, J.; Zhu, L.; Guo, L. Gold nanoparticles electrodeposited on ordered mesoporous carbon as an enhanced material for nonenzymatic hydrogen peroxide sensor. *Electroanalysis* **2010**, *22*, 2536–2542.

64. Jirkovsky, J.S.; Halasa, M.; Schiffrin, D.J. Kinetics of electrocatalytic reduction of oxygen and hydrogen peroxide on dispersed gold nanoparticles. *Phys. Chem. Chem. Phys.* **2010**, *12*, 8042–8052.

65. Molaei Rad, A.; Ghourchian, H.; Moosavi-Movahedi, A.A.; Hong, J.; Nazari, K. Spectrophotometric assay for horseradish peroxidase activity based on pyrocatechol–aniline coupling hydrogen donor. *Anal. Biochem.* **2007**, *362*, 38–43.

66. Buck, R.P.; Linder, E. Recommendations for nomenclature of ion selective electrodes. *Pure Appl. Chem.* **1994**, *66*, 2527–2536.

67. Chen, S.; Yuan, R.; Chai, Y.; Zhang, L.; Wang, N.; Li, X. Amperometric third-generation hydrogen peroxide biosensor based on the immobilization of hemoglobin on multiwall carbon nanotubes and gold colloidal nanoparticles. *Biosens. Bioelectron.* **2007**, *22*, 1268–1274.

68. Shamsipur, M.; Asgari, M.; Maragheh, M.G.; Moosavi-Movahedi, A.A. A novel impedimetric nanobiosensor for low level determination of hydrogen peroxide based on biocatalysis of catalase. *Bioelectrochemistry* **2012**, *83*, 31–37.

INDEX

This page left intentionally blank.